Short Courses

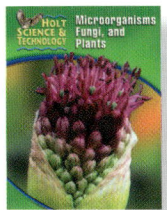 Microorganisms, Fungi, and Plants

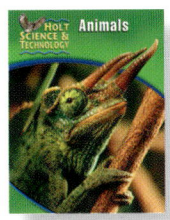 Animals

 Cells, Heredity, and Classification

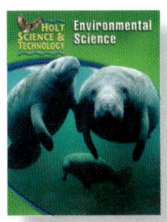 Environmental Science

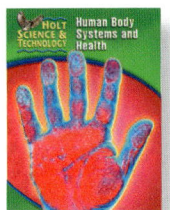 Human Body Systems and Health

 Inside the Restless Earth

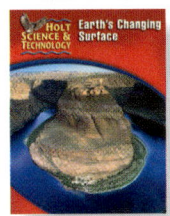 Earth's Changing Surface

 Water on Earth

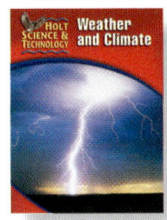 Weather and Climate

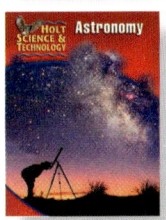 Astronomy

 Introduction to Matter

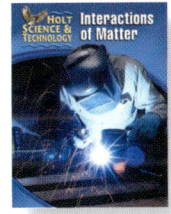 Interactions of Matter

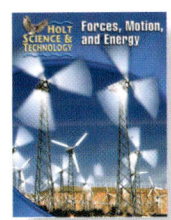 Forces, Motion, and Energy

 Electricity and Magnetism

 Sound and Light

Teacher Edition WALK-THROUGH

Student Edition CONTENTS IN BRIEF

HOLT, RINEHART AND WINSTON

A Harcourt Education Company

Orlando • **Austin** • New York • San Diego • Toronto • London

Designed to meet the needs of all students

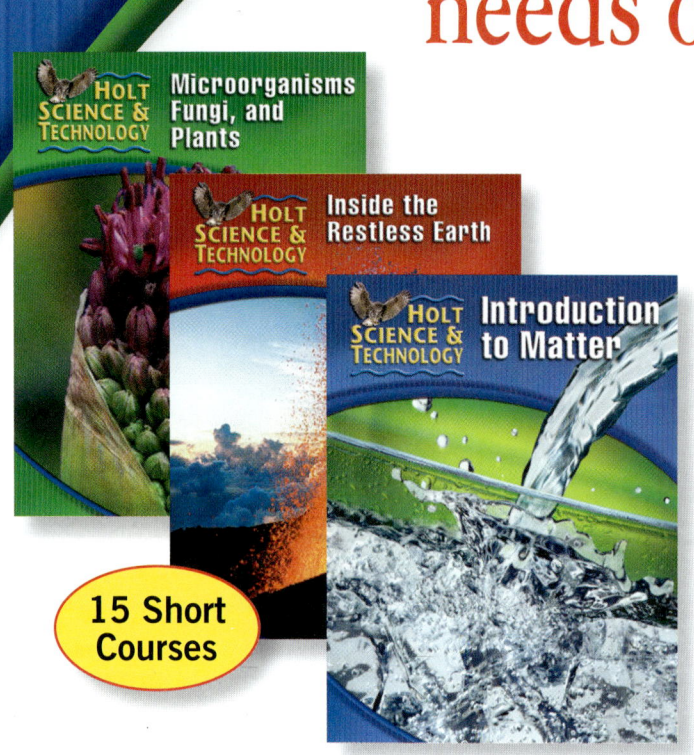

15 Short Courses

Holt Science & Technology: Short Course Series allows you to match your curriculum by choosing from 15 books covering life, earth, and physical sciences. The program reflects current curriculum developments and includes the strongest skills-development strand of any middle school science series. Students of all abilities will develop skills that they can use both in science as well as in other courses.

STUDENTS OF ALL ABILITIES RECEIVE THE READING HELP AND TAILORED INSTRUCTION THEY NEED.

- The *Student Edition* is accessible with a clean, easy-to-follow design and highlighted vocabulary words.
- Inclusion strategies and different learning styles help support all learners.
- Comprehensive **Section** and **Chapter Reviews** and **Standardized Test Preparation** allow students to practice their test-taking skills.
- **Reading Comprehension Guide** and **Guided Reading Audio CDs** help students better understand the content.

CROSS-DISCIPLINARY CONNECTIONS LET STUDENTS SEE HOW SCIENCE RELATES TO OTHER DISCIPLINES.

- **Mathematics, reading,** and **writing skills** are integrated throughout the program.
- Cross-discipline **Connection To** features show students how science relates to language arts, social studies, and other sciences.

A FLEXIBLE LABORATORY PROGRAM HELPS STUDENTS BUILD IMPORTANT INQUIRY AND CRITICAL-THINKING SKILLS.

- The laboratory program includes labs in each chapter, labs in the **LabBook** at the end of the text, six different lab books, and **Video Labs.**
- All labs are teacher-tested and rated by difficulty in the *Teacher Edition,* so you can be sure the labs will be appropriate for your students.
- A variety of labs, from **Inquiry Labs** to **Skills Practice Labs,** helps you meet the needs of your curriculum and work within the time constraints of your teaching schedule.

INTEGRATED TECHNOLOGY AND ONLINE RESOURCES EXPAND LEARNING BEYOND CLASSROOM WALLS.

- An **Enhanced Online Edition** or **CD-ROM Version** of the student text lightens your students' load.

- **SciLinks,** a Web service developed and maintained by the National Science Teachers Association (NSTA), contains current prescreened links directly related to the textbook.

- **Brain Food Video Quizzes** on videotape and DVD are game-show style quizzes that assess students' progress and motivate them to study.

- The **One-stop Planner® CD-ROM** with **Exam View® Test Generator** contains all of the resources you need including an *Interactive Teacher Edition,* worksheets, customizable lesson plans, **Holt Calendar Planner,** a powerful test generator, **Lab Materials QuickList Software,** and more.

- Spanish Resources include **Guided Reading Audio CD** in Spanish.

CHAPTER RESOURCE FILES FOR

Inside the Restless Earth

Skills Worksheets
- Directed Reading A
- Directed Reading B
- Vocabulary & Notes
- Section Reviews
- Chapter Review
- Reinforcement
- Critical Thinking

Assessments
- Section Quizzes
- Chapter Test A
- Chapter Test B
- Chapter Test C
- Performance-Based Assessment
- Standardized Test Preparation

Labs and Activities
- Datasheets for In-Text Labs
- Datasheets for Quick Labs
- Datasheets for LabBook
- Vocabulary Activity
- SciLinks® Activity

Teacher Resources
- Teacher Notes for Performance-Based Assessment
- Lab Notes and Answers
- Answer Keys
- Lesson Plans
- Test Item Listing for ExamView® Test Generator
- Teaching Transparencies
- Chapter Starter Transparencies
- Bellringer Transparencies
- Concept Mapping Transparencies

Life Science

PROGRAM SCOPE AND SEQUENCE

Selecting the right books for your course is easy. Just review the topics presented in each book to determine the best match to your district curriculum.

C CELLS, HEREDITY, & CLASSIFICATION

Cells: The Basic Units of Life
- Cells, tissues, and organs
- Populations, communities, and ecosystems
- Cell theory
- Surface-to-volume ratio
- Prokaryotic versus eukaryotic cells
- Cell organelles

The Cell in Action
- Diffusion and osmosis
- Passive versus active transport
- Endocytosis versus exocytosis
- Photosynthesis
- Cellular respiration and fermentation
- Cell cycle

Heredity
- Dominant versus recessive traits
- Genes and alleles
- Genotype, phenotype, the Punnett square and probability
- Meiosis
- Determination of sex

Genes and Gene Technology
- Structure of DNA
- Protein synthesis
- Mutations
- Heredity disorders and genetic counseling

The Evolution of Living Things
- Adaptations and species
- Evidence for evolution
- Darwin's work and natural selection
- Formation of new species

The History of Life on Earth
- Geologic time scale and extinctions
- Plate tectonics
- Human evolution

Classification
- Levels of classification
- Cladistic diagrams
- Dichotomous keys
- Characteristics of the six kingdoms

D HUMAN BODY SYSTEMS & HEALTH

Body Organization and Structure
- Homeostasis
- Types of tissue
- Organ systems
- Structure and function of the skeletal system, muscular system, and integumentary system

Circulation and Respiration
- Structure and function of the cardiovascular system, lymphatic system, and respiratory system
- Respiratory disorders

The Digestive and Urinary Systems
- Structure and function of the digestive system
- Structure and function of the urinary system

Communication and Control
- Structure and function of the nervous system and endocrine system
- The senses
- Structure and function of the eye and ear

Reproduction and Development
- Asexual versus sexual reproduction
- Internal versus external fertilization
- Structure and function of the human male and female reproductive systems
- Fertilization, placental development, and embryo growth
- Stages of human life

Body Defenses and Disease
- Types of diseases
- Vaccines and immunity
- Structure and function of the immune system
- Autoimmune diseases, cancer, and AIDS

Staying Healthy
- Nutrition and reading food labels
- Alcohol and drug effects on the body
- Hygiene, exercise, and first aid

 E ENVIRONMENTAL SCIENCE

Interactions of Living Things
- Biotic versus abiotic parts of the environment
- Producers, consumers, and decomposers
- Food chains and food webs
- Factors limiting population growth
- Predator-prey relationships
- Symbiosis and coevolution

Cycles in Nature
- Water cycle
- Carbon cycle
- Nitrogen cycle
- Ecological succession

The Earth's Ecosystems
- Kinds of land and water biomes
- Marine ecosystems
- Freshwater ecosystems

Environmental Problems and Solutions
- Types of pollutants
- Types of resources
- Conservation practices
- Species protection

Energy Resources
- Types of resources
- Energy resources and pollution
- Alternative energy resources

Earth Science

WATER ON EARTH

 WEATHER AND CLIMATE

 ASTRONOMY

The Flow of Fresh Water
- Water cycle
- River systems
- Stream erosion
- Life cycle of rivers
- Deposition
- Aquifers, springs, and wells
- Ground water
- Water treatment and pollution

The Atmosphere
- Structure of the atmosphere
- Air pressure
- Radiation, convection, and conduction
- Greenhouse effect and global warming
- Characteristics of winds
- Types of winds
- Air pollution

Studying Space
- Astronomy
- Keeping time
- Types of telescope
- Radioastronomy
- Mapping the stars
- Scales of the universe

Exploring the Oceans
- Properties and characteristics of the oceans
- Features of the ocean floor
- Ocean ecology
- Ocean resources and pollution

Understanding Weather
- Water cycle
- Humidity
- Types of clouds
- Types of precipitation
- Air masses and fronts
- Storms, tornadoes, and hurricanes
- Weather forecasting
- Weather maps

Stars, Galaxies, and the Universe
- Composition of stars
- Classification of stars
- Star brightness, distance, and motions
- H-R diagram
- Life cycle of stars
- Types of galaxies
- Theories on the formation of the universe

The Movement of Ocean Water
- Types of currents
- Characteristics of waves
- Types of ocean waves
- Tides

Climate
- Weather versus climate
- Seasons and latitude
- Prevailing winds
- Earth's biomes
- Earth's climate zones
- Ice ages
- Global warming
- Greenhouse effect

Formation of the Solar System
- Birth of the solar system
- Structure of the sun
- Fusion
- Earth's structure and atmosphere
- Planetary motion
- Newton's Law of Universal Gravitation

A Family of Planets
- Properties and characteristics of the planets
- Properties and characteristics of moons
- Comets, asteroids, and meteoroids

Exploring Space
- Rocketry and artificial satellites
- Types of Earth orbit
- Space probes and space exploration

Physical Science

	K INTRODUCTION TO MATTER	**L** INTERACTIONS OF MATTER
CHAPTER 1	**The Properties of Matter** • Definition of matter • Mass and weight • Physical and chemical properties • Physical and chemical change • Density	**Chemical Bonding** • Types of chemical bonds • Valence electrons • Ions versus molecules • Crystal lattice
CHAPTER 2	**States of Matter** • States of matter and their properties • Boyle's and Charles's laws • Changes of state	**Chemical Reactions** • Writing chemical formulas and equations • Law of conservation of mass • Types of reactions • Endothermic versus exothermic reactions • Law of conservation of energy • Activation energy • Catalysts and inhibitors
CHAPTER 3	**Elements, Compounds, and Mixtures** • Elements and compounds • Metals, nonmetals, and metalloids (semiconductors) • Properties of mixtures • Properties of solutions, suspensions, and colloids	**Chemical Compounds** • Ionic versus covalent compounds • Acids, bases, and salts • pH • Organic compounds • Biomolecules
CHAPTER 4	**Introduction to Atoms** • Atomic theory • Atomic model and structure • Isotopes • Atomic mass and mass number	**Atomic Energy** • Properties of radioactive substances • Types of decay • Half-life • Fission, fusion, and chain reactions
CHAPTER 5	**The Periodic Table** • Structure of the periodic table • Periodic law • Properties of alkali metals, alkaline-earth metals, halogens, and noble gases	
CHAPTER 6		

M FORCES, MOTION, AND ENERGY

Matter in Motion
- Speed, velocity, and acceleration
- Measuring force
- Friction
- Mass versus weight

Forces in Motion
- Terminal velocity and free fall
- Projectile motion
- Inertia
- Momentum

Forces in Fluids
- Properties in fluids
- Atmospheric pressure
- Density
- Pascal's principle
- Buoyant force
- Archimedes' principle
- Bernoulli's principle

Work and Machines
- Measuring work
- Measuring power
- Types of machines
- Mechanical advantage
- Mechanical efficiency

Energy and Energy Resources
- Forms of energy
- Energy conversions
- Law of conservation of energy
- Energy resources

Heat and Heat Technology
- Heat versus temperature
- Thermal expansion
- Absolute zero
- Conduction, convection, radiation
- Conductors versus insulators
- Specific heat capacity
- Changes of state
- Heat engines
- Thermal pollution

N ELECTRICITY AND MAGNETISM

Introduction to Electricity
- Law of electric charges
- Conduction versus induction
- Static electricity
- Potential difference
- Cells, batteries, and photocells
- Thermocouples
- Voltage, current, and resistance
- Electric power
- Types of circuits

Electromagnetism
- Properties of magnets
- Magnetic force
- Electromagnetism
- Solenoids and electric motors
- Electromagnetic induction
- Generators and transformers

Electronic Technology
- Properties of semiconductors
- Integrated circuits
- Diodes and transistors
- Analog versus digital signals
- Microprocessors
- Features of computers

O SOUND AND LIGHT

The Energy of Waves
- Properties of waves
- Types of waves
- Reflection and refraction
- Diffraction and interference
- Standing waves and resonance

The Nature of Sound
- Properties of sound waves
- Structure of the human ear
- Pitch and the Doppler effect
- Infrasonic versus ultrasonic sound
- Sound reflection and echolocation
- Sound barrier
- Interference, resonance, diffraction, and standing waves
- Sound quality of instruments

The Nature of Light
- Electromagnetic waves
- Electromagnetic spectrum
- Law of reflection
- Absorption and scattering
- Reflection and refraction
- Diffraction and interference

Light and Our World
- Luminosity
- Types of lighting
- Types of mirrors and lenses
- Focal point
- Structure of the human eye
- Lasers and holograms

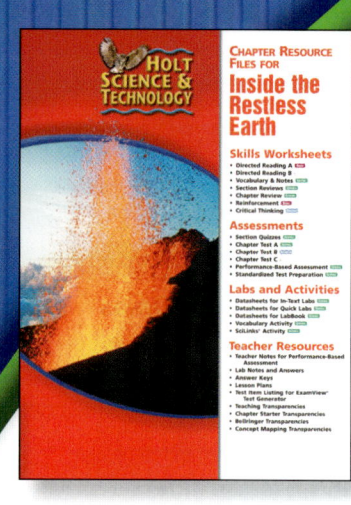

Program resources make teaching and learning easier.

CHAPTER RESOURCES

A *Chapter Resources book* accompanies each of the 15 *Short Courses*. Here you'll find everything you need to make sure your students are getting the most out of learning science—all in one book.

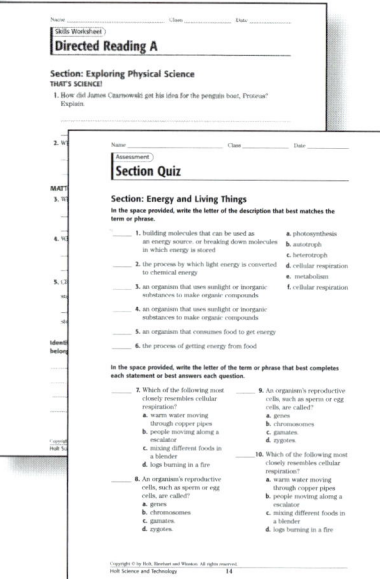

Skills Worksheets

- Directed Reading A: Basic
- Directed Reading B: Special Needs
- Vocabulary and Chapter Summary
- Section Reviews
- Chapter Reviews
- Reinforcement
- Critical Thinking

Labs & Activities

- Datasheets for Chapter Labs
- Datasheets for Quick Labs
- Datasheets for LabBook
- Vocabulary Activity
- SciLinks® Activity

Assessments

- Section Quizzes
- Chapter Tests A: General
- Chapter Tests B: Advanced
- Chapter Tests C: Special Needs
- Performance-Based Assessments
- Standardized Test Preparation

Teacher Resources

- Lab Notes and Answers
- Teacher Notes for Performance-Based Assessment
- Answer Keys
- Lesson Plans
- Test Item Listing for ExamView® Test Generator
- Full-color **Teaching Transparencies**, plus section **Bellringers, Concept Mapping,** and **Chapter Starter Transparencies.**

SPANISH RESOURCES

Spanish materials are available for each *Short Course:*

- *Student Edition*
- *Spanish Resources* booklet contains worksheets and assessments translated into Spanish with an English **Answer Key.**
- **Guided Reading Audio CD Program**

ONLINE RESOURCES

- *Enhanced Online Editions* engage students and assist teachers with a host of interactive features that are available anytime and anywhere you can connect to the Internet.
- **CNNStudentNews.com** provides award-winning news and information for both teachers and students.
- **SciLinks**—a Web service developed and maintained by the National Science Teachers Association—links you and your students to up-to-date online resources directly related to chapter topics.
- **go.hrw.com** links you and your students to online chapter activities and resources.
- **Current Science** articles relate to students' lives.

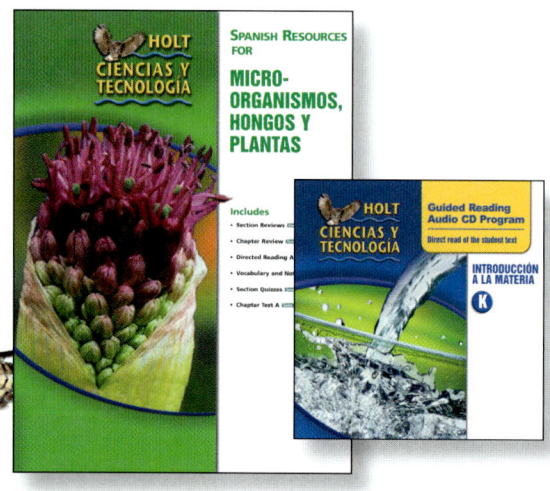

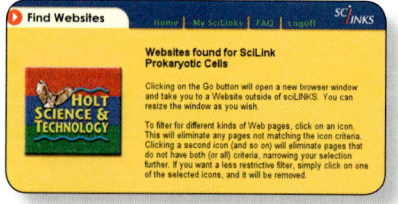

ADDITIONAL LAB AND SKILLS RESOURCES

- *Calculator-Based Labs* incorporates scientific instruments, offering students insight into modern scientific investigation.
- *EcoLabs & Field Activities* develops awareness of the natural world.
- *Holt Science Skills Workshop: Reading in the Content Area* contains exercises that target reading skills key.
- *Inquiry Labs* taps students' natural curiosity and creativity with a focus on the process of discovery.
- *Labs You Can Eat* safely incorporates edible items into the classroom.
- *Long-Term Projects & Research Ideas* extends and enriches lessons.
- *Math Skills for Science* provides additional explanations, examples, and math problems so students can develop their skills.
- *Science Skills Worksheets* helps your students hone important learning skills.
- *Whiz-Bang Demonstrations* gets your students' attention at the beginning of a lesson.

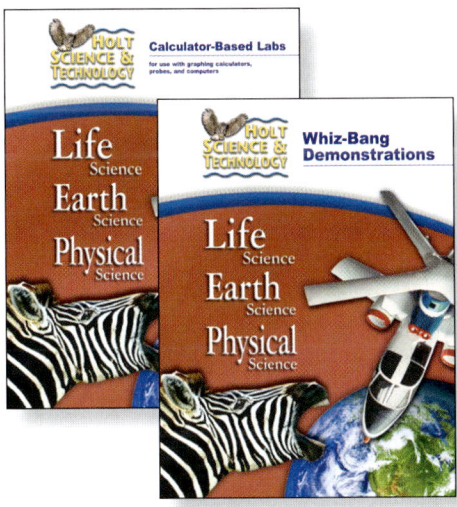

ADDITIONAL RESOURCES

- *Assessment Checklists & Rubrics* gives you guidelines for evaluating students' progress.
- *Holt Anthology of Science Fiction* sparks your students' imaginations with thought-provoking stories.
- *Holt Science Posters* visually reinforces scientific concepts and themes with seven colorful posters including **The Periodic Table of the Elements.**

- *Professional Reference for Teachers* contains professional articles that discuss a variety of topics, such as classroom management.
- *Program Introduction Resource File* explains the program and its features and provides several additional references, including lab safety, scoring rubrics, and more.
- *Science Fair Guide* gives teachers, students, and parents tips for planning and assisting in a science fair.
- *Science Puzzlers, Twisters & Teasers* activities challenge students to think about science concepts in different ways.

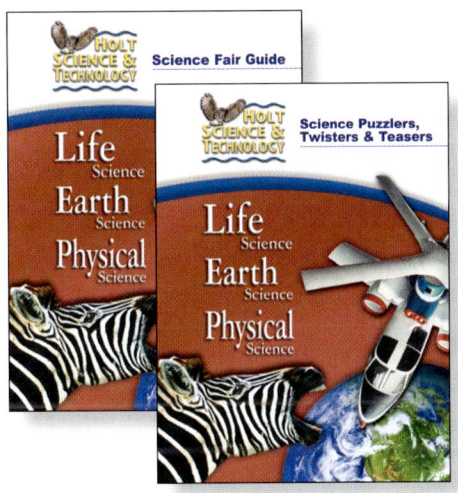

TECHNOLOGY RESOURCES

- *CNN Presents Science in the News: Video Library* helps students see the impact of science on their everyday lives with actual news video clips.
  - Multicultural Connections
  - Science, Technology & Society
  - Scientists in Action
  - Eye on the Environment
- *Guided Reading Audio CD Program*, available in English and Spanish, provides students with a direct read of each section.
- *HRW Earth Science Videotape* takes your students on a geology "field trip" with full-motion video.
- *Interactive Explorations CD-ROM Program* develops students' inquiry and decision-making skills as they investigate science phenomena in a virtual lab setting.

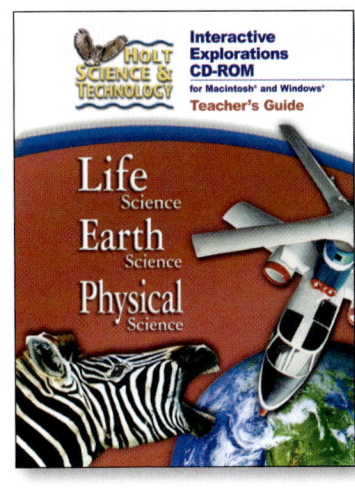

- *One-Stop Planner CD-ROM*® organizes everything you need on one disc, including printable worksheets, customizable lesson plans, a powerful test generator, **PowerPoint® LectureNotes, Lab Materials QuickList Software, Holt Calendar Planner, Interactive Teacher Edition,** and more.
- *Science Tutor CD-ROMs* help students practice what they learn with immediate feedback.
- *Lab Videos* make it easier to integrate more experiments into your lessons without the preparation time and costs. Available on DVD and VHS.
- **Brain Food Video Quizzes** are game-show style quizzes that assess students' progress. Available on DVD and VHS.
- *Visual Concepts CD-ROMs* include graphics, animations, and movie clips that demonstrate key chapter concepts.

Science and Math Worksheets

The **Holt Science & Technology** program helps you meet the needs of a wide variety of students, regardless of their skill level. The following pages provide examples of the worksheets available to improve your students' science and math skills whether they already have a strong science and math background or are weak in these areas. Samples of assessment checklists and rubrics are also provided.

In addition to the skills worksheets represented here, **Holt Science & Technology** provides a variety of worksheets that are correlated directly with each chapter of the program. Representations of these worksheets are found at the beginning of each chapter in this *Teacher Edition*.

Many worksheets are also available on the Holt Web site. The address is **go.hrw.com**.

Science Skills Worksheets: Thinking Skills

BEING FLEXIBLE
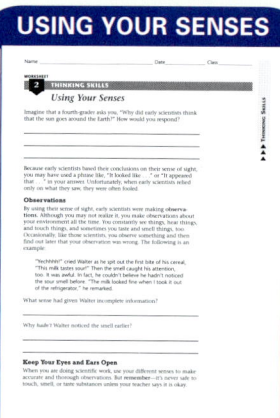

USING YOUR SENSES
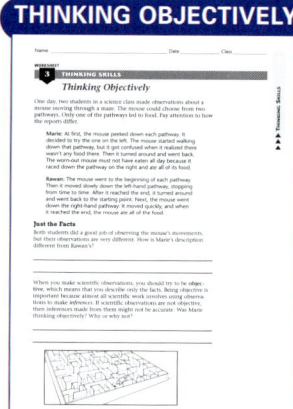

THINKING OBJECTIVELY
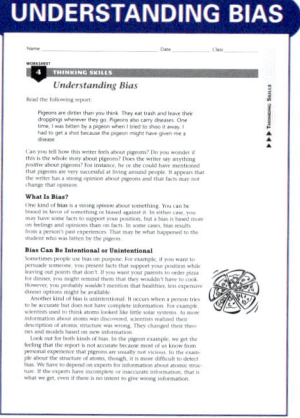

UNDERSTANDING BIAS

USING LOGIC
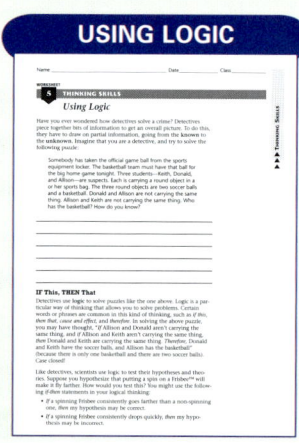

BOOSTING YOUR MEMORY
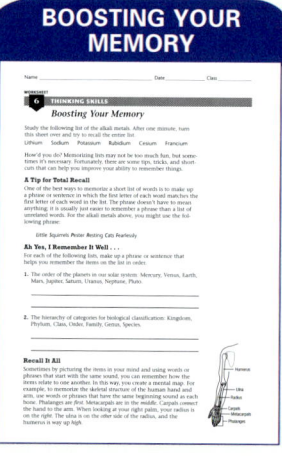

IMPROVING YOUR STUDY HABITS
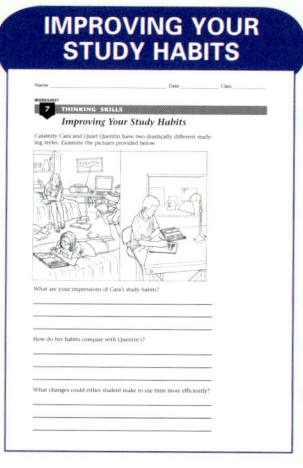

READING A SCIENCE TEXTBOOK
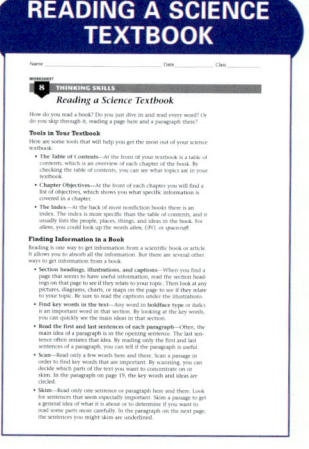

Science Skills Worksheets: Experimenting Skills

SAFETY RULES!

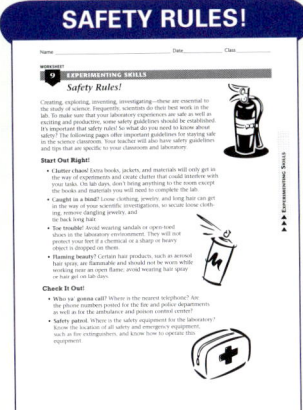

DOING A LAB WRITE-UP

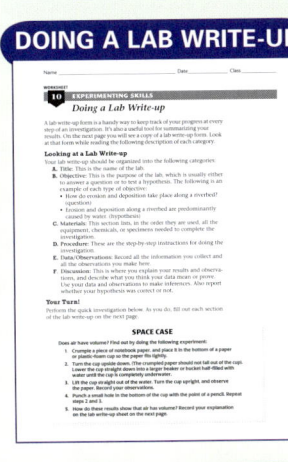

UNDERSTANDING VARIABLES

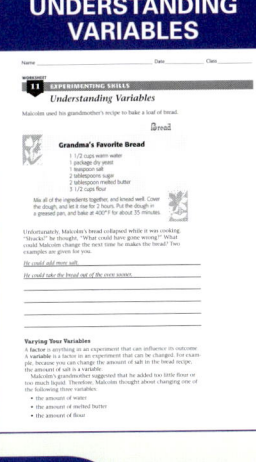

WORKING WITH HYPOTHESES

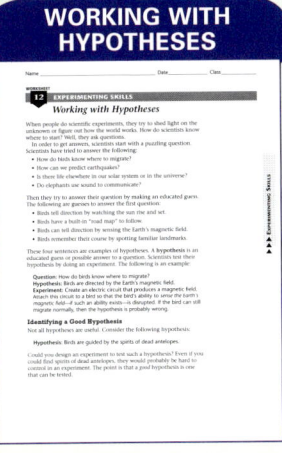

DESIGNING AN EXPERIMENT

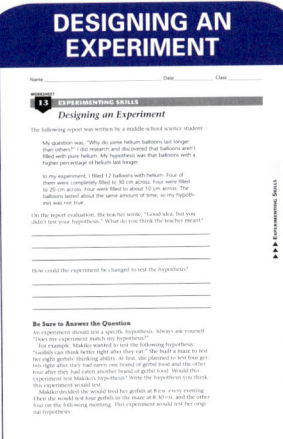

USING THE INTERNATIONAL SYSTEM OF UNITS (SI)

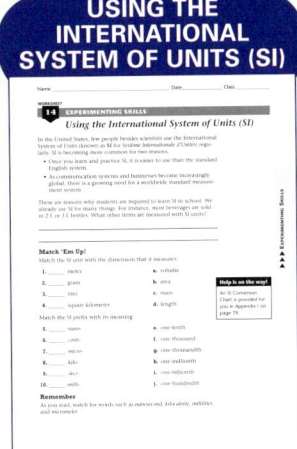

MEASURING

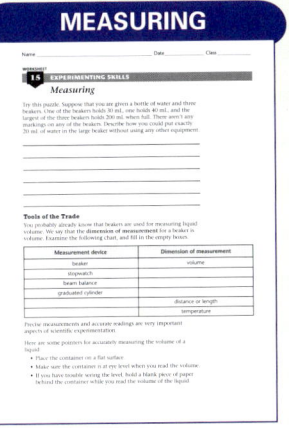

Science Skills Worksheets: Researching Skills

CHOOSING YOUR TOPIC

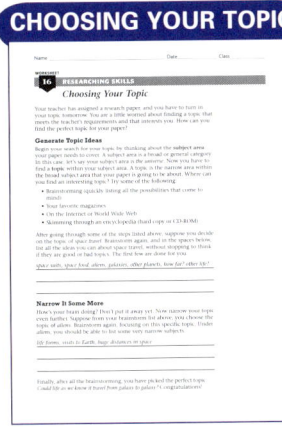

ORGANIZING YOUR RESEARCH

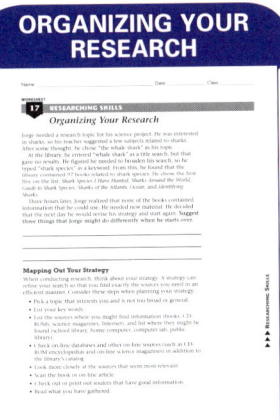

FINDING USEFUL SOURCES

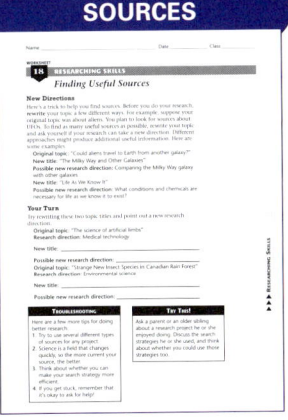

RESEARCHING ON THE WEB

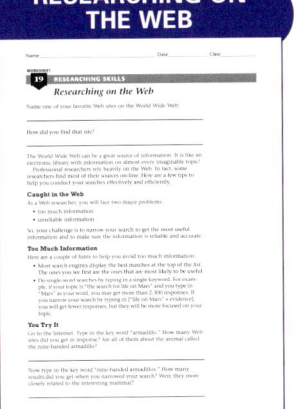

Science Skills Worksheets: Researching Skills (continued)

IDENTIFYING BIAS

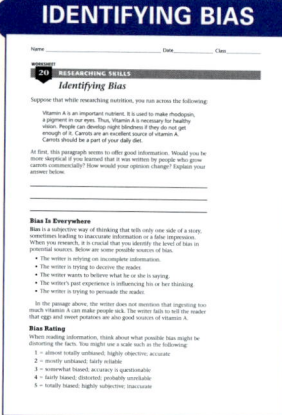

TAKING NOTES

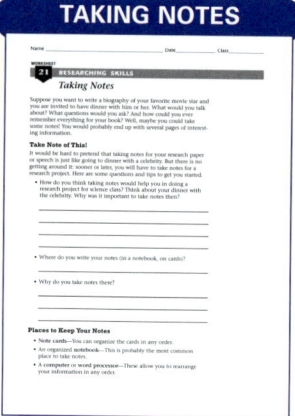

Science Skills Worksheets: Communicating Skills

SCIENCE WRITING

SCIENCE DRAWING

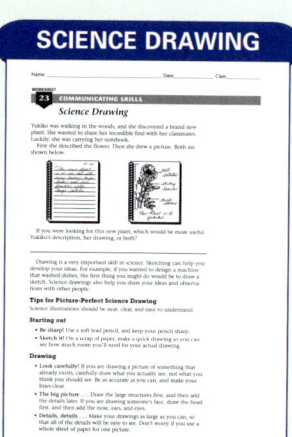

USING MODELS TO COMMUNICATE

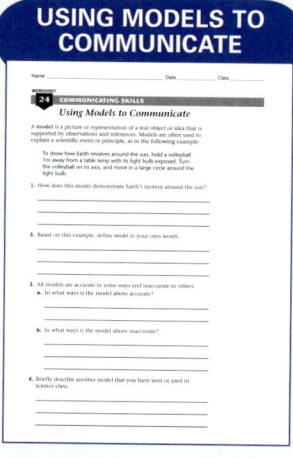

INTRODUCTION TO GRAPHS

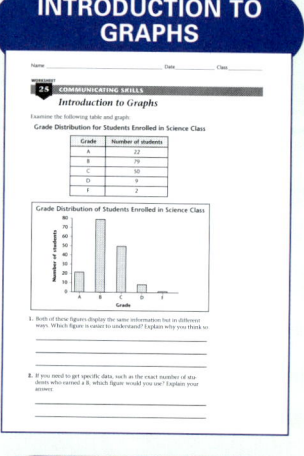

GRASPING GRAPHING

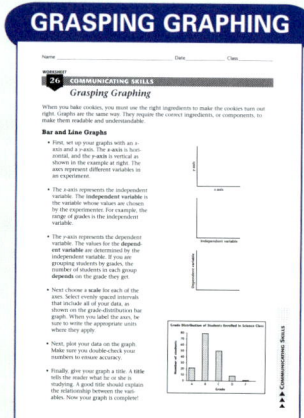

INTERPRETING YOUR DATA

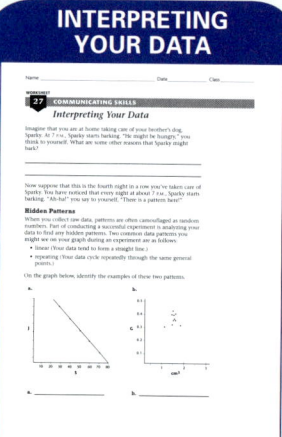

RECOGNIZING BIAS IN GRAPHS

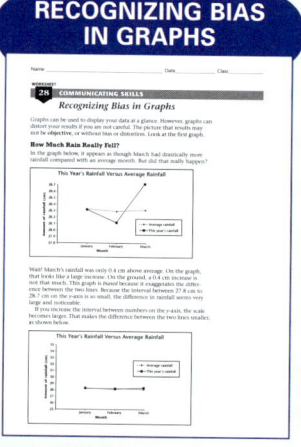

MAKING DATA MEANINGFUL

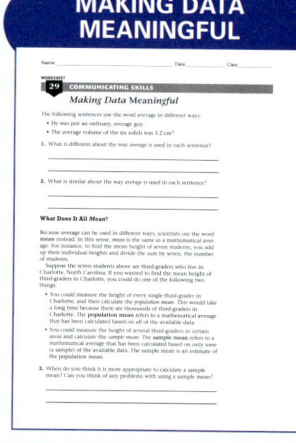

HINTS FOR ORAL PRESENTATIONS

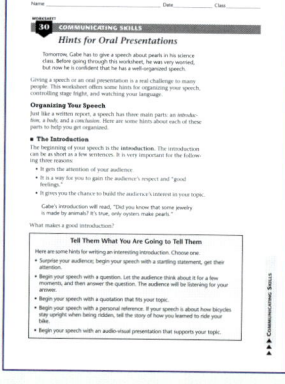

Math Skills for Science

ADDITION AND SUBTRACTION

MULTIPLICATION

DIVISION

AVERAGES

POSITIVE AND NEGATIVE NUMBERS

FRACTIONS

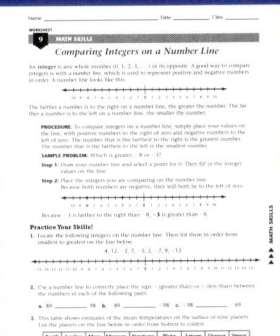

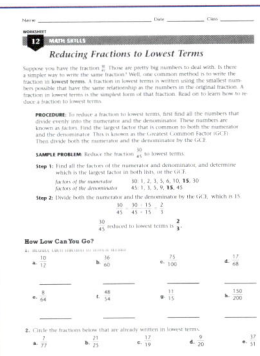

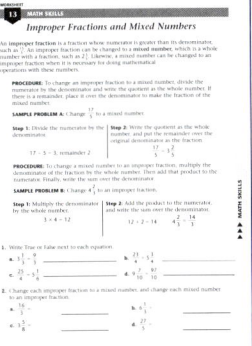

Math Skills for Science (continued)

RATIOS AND PROPORTIONS

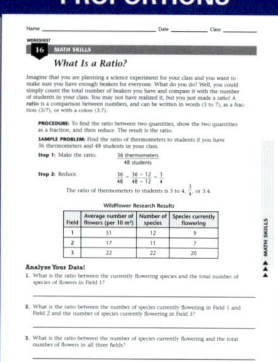

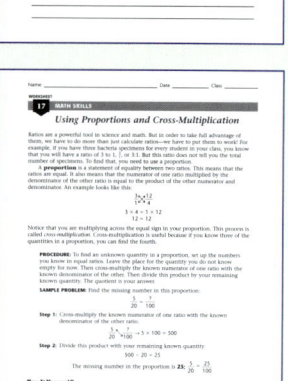

DECIMALS

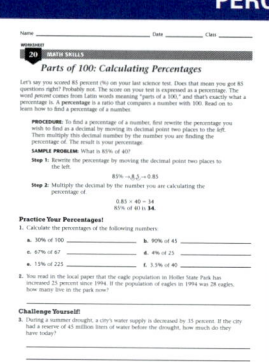

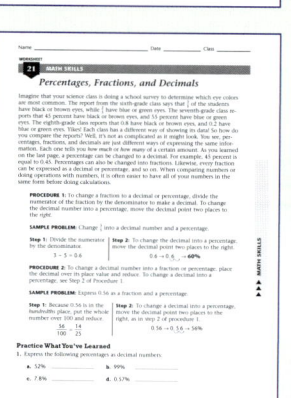

PERCENTAGES

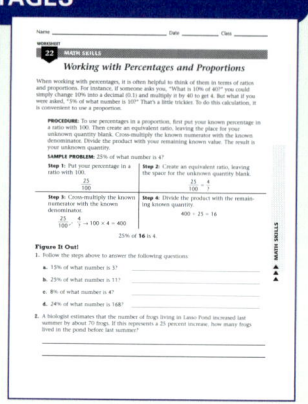

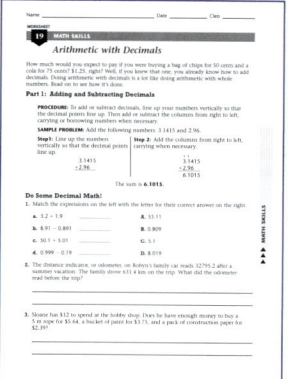

POWERS OF 10

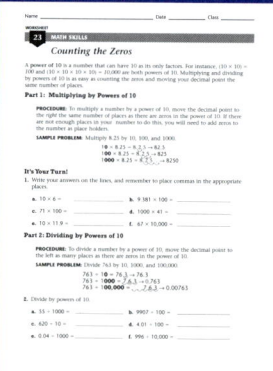

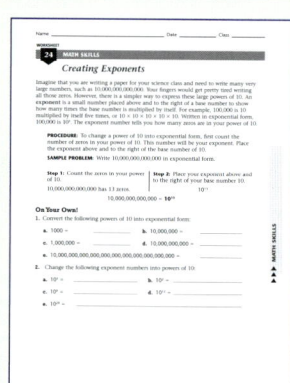

SCIENTIFIC NOTATION

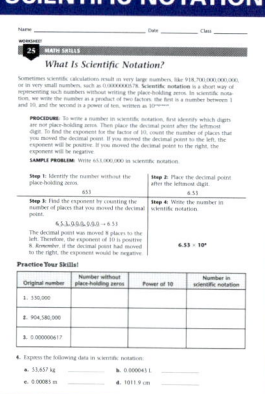

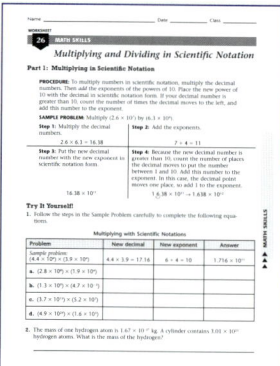

SI MEASUREMENT AND CONVERSION

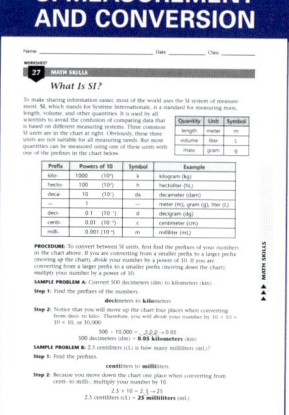

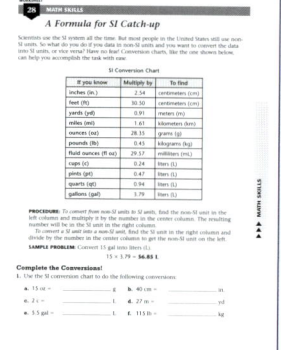

Math Skills for Science (continued)

GEOMETRY

THE UNIT FACTOR AND DIMENSIONAL ANALYSIS

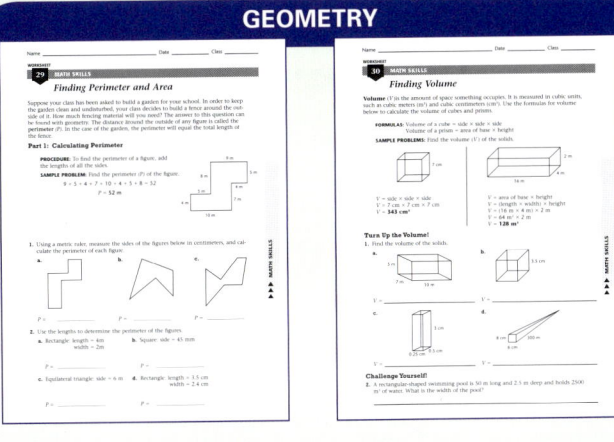

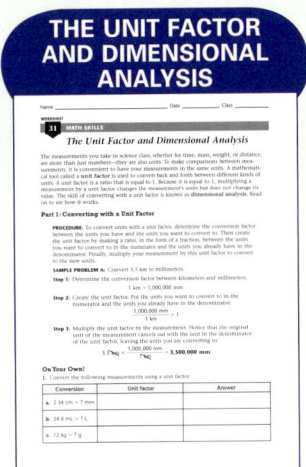

MATH IN SCIENCE: INTEGRATED SCIENCE

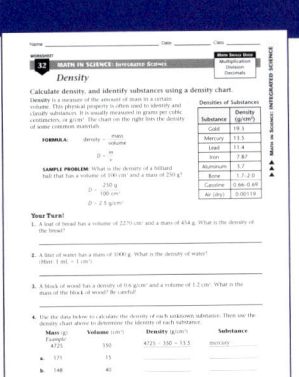

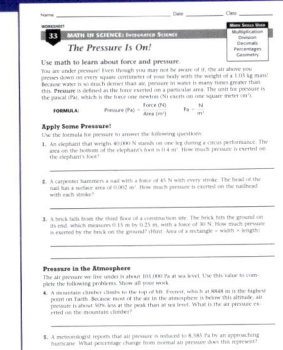

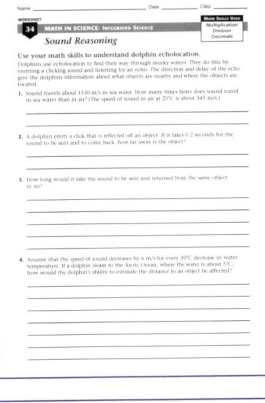

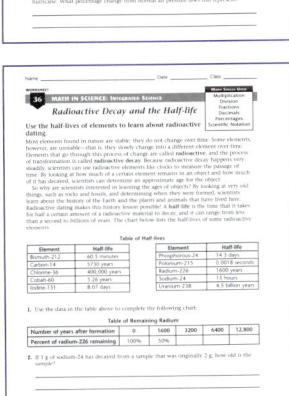

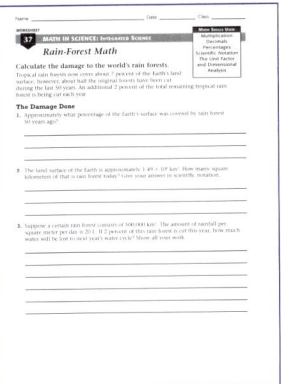

Math Skills for Science (continued)

MATH IN SCIENCE: LIFE SCIENCE

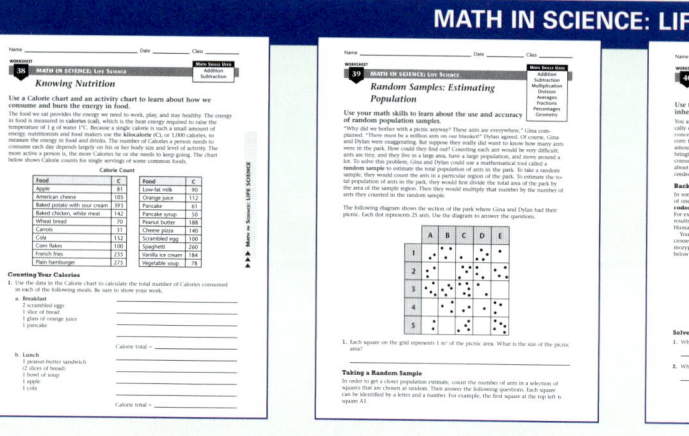
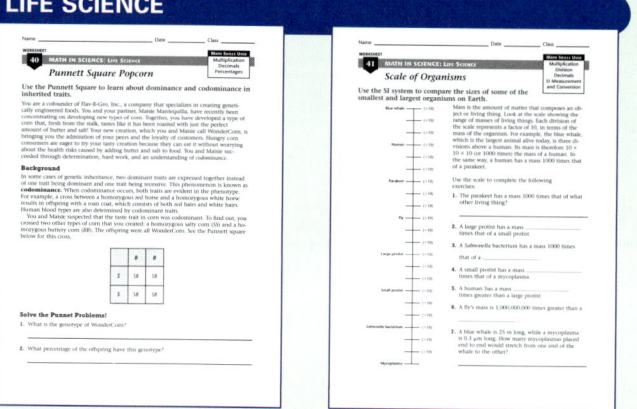

38 — Knowing Nutrition

39 — Random Samples: Estimating Population

40 — Punnett Square Popcorn

41 — Scale of Organisms

MATH IN SCIENCE: EARTH SCIENCE

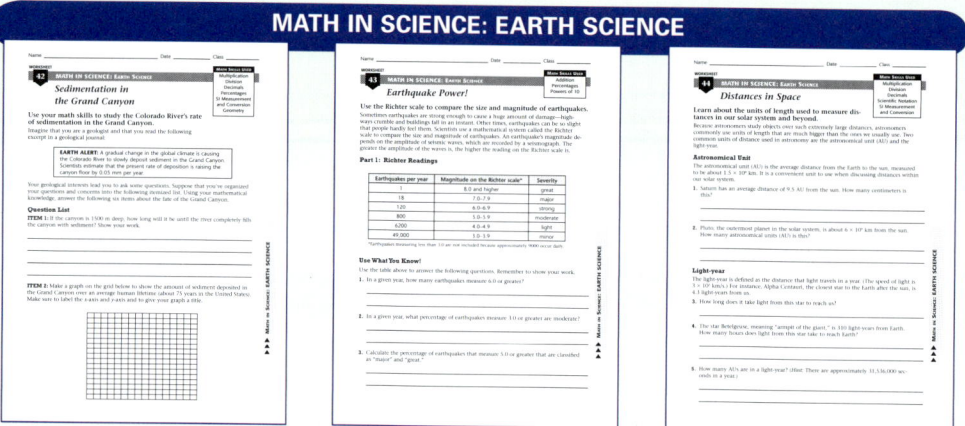

42 — Sedimentation in the Grand Canyon

43 — Earthquake Power!

44 — Distances in Space

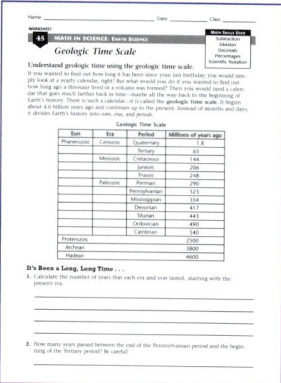
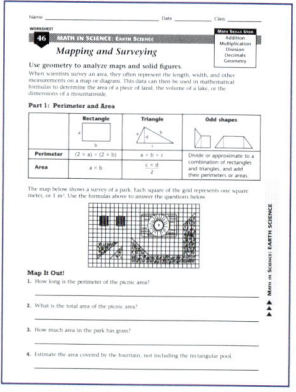

45 — Geologic Time Scale

46 — Mapping and Surveying

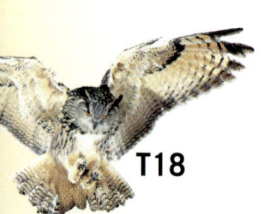

Math Skills for Science (continued)

MATH IN SCIENCE: PHYSICAL SCIENCE

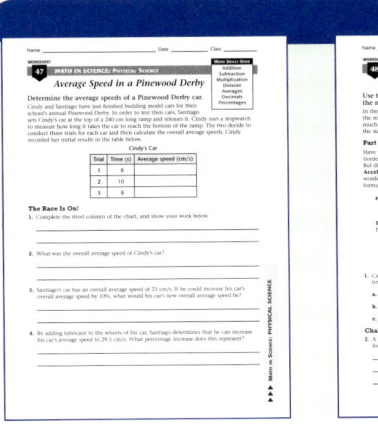
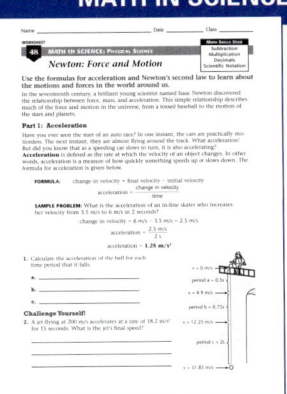
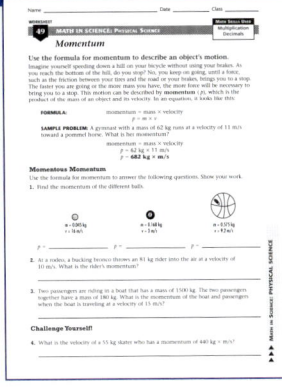
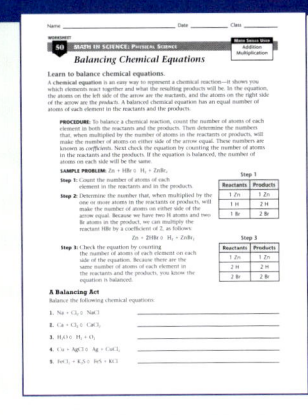

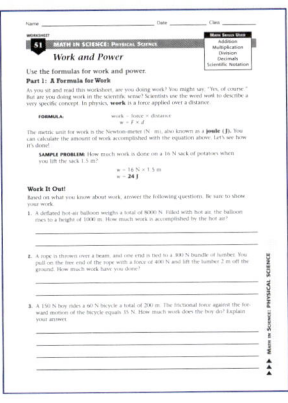
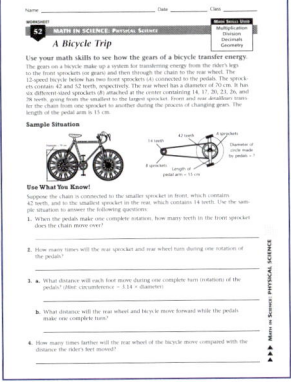
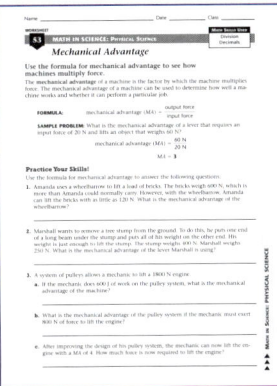
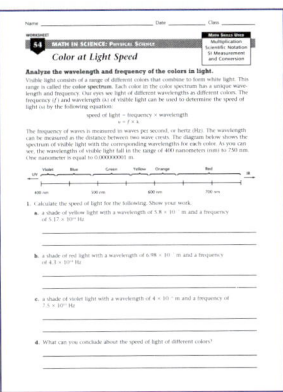

Assessment Checklist & Rubrics

The following is just a sample of over 50 checklists and rubrics contained in this booklet.

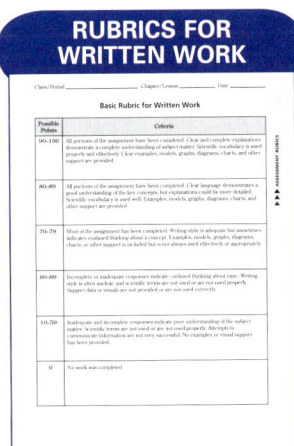
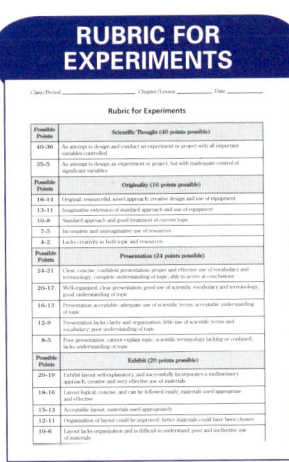
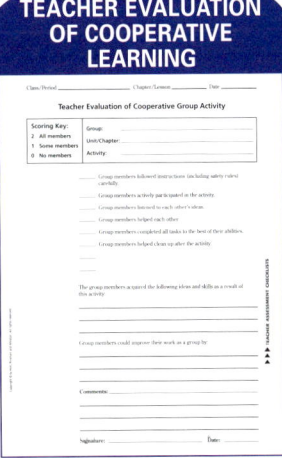

RUBRICS FOR WRITTEN WORK

RUBRIC FOR EXPERIMENTS

TEACHER EVALUATION OF COOPERATIVE LEARNING

TEACHER EVALUATION OF STUDENT PROGRESS

National Science Education Standards

The following lists show the chapter correlation of *Holt Science & Technology: Interactions of Matter* with the *National Science Education Standards* (grades 5–8).

Unifying Concepts and Processes

Standard	Chapter Correlation	
Systems, order, and organization Code: UCP 1	Chapter 1	1.1, 1.2, 1.3
	Chapter 3	3.1, 3.2, 3.3, 3.4
	Chapter 4	4.1
Evidence, models, and explanation Code: UCP 2	Chapter 1	1.1, 1.2, 1.3
	Chapter 3	3.3
	Chapter 4	4.1, 4.2
Change, constancy, and measurement Code: UCP 3	Chapter 1	1.2, 1.3
	Chapter 2	2.1, 2.2, 2.3, 2.4
	Chapter 4	4.1, 4.2

Science as Inquiry

Standard	Chapter Correlation	
Abilities necessary to do scientific inquiry Code: SAI 1	Chapter 1	1.3
	Chapter 2	2.2, 2.4
	Chapter 3	3.2, 3.3, 3.4
Understandings about scientific inquiry Code: SAI 2	Chapter 3	3.1
	Chapter 4	4.1

Science and Technology

Standard	Chapter Correlation	
Abilities of technological design Code: ST 1	Chapter 4	4.2
Understandings about science and technology Code: ST 2	Chapter 3	3.3, 3.4
	Chapter 4	4.1, 4.2

Science in Personal and Social Perspectives

Standard	Chapter Correlation	
Personal health Code: SPSP 1	Chapter 3	3.2, 3.3
	Chapter 4	4.1
Natural hazards Code: SPSP 3	Chapter 3	3.2
	Chapter 4	4.2
Risks and benefits Code: SPSP 4	Chapter 3	3.2
	Chapter 4	4.1, 4.2
Science and technology in society Code: SPSP 5	Chapter 4	4.1, 4.2

History and Nature of Science

Standard	Chapter Correlation	
Science as a human endeavor Code: HNS 1	Chapter 4	4.1, 4.2
Nature of science Code: HNS 2	Chapter 4	4.1, 4.2
History of science Code: HNS 3	Chapter 4	4.1, 4.2

Physical Science Content Standards

Properties and changes of properties in matter

Standard	Chapter Correlation	
A substance has characteristic properties, such as density, a boiling point, and solubility, all of which are independent of the amount of the sample. A mixture of substances often can be separated into the original substances using one or more of the characteristic properties. Code: PS 1a	**Chapter 3**	3.1, 3.2, 3.4
Substances react chemically in characteristic ways with other substances to form new substances (compounds) with different characteristic properties. In chemical reactions, the total mass is conserved. Substances often are placed in categories or groups if they react in similar ways; metals is an example of such a group. Code: PS 1b	**Chapter 1** **Chapter 2** **Chapter 3**	1.1, 1.2, 1.3 2.1, 2.2, 2.3 3.1, 3.2, 3.3, 3.4
Chemical elements do not break down during normal laboratory reactions involving such treatments as heating, exposure to electric current, or reaction with acids. There are more than 100 known elements that combine in a multitude of ways to produce compounds, which account for the living and nonliving substances that we encounter. Code: PS 1c	**Chapter 1** **Chapter 3**	1.1 3.4

Transfer of energy

Standard	Chapter Correlation	
Energy is a property of many substances and is associated with heat, light, electricity, mechanical motion, sound, nuclei, and the nature of a chemical. Energy is transferred in many ways. Code: PS 3a	**Chapter 1** **Chapter 2** **Chapter 4**	1.2 2.1, 2.4 4.1, 4.2
In most chemical and nuclear reactions, energy is transferred into or out of a system. Heat, light, mechanical motion, or electricity might all be involved in such transfers. Code: PS 3e	**Chapter 1** **Chapter 2** **Chapter 4**	1.2 2.1, 2.4 4.1, 4.2

HOLT SCIENCE & TECHNOLOGY

Interactions of Matter

HOLT, RINEHART AND WINSTON

A Harcourt Education Company

Orlando • Austin • New York • San Diego • Toronto • London

Acknowledgments

Contributing Authors

Christie Borgford, Ph.D.
Assistant Professor of Chemistry
Department of Chemistry
The University of Alabama
Birmingham, Alabama

Sally Ann Vonderbrink, Ph.D.
Chemistry Teacher (retired)
Cincinnati, Ohio

Inclusion Specialist

Ellen McPeek Glisan
Special Needs Consultant
San Antonio, Texas

Safety Reviewer

Jack Gerlovich, Ph.D.
Associate Professor
School of Education
Drake University
Des Moines, Iowa

Academic Reviewers

Scott Darveau, Ph.D.
Associate Professor of Chemistry
Chemistry Department
University of Nebraska at Kearney
Kearney, Nebraska

Richard F. Niedziela, Ph.D.
Assistant Professor of Chemistry
Department of Chemistry
DePaul University
Chicago, Illinois

Kate Queeney, Ph.D.
Assistant Professor of Chemistry
Chemistry Department
Smith College
Northampton, Massachusetts

Fred Seaman, Ph.D.
Retired Research Associate
College of Pharmacy
The University of Texas at Austin
Austin, Texas

Richard S. Treptow, Ph.D.
Professor of Chemistry
Department of Chemistry and Physics
Chicago State University
Chicago, Illinois

Dale Wheeler
Assistant Professor of Chemistry
A. R. Smith Department of Chemistry
Appalachian State University
Boone, North Carolina

Lab Testing

Paul Boyle
Science Teacher
Perry Heights Middle School
Evansville, Indiana

Rebecca Ferguson
Science Teacher
North Ridge Middle School
North Richland Hills, Texas

Laura Fleet
Science Teacher
Alice B. Landrum Middle School
Ponte Verde Beach, Florida

Dennis Hanson
Science Teacher and Dept. Chair
Big Bear Middle School
Big Bear Lake, California

Tracy Jahn
Science Teacher
Berkshire Junior-Senior High School
Canaan, New York

Rodney A. Sandefur
Science Teacher
Naturita Middle School
Naturita, Colorado

Larry Tackett
Science Teacher and Dept. Chair
R. H. Terrell Junior High School
Washington, D.C.

L Interactions of Matter

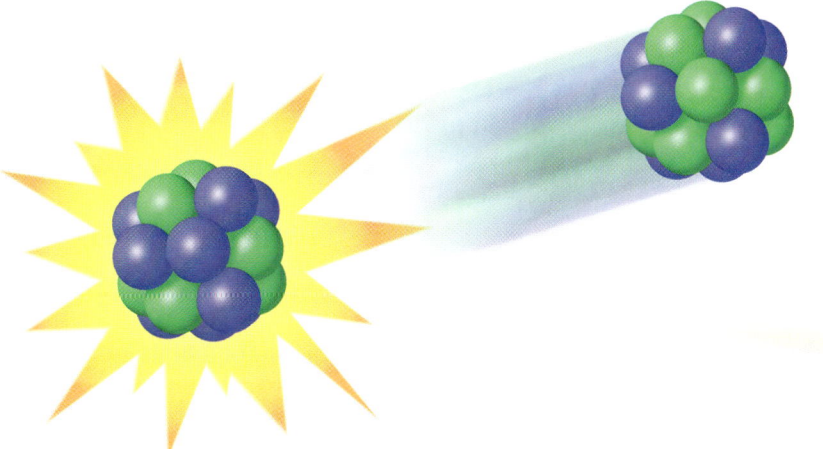

Skills Development

Connection to . . .

Science in Action

How to Use Your Textbook

Your Roadmap for Success with Holt Science and Technology

Reading Warm-Up

A Reading Warm-Up at the beginning of every section provides you with the section's objectives and key terms. The objectives tell you what you'll need to know after you finish reading the section.

Key terms are listed for each section. Learn the definitions of these terms because you will most likely be tested on them. Each key term is highlighted in the text and is defined at point of use and in the margin. You can also use the glossary to locate definitions quickly.

STUDY TIP Reread the objectives and the definitions to the key terms when studying for a test to be sure you know the material.

Get Organized

A Reading Strategy at the beginning of every section provides tips to help you organize and remember the information covered in the section. Keep a science notebook so that you are ready to take notes when your teacher reviews the material in class. Keep your assignments in this notebook so that you can review them when studying for the chapter test.

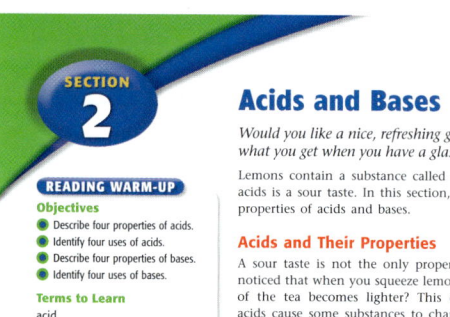

SECTION 2

Acids and Bases

Would you like a nice, refreshing glass of acid? This is just what you get when you have a glass of lemonade.

Lemons contain a substance called an *acid*. One property of acids is a sour taste. In this section, you will learn about the properties of acids and bases.

READING WARM-UP

Objectives
- Describe four properties of acids.
- Identify four uses of acids.
- Describe four properties of bases.
- Identify four uses of bases.

Terms to Learn
acid
indicator
base

READING STRATEGY

Reading Organizer As you read this section, make a table comparing acids and bases.

acid any compound that increases the number of hydronium ions when dissolved in water

Acids and Their Properties

A sour taste is not the only property of an acid. Have you noticed that when you squeeze lemon juice into tea, the color of the tea becomes lighter? This change happens because acids cause some substances to change color. An **acid** is any compound that increases the number of hydronium ions, H_3O^+, when dissolved in water. Hydronium ions form when a hydrogen ion, H^+, separates from the acid and bonds with a water molecule, H_2O, to form a hydronium ion, H_3O^+.

✓ **Reading Check** How is a hydronium ion formed? (*See the Appendix for answers to Reading Checks.*)

Acids Have a Sour Flavor

Have you ever taken a bite of a lemon or lime? If so, like the boy in **Figure 1**, you know the sour taste of an acid. The taste of lemons, limes, and other citrus fruits is a result of citric acid. However, taste, touch, or smell should NEVER be used to identify an unknown chemical. Many acids are *corrosive*, which means that they destroy body tissue, clothing, and many other things. Most acids are also poisonous.

Figure 1 Foods that have a sour taste usually contain acids.

NEVER touch or taste a concentrated solution of a strong acid.

422   Chapter 15

Be Resourceful—Use the Web

Internet Connect

boxes in your textbook take you to resources that you can use for science projects, reports, and research papers. Go to scilinks.org, and type in the SciLinks code to get information on a topic.

Visit go.hrw.com

Find worksheets, **Current Science**® magazine articles online, and other materials that go with your textbook at **go.hrw.com.** Click on the textbook icon and the table of contents to see all of the resources for each chapter.

Figure 2 Detecting Acids with Indicators

The indicator, bromthymol blue, is pale blue in water.

When acid is added, the color changes to yellow because of the presence of the indicator.

indicator a compound that can reversibly change color depending on the pH of the solution or other chemical change

Acids Change Colors in Indicators

A substance that changes color in the presence of an acid or base is an **indicator**. Look at **Figure 2**. The flask on the left contains water and an indicator called *bromthymol blue* (BROHM THIE MAWL BLOO). Acid has been added to the flask on the right. The color changes from pale blue to yellow because the indicator detects the presence of an acid.

Another indicator commonly used in the lab is litmus. Paper strips containing litmus are available in both blue and red. When an acid is added to blue litmus paper, the color of the litmus changes to red.

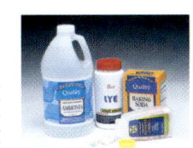

Acids React with Metals

Acids react with some metals to produce hydrogen gas. For example, hy
hydrogen ga
reaction is t

In this re
hydrochlorit
is an active
hydrochlorit
active metal

Uses of Bases

Like acids, bases have many uses. Sodium hydroxide is a base used to make soap and paper. It is also used in oven cleaners and in products that unclog drains. Calcium hydroxide, $Ca(OH)_2$, is used to make cement and plaster. Ammonia is found in many household cleaners and is used to make fertilizers. And magnesium hydroxide and aluminum hydroxide are used in antacids to treat heartburn. **Figure 7** shows some of the many products that contain bases. Carefully follow the safety instructions when using these products. Remember that bases can harm your skin.

Figure 7 Bases are common around the house. They are useful as cleaning agents, as cooking aids, and as medicines.

Reading Check What three ways can bases be used at home?

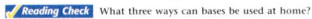

SECTION Review

Summary

- An acid is a compound that increases the number of hydronium ions in solution.
- Acids taste sour, turn blue litmus paper red, react with metals to produce hydrogen gas, and may conduct an electric current when in solution.
- Acids are used for industrial purposes and in household products.
- A base is a compound that increases the number of hydroxide ions in solution.
- Bases taste bitter, feel slippery, and turn red litmus paper blue. Most solutions of bases conduct an electric current.
- Bases are used in cleaning products and acid neutralizers.

Using Key Terms

1. In your own words, write a definition for each of the following terms: *acid, base,* and *indicator*.

Understanding Key Ideas

2. A base is a substance that
   a. feels slippery.
   b. tastes sour.
   c. reacts with metals to produce hydrogen gas.
   d. turns blue litmus paper red.

3. Acids are important in
   a. making antacids.
   b. preparing detergents.
   c. keeping algae out of swimming pools.
   d. manufacturing cement.

4. What happens to red litmus paper when it touches a base?

Math Skills

5. A cake recipe calls for 472 mL of milk. You don't have a metric measuring cup at home, so you need to convert milliliters to cups. You know that 1 L equals 1.06 quarts and that there are 4 cups in 1 quart. How many cups of milk will you need to use?

Critical Thinking

6. **Making Comparisons** Compare the properties of acids and bases.

7. **Applying Concepts** Why would it be useful for a gardener or a vegetable farmer to use litmus paper to test soil samples?

8. **Analyzing Processes** Suppose that your teacher gives you a solution of an unknown chemical. The chemical is either an acid or a base. You know that touching or tasting acids and bases is not safe. What two tests could you perform on the chemical to determine whether it is an acid or a base? What results would help you decide if the chemical was an acid or a base?

SCiLINKS.

For a variety of links related to this chapter, go to www.scilinks.org

Topic: Acids and Bases
SciLinks code: HSM0013

427

Use the Illustrations and Photos

Art shows complex ideas and processes. Learn to analyze the art so that you better understand the material you read in the text.

Tables and graphs display important information in an organized way to help you see relationships.

A picture is worth a thousand words. Look at the photographs to see relevant examples of science concepts that you are reading about.

Answer the Section Reviews

Section Reviews test your knowledge of the main points of the section. Critical Thinking items challenge you to think about the material in greater depth and to find connections that you infer from the text.

STUDY TIP When you can't answer a question, reread the section. The answer is usually there.

Do Your Homework

Your teacher may assign worksheets to help you understand and remember the material in the chapter.

STUDY TIP Don't try to answer the questions without reading the text and reviewing your class notes. A little preparation up front will make your homework assignments a lot easier. Answering the items in the Chapter Review will help prepare you for the chapter test.

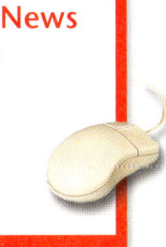

Holt Online Learning

Visit Holt Online Learning

If your teacher gives you a special password to log onto the Holt Online Learning site, you'll find your complete textbook on the Web. In addition, you'll find some great learning tools and practice quizzes. You'll be able to see how well you know the material from your textbook.

CNN student News

Visit CNN Student News

You'll find up-to-date events in science at cnnstudentnews.com.

SAFETY FIRST!

Exploring, inventing, and investigating are essential to the study of science. However, these activities can also be dangerous. To make sure that your experiments and explorations are safe, you must be aware of a variety of safety guidelines. You have probably heard of the saying, "It is better to be safe than sorry." This is particularly true in a science classroom where experiments and explorations are being performed. Being uninformed and careless can result in serious injuries. Don't take chances with your own safety or with anyone else's.

The following pages describe important guidelines for staying safe in the science classroom. Your teacher may also have safety guidelines and tips that are specific to your classroom and laboratory. Take the time to be safe.

Safety Rules!

Start Out Right

Always get your teacher's permission before attempting any laboratory exploration. Read the procedures carefully, and pay particular attention to safety information and caution statements. If you are unsure about what a safety symbol means, look it up or ask your teacher. You cannot be too careful when it comes to safety. If an accident does occur, inform your teacher immediately regardless of how minor you think the accident is.

Safety Symbols

All of the experiments and investigations in this book and their related worksheets include important safety symbols to alert you to particular safety concerns. Become familiar with these symbols so that when you see them, you will know what they mean and what to do. It is important that you read this entire safety section to learn about specific dangers in the laboratory.

If you are instructed to note the odor of a substance, wave the fumes toward your nose with your hand. Never put your nose close to the source.

Eye protection

Clothing protection

Hand safety

Heating safety

Electric safety

Chemical safety

Animal safety

Sharp object

Plant safety

Eye Safety

Wear safety goggles when working around chemicals, acids, bases, or any type of flame or heating device. Wear safety goggles any time there is even the slightest chance that harm could come to your eyes. If any substance gets into your eyes, notify your teacher immediately and flush your eyes with running water for at least 15 minutes. Treat any unknown chemical as if it were a dangerous chemical. Never look directly into the sun. Doing so could cause permanent blindness.

Avoid wearing contact lenses in a laboratory situation. Even if you are wearing safety goggles, chemicals can get between the contact lenses and your eyes. If your doctor requires that you wear contact lenses instead of glasses, wear eye-cup safety goggles in the lab.

Safety Equipment

Know the locations of the nearest fire alarms and any other safety equipment, such as fire blankets and eyewash fountains, as identified by your teacher, and know the procedures for using the equipment.

Neatness

Keep your work area free of all unnecessary books and papers. Tie back long hair, and secure loose sleeves or other loose articles of clothing, such as ties and bows. Remove dangling jewelry. Don't wear open-toed shoes or sandals in the laboratory. Never eat, drink, or apply cosmetics in a laboratory setting. Food, drink, and cosmetics can easily become contaminated with dangerous materials.

Certain hair products (such as aerosol hair spray) are flammable and should not be worn while working near an open flame. Avoid wearing hair spray or hair gel on lab days.

Sharp/Pointed Objects

Use knives and other sharp instruments with extreme care. Never cut objects while holding them in your hands. Place objects on a suitable work surface for cutting.

Be extra careful when using any glassware. When adding a heavy object to a graduated cylinder, tilt the cylinder so that the object slides slowly to the bottom.

Heat

Wear safety goggles when using a heating device or a flame. Whenever possible, use an electric hot plate as a heat source instead of using an open flame. When heating materials in a test tube, always angle the test tube away from yourself and others. To avoid burns, wear heat-resistant gloves whenever instructed to do so.

Electricity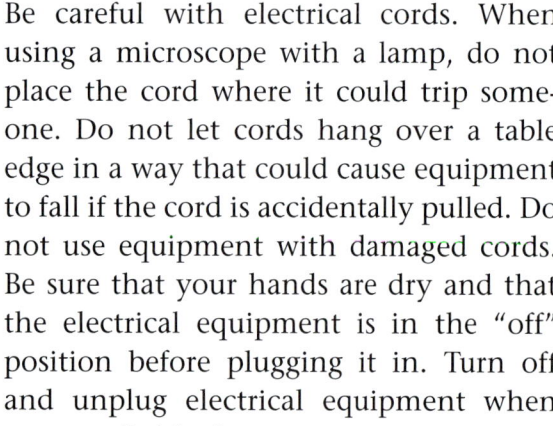

Be careful with electrical cords. When using a microscope with a lamp, do not place the cord where it could trip someone. Do not let cords hang over a table edge in a way that could cause equipment to fall if the cord is accidentally pulled. Do not use equipment with damaged cords. Be sure that your hands are dry and that the electrical equipment is in the "off" position before plugging it in. Turn off and unplug electrical equipment when you are finished.

Chemicals

Wear safety goggles when handling any potentially dangerous chemicals, acids, or bases. If a chemical is unknown, handle it as you would a dangerous chemical. Wear an apron and protective gloves when you work with acids or bases or whenever you are told to do so. If a spill gets on your skin or clothing, rinse it off immediately with water for at least 5 minutes while calling to your teacher.

Never mix chemicals unless your teacher tells you to do so. Never taste, touch, or smell chemicals unless you are specifically directed to do so. Before working with a flammable liquid or gas, check for the presence of any source of flame, spark, or heat.

Animal Safety

Always obtain your teacher's permission before bringing any animal into the school building. Handle animals only as your teacher directs. Always treat animals carefully and respectfully. Wash your hands thoroughly after handling any animal.

Plant Safety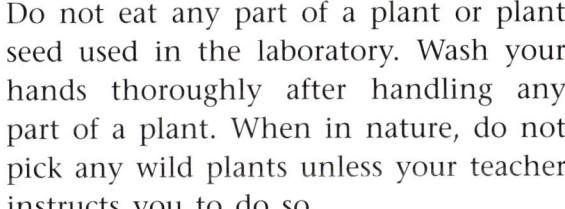

Do not eat any part of a plant or plant seed used in the laboratory. Wash your hands thoroughly after handling any part of a plant. When in nature, do not pick any wild plants unless your teacher instructs you to do so.

Glassware

Examine all glassware before use. Be sure that glassware is clean and free of chips and cracks. Report damaged glassware to your teacher. Glass containers used for heating should be made of heat-resistant glass.

Compression guide:
To shorten instruction because of time limitations, omit the Chapter Lab.

OBJECTIVES	LABS, DEMONSTRATIONS, AND ACTIVITIES	TECHNOLOGY RESOURCES
PACING • 90 min pp. 2–7 **Chapter Opener**	**SE** Start-up Activity, p. 3 `GENERAL`	**OSP** Parent Letter ■ `GENERAL` **CD** Student Edition on CD-ROM **CD** Guided Reading Audio CD ■ **TR** Chapter Starter Transparency* **VID** Brain Food Video Quiz
Section 1 **Electrons and Chemical Bonding** • Describe chemical bonding. • Identify the number of valence electrons in an atom. • Predict whether an atom is likely to form bonds.	**TE** Demonstration Breaking Bonds, p. 4 `GENERAL` **SE** Science in Action Math, Social Studies, and Language Arts Activities, pp. 24–25 `GENERAL`	**CRF** Lesson Plans* **TR** Bellringer Transparency* **TR** Electron Arrangement in an Atom* **TR** Determining the Number of Valence Electrons*
PACING • 45 min pp. 8–11 **Section 2** **Ionic Bonds** • Explain how ionic bonds form. • Describe how positive ions form. • Describe how negative ions form. • Explain why ionic compounds are neutral.	**SE** School-to-Home Activity Studying Salt, p. 9 `GENERAL`	**CRF** Lesson Plans* **TR** Bellringer Transparency* **TR** Forming Positive and Negative Ions* **CRF** SciLinks Activity* `GENERAL`
PACING • 90 min pp. 12–17 **Section 3** **Covalent and Metallic Bonds** • Explain how covalent bonds form. • Describe molecules. • Explain how metallic bonds form. • Describe the properties of metals.	**TE** Activity Cereal-Dot Diagrams, p. 12 `GENERAL` **TE** Activity Hydrogen Bonds, p. 13 `ADVANCED` **TE** Making Models Three-Dimensional Models, p. 13 `GENERAL` **TE** Activity Drawing Diagrams, p. 14 `GENERAL` **SE** Connection to Biology Proteins, p. 15 `GENERAL` **TE** Connection Activity Art, p. 15 `GENERAL` **SE** Quick Lab Bending with Bonds, p. 16 `GENERAL` **CRF** Datasheet for Quick Lab* **SE** Skills Practice Lab Covalent Marshmallows, p. 18 `GENERAL` **CRF** Datasheet for Chapter Lab* **LB** Long-Term Projects & Research Ideas The Wonders of Water* `ADVANCED`	**CRF** Lesson Plans* **TR** Bellringer Transparency* **TR** Covalent Bond* **TR** Covalent Bonds in a Water Molecule* **TR** **LINK TO LIFE SCIENCE** Making of a Protein A and B* **SE** Internet Activity, p. 14 `GENERAL` **VID** Lab Videos for Physical Science

PACING • 90 min

CHAPTER REVIEW, ASSESSMENT, AND STANDARDIZED TEST PREPARATION

CRF Vocabulary Activity* `GENERAL`
SE Chapter Review, pp. 20–21 `GENERAL`
CRF Chapter Review* ■ `GENERAL`
CRF Chapter Tests A* ■ `GENERAL`, B* `ADVANCED`, C* `SPECIAL NEEDS`
SE Standardized Test Preparation, pp. 22–23 `GENERAL`
CRF Standardized Test Preparation* `GENERAL`
CRF Performance-Based Assessment* `GENERAL`
OSP Test Generator `GENERAL`
CRF Test Item Listing* `GENERAL`

Online and Technology Resources

Visit **go.hrw.com** for a variety of free resources related to this textbook. Enter the keyword **HP5BND**.

Holt Online Learning

Students can access interactive problem-solving help and active visual concept development with the *Holt Science and Technology* Online Edition available at **www.hrw.com**.

Guided Reading Audio CD

A direct reading of each chapter using instructional visuals as guideposts. For auditory learners, reluctant readers, and Spanish-speaking students. Available in English and Spanish.

SKILLS DEVELOPMENT RESOURCES	SECTION REVIEW AND ASSESSMENT	STANDARDS CORRELATIONS
SE Pre-Reading Activity, p. 2 `GENERAL` **OSP** Science Puzzlers, Twisters & Teasers* ■ `GENERAL`		National Science Education Standards SAI 1; PS 1b
CRF Directed Reading A* ■ `BASIC`, B* `SPECIAL NEEDS` **CRF** Vocabulary and Section Summary* ■ `GENERAL` **SE** Reading Strategy Discussion, p. 4 `GENERAL` **SE** Connection to Social Studies History of a Noble Gas, p. 6 `GENERAL` **TE** Inclusion Strategies, p. 6	**SE** Reading Checks, pp. 5, 6 `GENERAL` **TE** Homework, p. 5 `ADVANCED` **TE** Reteaching, p. 6 `BASIC` **TE** Quiz, p. 6 `GENERAL` **TE** Alternative Assessment, p. 6 `GENERAL` **SE** Section Review,* p. 7 `GENERAL` **CRF** Section Quiz* ■ `GENERAL`	UCP 1, 2; PS 1b, 1c
CRF Directed Reading A* ■ `BASIC`, B* `SPECIAL NEEDS` **CRF** Vocabulary and Section Summary* ■ `GENERAL` **SE** Reading Strategy Paired Summarizing, p. 8 `GENERAL` **TE** Reading Strategy Graphic Organizer, p. 9 `GENERAL` **SE** Math Practice Calculating Charge, p. 10 `GENERAL` **TE** Inclusion Strategies, p. 10 **MS** Math Skills for Science Comparing Integers on a Number Line* `GENERAL` **MS** Math Skills for Science Arithmetic with Positive and Negative Numbers* `GENERAL` **CRF** Reinforcement Worksheet Is It an Ion?* `BASIC`	**SE** Reading Checks, pp. 8, 10 `GENERAL` **TE** Reteaching, p. 10 `BASIC` **TE** Quiz, p. 10 `GENERAL` **TE** Alternative Assessment, p. 10 `GENERAL` **SE** Section Review,* p. 11 `GENERAL` **CRF** Section Quiz* ■ `GENERAL`	UCP 1, 2, 3; PS 1b, 3a, 3e
CRF Directed Reading A* ■ `BASIC`, B* `SPECIAL NEEDS` **CRF** Vocabulary and Section Summary* ■ `GENERAL` **SE** Reading Strategy Reading Organizer, p. 12 `GENERAL` **CRF** Reinforcement Worksheet Interview with an Electron* `BASIC` **CRF** Critical Thinking The Road to Knowledge* `ADVANCED`	**SE** Reading Checks, pp. 12, 14, 16 `GENERAL` **TE** Homework, p. 15 `GENERAL` **TE** Reteaching, p. 16 `BASIC` **TE** Quiz, p. 16 `GENERAL` **TE** Alternative Assessment, p. 16 `GENERAL` **SE** Section Review,* p. 17 ■ `GENERAL` **CRF** Section Quiz* ■ `GENERAL`	UCP 1, 2, 3; SAI 1; PS 1b; *Chapter Lab:* UCP 2

One-Stop Planner® CD-ROM

This convenient CD-ROM includes:
- Lab Materials QuickList Software
- Holt Calendar Planner
- Customizable Lesson Plans
- Printable Worksheets
- ExamView® Test Generator

cnnstudentnews.com

Find the latest news, lesson plans, and activities related to important scientific events.

www.scilinks.org

Maintained by the **National Science Teachers Association.** See Chapter Enrichment pages for a complete list of topics.

Check out *Current Science* articles and activities by visiting the HRW Web site at **go.hrw.com.** Just type in the keyword **HP5CS13T.**

Classroom Videos
- **Lab Videos** demonstrate the chapter lab.
- **Brain Food Video Quizzes** help students review the chapter material.

Visual Resources

CHAPTER STARTER TRANSPARENCY

BELLRINGER TRANSPARENCIES

TEACHING TRANSPARENCIES

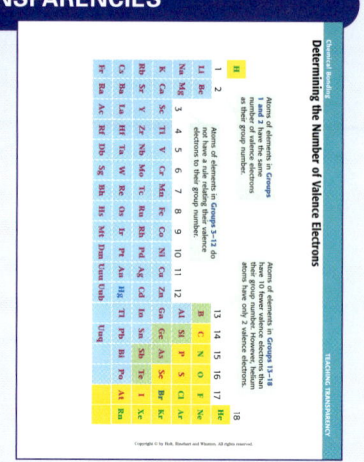

TEACHING TRANSPARENCIES

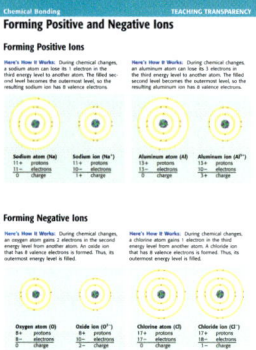

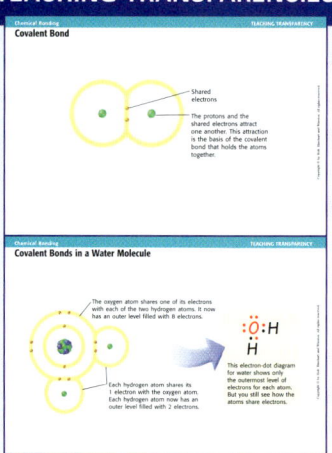

CONCEPT MAPPING TRANSPARENCY

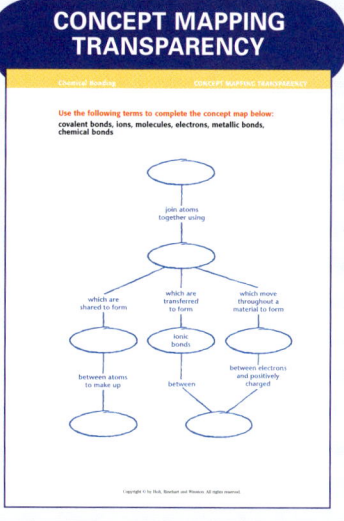

Planning Resources

LESSON PLANS

PARENT LETTER

TEST ITEM LISTING

One-Stop
Planner® CD-ROM

This CD-ROM includes all of the resources shown here and the following time-saving tools:

- *Lab Materials QuickList Software*
- *Customizable lesson plans*
- *Holt Calendar Planner*
- *The powerful ExamView® Test Generator*

Meeting Individual Needs

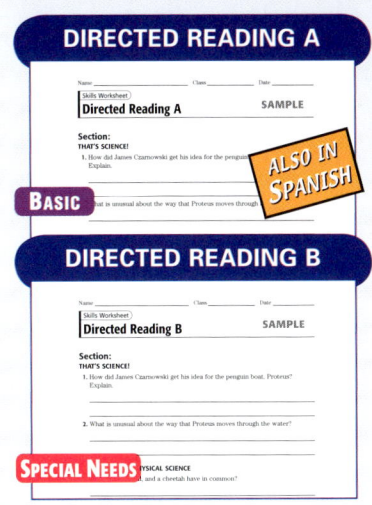

DIRECTED READING A

Skills Worksheet
Directed Reading A — SAMPLE

Section:
THAT'S SCIENCE!

1. How did James Czarnowski get his idea for the penguin boat, Proteus? Explain.

What is unusual about the way that Proteus moves through the water?

BASIC

ALSO IN SPANISH

DIRECTED READING B

Skills Worksheet
Directed Reading B — SAMPLE

Section:
THAT'S SCIENCE!

1. How did James Czarnowski get his idea for the penguin boat, Proteus? Explain.

2. What is unusual about the way that Proteus moves through the water?

SPECIAL NEEDS   PHYSICAL SCIENCE
and a cheetah have in common?

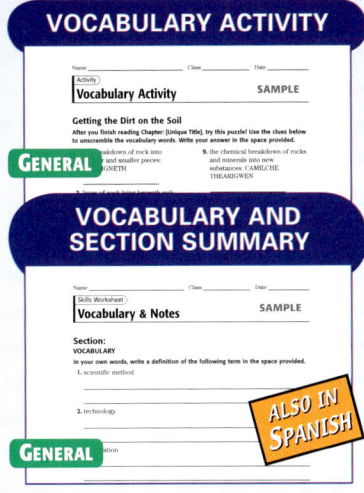

VOCABULARY ACTIVITY

Activity
Vocabulary Activity — SAMPLE

Getting the Dirt on the Soil
After you finish reading the vocabulary words, try this puzzle! Use the clues below to unscramble the vocabulary words. Write your answer in the space provided.

9. the chemical breakdown of rocks
GNETH   THEARDGWEN
substances: CAMILCHE

GENERAL

VOCABULARY AND SECTION SUMMARY

Skills Worksheet
Vocabulary & Notes — SAMPLE

Section:
VOCABULARY
In your own words, write a definition of the following term in the space provided.

1. scientific method

2. technology

GENERAL

ALSO IN SPANISH

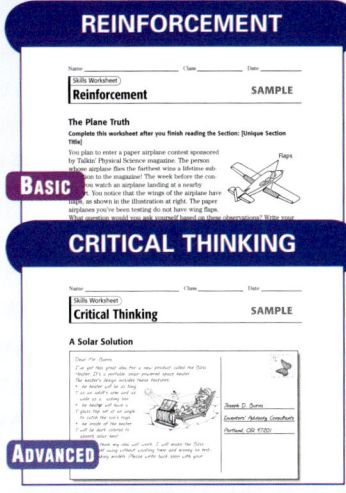

REINFORCEMENT

Skills Worksheet
Reinforcement — SAMPLE

The Plane Truth
Complete this worksheet after you finish reading the Section: [Unique Section Title]

You plan to enter a paper airplane contest sponsored by Talkin' Physical Science magazine. The person whose airplane flies the farthest wins a lifetime subscription to the magazine! The week before the contest, you watch an airplane landing at a nearby airport. You notice that the wings of the airplane have flaps, as shown in the illustration at right. The paper airplanes you've been testing do not have wing flaps. What question would you ask yourself based on these observations? Write your

BASIC

CRITICAL THINKING

Skills Worksheet
Critical Thinking — SAMPLE

A Solar Solution

Joseph D. Burns

ADVANCED

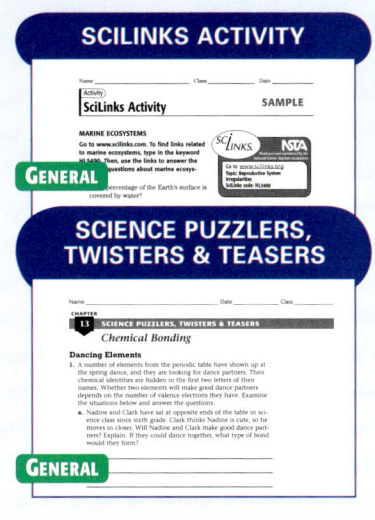

SCILINKS ACTIVITY

Activity
SciLinks Activity — SAMPLE

MARINE ECOSYSTEMS
Go to www.scilinks.com. To find links related to marine ecosystems, type in the keyword HL5400. Then, use the links to answer the questions.

percentage of the Earth's surface is covered by water?

GENERAL

SCIENCE PUZZLERS, TWISTERS & TEASERS

CHAPTER 13   SCIENCE PUZZLERS, TWISTERS & TEASERS
Chemical Bonding

Dancing Elements
1. A number of elements from the periodic table have shown up at the spring dance, and they are looking for dance partners. Their chemical identities are hidden in the first two letters of their names. Whether two elements will make good dance partners depends on the number of valence electrons they have. Examine the situations below and answer the questions.

a. Nadine and Clark have sat at opposite ends of the table in science class since sixth grade. Clark thinks Nadine is cute, so he moves in closer. Will Nadine and Clark make good dance partners? Explain. If they could dance together, what type of bond would they form?

GENERAL

Labs and Activities

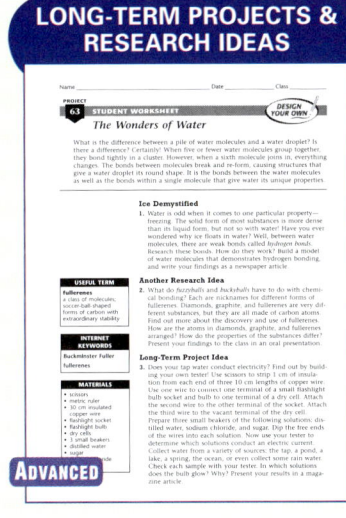

LONG-TERM PROJECTS & RESEARCH IDEAS

PROJECT 6.3   STUDENT WORKSHEET   DESIGN YOUR OWN
The Wonders of Water

What is the difference between a pile of water molecules and a water droplet? Is there a difference? Certainly! When five or fewer water molecules group together, they bond tightly in a cluster. However, when a sixth molecule joins in, everything changes. The bonds between molecules break and re-form, causing structures that give a water droplet its round shape. It is the bonds between the water molecules as well as the bonds within a single molecule that give water its unique properties.

Ice Demystified
1. Water is odd when it comes to one particular property—freezing. The solid form of most substances is more dense than its liquid form, but not so with water! Have you ever wondered why ice floats in water? Well, between water molecules, there are weak bonds called hydrogen bonds. Research these bonds. How do they work? Build a model of water molecules that demonstrates hydrogen bonding, and write your findings as a newspaper article.

Another Research Idea
2. What do buckyballs and buckballs have to do with chemical bonding? Each are nicknames for different forms of fullerenes. Diamonds, graphite, and fullerenes are very different substances, but they are all made of carbon atoms. Find out more about the discovery and use of fullerenes. How are the atoms in diamonds, graphite, and fullerenes arranged? How do the properties of the substances differ? Present your findings to the class in an oral presentation.

Long-Term Project Idea
3. Does your tap water conduct electricity? Find out by building your own tester! Use scissors to strip 1 cm of insulation from each end of three 10 cm lengths of copper wire. Use one wire to connect one terminal of a small flashlight bulb socket and bulb to one terminal of a dry cell. Attach the second wire to the other terminal of the dry cell. The third wire to the vacant terminal of the dry cell. Prepare three small beakers of the following solutions: distilled water, sodium chloride, and sugar. Dip the free ends of the wires into each solution. Now use your tester to determine which solutions conduct an electric current. Collect water from a variety of sources: the tap, a pond, a lake, a spring, the ocean, or even collect some rain water. Check each sample with your tester. In which solutions does the bulb glow? Why? Present your results in a magazine article.

USEFUL TERM
fullerenes
a class of molecules, soccer-ball-shaped forms of carbon with extraordinary stability

INTERNET KEYWORDS
Buckminster Fuller
fullerenes

MATERIALS
• scissors
• metric ruler
• 30 cm insulated copper wire
• flashlight socket
• dry cells
• 3 small beakers
• distilled water
• sugar

ADVANCED

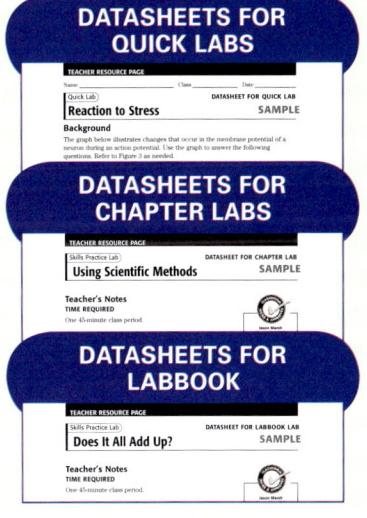

DATASHEETS FOR QUICK LABS

TEACHER RESOURCE PAGE

Quick Lab   DATASHEET FOR QUICK LAB
Reaction to Stress — SAMPLE

Background
The graph below illustrates changes that occur in the membrane potential of a neuron during an action potential. Use the graph to answer the following questions. Refer to Figure 3 as needed.

DATASHEETS FOR CHAPTER LABS

TEACHER RESOURCE PAGE

Skills Practice Lab   DATASHEET FOR CHAPTER LAB
Using Scientific Methods — SAMPLE

Teacher's Notes
TIME REQUIRED
One 45-minute class period.

DATASHEETS FOR LABBOOK

TEACHER RESOURCE PAGE

Skills Practice Lab   DATASHEET FOR LABBOOK LAB
Does It All Add Up? — SAMPLE

Teacher's Notes
TIME REQUIRED
One 45-minute class period.

Review and Assessments

SECTION QUIZ

Assessment
Section Quiz — SAMPLE

Section:
In the space provided, write the letter of the description that best matches the term or phrase.

_____ 1. building molecules that can be used in an energy source or breaking down molecules in which energy is stored

_____ the process by which light energy is converted to chemical energy

_____ an organism that uses sunlight or inorganic substances to make organic compounds

a.
b.
c.
f. cellular respiration

GENERAL

ALSO IN SPANISH

SECTION REVIEW

Skills Worksheet
Section Review — SAMPLE

Section:
KEY TERMS
1. What do paleontologist study?

2. How does a trace fossil differ from petrified wood?

UNDERSTANDING KEY IDEAS

GENERAL

ALSO IN SPANISH

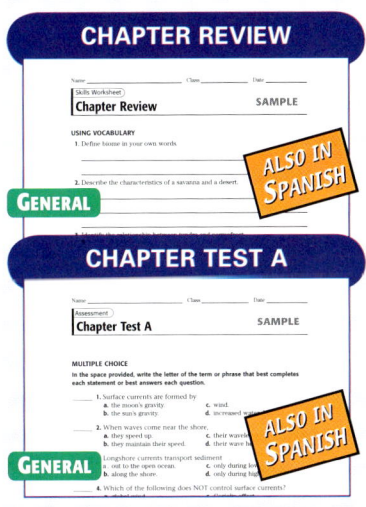

CHAPTER REVIEW

Skills Worksheet
Chapter Review — SAMPLE

USING VOCABULARY
1. Define biome in your own words.

2. Describe the characteristics of a savanna and a desert.

GENERAL

ALSO IN SPANISH

CHAPTER TEST A

Assessment
Chapter Test A — SAMPLE

MULTIPLE CHOICE
In the space provided, write the letter of the term or phrase that best completes each statement or best answers each question.

_____ 1. Surface currents are formed by
a. the moon's gravity.   c. wind.
b. the sun's gravity.   d. increased water density.

_____ 2. When waves come near the shore,
a. they speed up.   c. their wavelength increases.
b. they maintain their speed.   d. their wave height increases.

Longshore currents transport sediment
a. out to the open ocean.   c. only during low tide.
b. along the shore.   d. only during high tide.

4. Which of the following does NOT control surface currents?

GENERAL

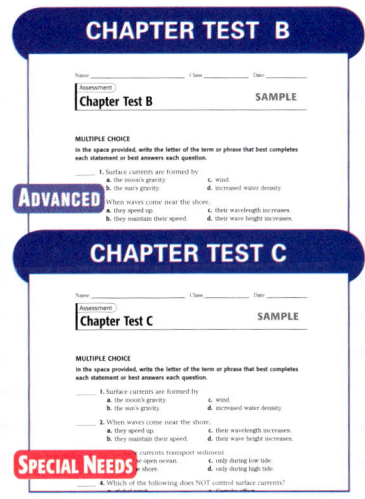

CHAPTER TEST B

Assessment
Chapter Test B — SAMPLE

MULTIPLE CHOICE
In the space provided, write the letter of the term or phrase that best completes each statement or best answers each question.

_____ 1. Surface currents are formed by
a. the moon's gravity.   c. wind.
b. the sun's gravity.   d. increased water density.

_____ When waves come near the shore,
a. they speed up.   c. their wavelength increases.
b. they maintain their speed.   d. their wave height increases.

ADVANCED

CHAPTER TEST C

Assessment
Chapter Test C — SAMPLE

MULTIPLE CHOICE
In the space provided, write the letter of the term or phrase that best completes each statement or best answers each question.

_____ 1. Surface currents are formed by
a. the moon's gravity.   c. wind.
b. the sun's gravity.   d. increased water density.

_____ 2. When waves come near the shore,
a. they speed up.   c. their wavelength increases.
b. they maintain their speed.   d. their wave height increases.

a. out to the open ocean.   c. only during low tide.
b. along the shore.   d. only during high tide.

4. Which of the following does NOT control surface currents?

SPECIAL NEEDS

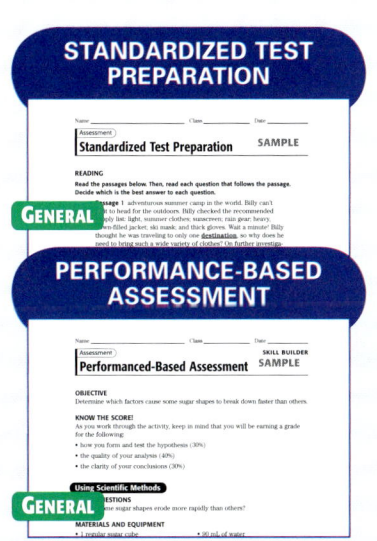

STANDARDIZED TEST PREPARATION

Assessment
Standardized Test Preparation — SAMPLE

READING
Read the passages below. Then, read each question that follows the passage. Decide which is the best answer to each question.

Passage 1 adventurous summer camp in the world. Billy can't wait to head for the outdoors. Billy checked the recommended supply list: light, summer clothes; sunscreen; rain gear; heavy, down-filled jacket; ski mask; and thick gloves. Wait a minute! Billy thought he was traveling to only one destination, so why does he need to bring such a wide variety of clothes? On further investiga-

GENERAL

PERFORMANCE-BASED ASSESSMENT

Assessment
Performanced-Based Assessment   SKILL BUILDER   SAMPLE

OBJECTIVE
Determine which factors cause some sugar shapes to break down faster than others.

KNOW THE SCORE!
As you work through the activity, keep in mind that you will be earning a grade for the following:
• how you form and test the hypothesis (30%)
• the quality of your analysis (40%)
• the clarity of your conclusions (30%)

Using Scientific Methods
QUESTIONS
ome sugar shapes erode more rapidly than others?

MATERIALS AND EQUIPMENT

GENERAL

Chapter Enrichment

This Chapter Enrichment provides relevant and interesting information to expand and enhance your presentation of the chapter material.

Section 1

Electrons and Chemical Bonding

Discovery of the Electron

- Electrons are important in all types of chemical bonding. English physicist Joseph John Thomson (1856–1940) discovered the electron in 1897 when studying cathode rays. Cathode rays are invisible beams emitted from negative electrodes when electrical energy is passed through a vacuum tube.

- Thomson believed that cathode rays were composed of particles of matter that he called *corpuscles*. He also theorized that the corpuscles—later renamed electrons—were negatively charged and were identical, no matter what type of gas or metal carried the electrical energy.

Is That a Fact!

- ◆ J. J. Thomson's model of the atom, which was eventually superseded by other models, was dubbed the plum-pudding model because some people visualized Thomson's model as a positively charged sphere of "pudding" interspersed with negatively charged "plums," or electrons.

Bohr's Theory of Atomic Structure

- In 1913, Danish physicist Niels Bohr (1885–1962) theorized that electrons occupy energy levels, which are at certain specified distances from an atom's nucleus.

- Energy is absorbed when an electron moves from a lower energy level to a higher energy level. Energy is released when an electron moves from a higher level to a lower one. Bohr's model of the atom has been compared to the structure of an onion, with the layers of onion corresponding to the energy levels occupied by electrons.

- Bohr's idea of fixed energy levels was eventually proved wrong. Scientists now describe electrons as particles with properties similar to those of waves. These properties led scientists to describe regions in the atom where electrons are likely to be found because the exact path of an electron cannot be predicted. These regions are called *electron clouds*.

Section 2

Ionic Bonds

Ionic or Covalent?

- Bonding between atoms of different elements is rarely purely ionic or purely covalent. Bonding is usually somewhere between these two extremes, depending on how strongly the atoms of each element attract electrons. Electronegativity is a measure of an atom's ability to attract electrons. In general, the electronegativies of atoms increase across the rows of the periodic table and decrease down the columns. The degree to which a bond between atoms of two elements is ionic or covalent can be estimated by calculating the difference in the elements' electronegativities.

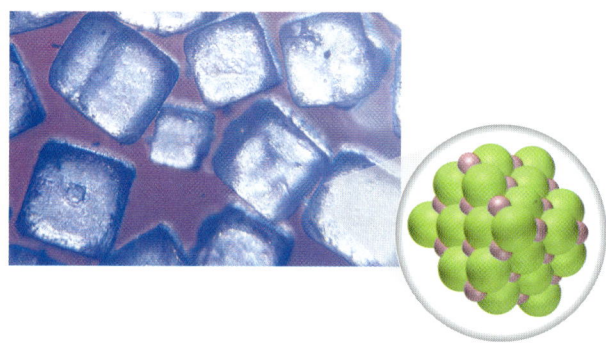

- In general, bonding between atoms with an electronegativity difference of 1.7 or less has an ionic character of 50% or less and is classified as covalent. Bonding between atoms with an electronegativity difference greater than 1.7 is classified as ionic. However, this rule has exceptions, and there are other ways to estimate the ionic or covalent character of a bond.

Section 3

Covalent and Metallic Bonds

G. N. Lewis and the Theory of the Electron Pair

- In the early 1900s, American chemist Gilbert Newton Lewis (1875–1946) noticed that elements with certain numbers of electrons seemed to be especially unreactive, and that other elements were highly reactive. Helium, which has 2 electrons, is inert, but hydrogen, which has 1 electron, is reactive. Lewis also noticed that the next nonreactive element on the periodic table, neon, had 8 more electrons than helium.

- Lewis theorized that atoms have "layers" of electrons and that a specific number of electrons—8 electrons, for example—is required to fill the outer layer. From this observation came Lewis's "octet rule." The octet rule states that ions or atoms with a filled outer layer of 8 electrons are unreactive.

- Lewis published his theory in 1916. It was the first explanation of the covalent bond and went a long way toward explaining the mechanism of many chemical reactions.

- Electron-dot diagrams, often called _Lewis structures,_ have helped several generations of chemistry students visualize molecular structures and bonding.

The Unique Bonding Properties of Carbon

- Carbon and carbon compounds are the basis of all living things. Because carbon atoms have 4 valence electrons, carbon atoms can combine with other carbon atoms to form molecules with high molecular weight. These large molecules may take many forms, such as rings or long chains.

- Carbon rings can form when carbon atoms bond to other carbon atoms. The carbon compound benzene, C_6H_6, for example, is a hexagonal ring consisting of six carbon atoms bonded to each other with a hydrogen atom bonded to each carbon. The benzene ring is the parent compound of many substances, including aromatics, such as vanilla, perfumes, and mothballs.

- Long chains of repeating molecular units are known as polymers. The wide variety of plastics we use, with all their different physical properties, are examples of polymers. These different properties are the result of the type of repeating unit and the way in which these units are bonded together.

Is That a Fact!

- To some people, the word _polymer_ is synonymous with _plastic._ However, many polymers exist in nature. Cellulose, a polymer chain containing repeating units of the molecule glucose, is the chief constituent of plant cells. Wood is about 50% cellulose, and cotton is 90% cellulose.

SciLINKS

NSTA
Developed and maintained by the
National Science Teachers Association

SciLinks is maintained by the National Science Teachers Association to provide you and your students with interesting, up-to-date links that will enrich your classroom presentation of the chapter.

Visit  www.scilinks.org and enter the SciLinks code for more information about the topic listed.

Topic: The Electron
SciLinks code: HSM0489

Topic: Types of Chemical Bonds
SciLinks code: HSM1565

Topic: Periodic Table
SciLinks code: HSM1125

Topic: Properties of Metals
SciLinks code: HSM1231

Overview

Tell students that this chapter will help them learn about chemical bonding. The chapter describes how the valence electrons of atoms are involved in forming chemical bonds and describes the three kinds of chemical bonds—ionic, covalent, and metallic.

Assessing Prior Knowledge

Students should be familiar with the following topics:
- the periodic table
- the structure of the atom

Identifying Misconceptions

As students learn the material in this chapter, some of them may think that ionic compounds are made up of molecules of the compounds. Tell students that because ionic compounds form crystal lattices, chemists usually do not refer to molecules when discussing ionic compounds. Instead, chemists refer to the smallest ratio of ions in an ionic compound as a *formula unit*. For example, the chemical formulas of sodium chloride, $NaCl$, and calcium fluoride, CaF_2, represent one formula unit of each compound.

1

Chemical Bonding

About the PHOTO

What looks like a fantastic "sculpture" is really a model of deoxyribonucleic acid (DNA). DNA is one of the most complex molecules in living things. In DNA, atoms are bonded together in two very long spiral strands. These strands join to form a double spiral. The DNA in living cells has all the coding for passing on the traits of that cell and that organism.

PRE-READING ACTIVITY

FOLDNOTES

Three-Panel Flip Chart
Before you read the chapter, create the FoldNote entitled "Three-Panel Flip Chart" described in the **Study Skills** section of the Appendix. Label the flaps of the three-panel flip chart with "Ionic bond," "Covalent bond," and "Metallic bond." As you read the chapter, write information you learn about each category under the appropriate flap.

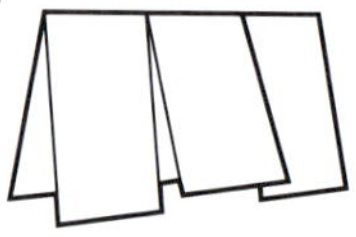

Standards Correlations

National Science Education Standards

The following codes indicate the National Science Education Standards that correlate to this chapter. The full text of the standards is at the front of the book.

Chapter Opener
SAI 1; PS 1b

Section 1 Electrons and Chemical Bonding
UCP 1, 2; PS 1b, 1c

Section 2 Ionic Bonds
UCP 1, 2, 3; PS 1b, 3a, 3e

Section 3 Covalent and Metallic Bonds
UCP 1, 2, 3; SAI 1; PS 1b

Chapter Lab
UCP 2

Chapter Review
PS 1b

Science in Action
ST 1; SPSP 3, 5; HNS 1

START-UP **ACTIVITY**
MATERIALS
FOR EACH STUDENT
- borax solution, 40 g/L (100 ml)
- cups, paper (2)
- glue, white (1 : 1 solution of glue and water)
- spoon, plastic (or craft stick, wooden)

Safety Caution: Caution students to wear safety goggles, gloves, and aprons during this lab activity. Caution students to keep their hands away from their eyes and face during this lab activity.

Ingestion of large amounts of borax can cause severe vomiting, diarrhea, and shock. Have the telephone number for your local poison control center available during this activity.

Be sure eyewash equipment is available and is working.

Caution students to wash their hands thoroughly when they are finished with this activity.

Use only nontoxic white glue.

Answers

1. The glue is a white liquid that flows easily. The new material is white and has properties that are more like those of a solid than like those of a liquid.

2. The properties of the material would be more like the properties of the glue. The material would flow more easily and would not hold its shape well.

START-UP **ACTIVITY**

From Glue to Goop

Particles of glue can bond to other particles and hold objects together. Different types of bonds create differences in the properties of substances. In this activity, you will see how the formation of bonds causes a change in the properties of white glue.

Procedure

1. Fill a **small paper cup** 1/4 full of **white glue**. Record the properties of the glue.

2. Fill a **second small paper cup** 1/4 full of **borax solution**.

3. Pour the borax solution into the cup of white glue, and stir well using a **plastic spoon** or a **wooden craft stick.**

4. When the material becomes too thick to stir, remove it from the cup and knead it with your fingers. Record the properties of the material.

Analysis

1. Compare the properties of the glue with those of the new material.

2. The properties of the material resulted from bonds between the borax and the glue. Predict the properties of the material if less borax is used.

Strange but True!

In 1987, pilots Richard Rutan and Jeana Yeager flew the *Voyager* aircraft, shown above, around the world without refueling. The record-breaking trip lasted just over 9 days. In order to carry enough fuel for the trip, the plane had to be as lightweight as possible. Using fewer bolts than usual to attach parts would make the airplane lighter. But without the bolts, what would hold the parts together? The designers decided to use glue!

Not just any glue would do. They used superglue. When superglue is applied, it combines with water from the air to form chemical bonds. The result– the materials stick together as if they were one material. Superglue is so strong that the amount of superglue it takes to cover the size of a postage stamp can hold over a ton.

of the stitching with superglue. And dentists can use superglue to hold a cracked tooth together.

Chemical bonding is responsible for the properties of materials. In this chapter, you will learn about the different types of bonds that hold atoms together and how those bonds affect the properties of the materials.

Superglue was discovered in the early 1950s by a scientist who was trying to develop a new plastic for the cockpit bubble of a jet plane.

Chapter Starter Transparency
Use this transparency to help students begin thinking about chemical bonding.

CHAPTER RESOURCES
Technology
Transparencies
- Chapter Starter Transparency

READING SKILLS

Student Edition on CD-ROM

Guided Reading Audio CD
- English or Spanish

Classroom Videos
- Brain Food Video Quiz

Workbooks
Science Puzzlers, Twisters & Teasers
- Chemical Bonding GENERAL

Focus

Overview

This section defines chemical bonding and describes the role of electrons in the formation of chemical bonds.

🔔 Bellringer

Display the following chemical formulas but not their identities:

$C_6H_{12}O_6$ (glucose, a sugar)

C_2H_5OH (ethyl alcohol)

$C_6H_8O_6$ (vitamin C)

$C_6H_8O_7$ (citric acid)

Ask students to identify the elements in these compounds and to predict whether the compounds are similar to each other and why. Identify and discuss the compounds.

Motivate

Demonstration — GENERAL

Breaking Bonds Heat sugar in a crucible with a Bunsen burner until a black goo remains. Ask students to identify the black goo. (carbon) Ask students to explain why the sugar turned into carbon. (Sugar is composed of carbon, hydrogen, and oxygen. The hydrogen and oxygen combine to form water. The thermal energy evaporates the water.) Explain that the energy broke the chemical bonds that were holding the atoms of the elements together.
LS Visual

READING WARM-UP

Objectives
- Describe chemical bonding.
- Identify the number of valence electrons in an atom.
- Predict whether an atom is likely to form bonds.

Terms to Learn
chemical bonding
chemical bond
valence electron

READING STRATEGY

Discussion Read this section silently. Write down questions that you have about this section. Discuss your questions in a small group.

CHAPTER RESOURCES

Chapter Resource File

- Lesson Plan
- Directed Reading A **BASIC**
- Directed Reading B **SPECIAL NEEDS**

Technology

Transparencies
- Bellringer
- Electron Arrangement in an Atom

Electrons and Chemical Bonding

Have you ever stopped to consider that by using only the 26 letters of the alphabet, you make all of the words you use every day?

Although the number of letters is limited, combining the letters in different ways allows you to make a huge number of words. In the same way that words can be formed by combining letters, substances can be formed by combining atoms.

Combining Atoms Through Chemical Bonding

Look at **Figure 1.** Now, look around the room. Everything you see—desks, pencils, paper, and even your friends—is made of atoms of elements. All substances are made of atoms of one or more of the approximately 100 elements. For example, the atoms of carbon, hydrogen, and oxygen combine in different patterns to form sugar, alcohol, and citric acid. **Chemical bonding** is the joining of atoms to form new substances. The properties of these new substances are different from the properties of the original elements. An interaction that holds two atoms together is called a **chemical bond.** When chemical bonds form, electrons are shared, gained, or lost.

Discussing Bonding Using Theories and Models

We cannot see atoms and chemical bonds with the unaided eye. For more than 150 years, scientists have done many experiments that have led to a theory of chemical bonding. Remember that a theory is an explanation for some phenomenon that is based on observation, experimentation, and reasoning. The use of models helps people discuss the theory of how and why atoms form bonds.

Figure 1 *Everything you see in this photo is formed by combining atoms.*

SCIENTISTS AT ODDS

Noble Gas Compounds Before 1962, most scientists believed that noble gases could not form compounds with other elements. After all, none were known to exist. In that year, though, chemists first created a compound of xenon and fluorine called xenon tetrafluoride, XeF_4. Much to their surprise, these chemists found that xenon and fluorine reacted quite easily to form the compound. Under the right conditions, krypton and radon can also form compounds.

Figure 2   Electron Arrangement in an Atom

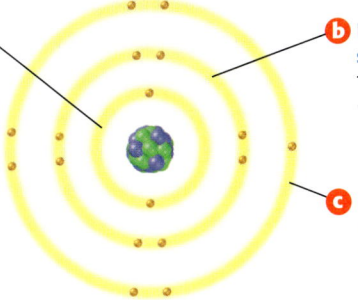

a The **first energy level** is closest to the nucleus and can hold up to 2 electrons.

b Electrons will begin filling the **second energy level** only after the first level is full. The second energy level can hold up to 8 electrons.

c The **third energy level** in this model of a chlorine atom has only 7 electrons, so the atom has a total of 17 electrons. This outer level of the atom is not full.

Electron Number and Organization

To understand how atoms form chemical bonds, you need to know about the electrons in an atom. The number of electrons in an atom can be determined from the atomic number of the element. The *atomic number* is the number of protons in an atom. But atoms have no charge. So, the atomic number also represents the number of electrons in the atom.

Electrons in an atom are organized in energy levels. **Figure 2** shows a model of the arrangement of electrons in a chlorine atom. This model and models like it are useful for counting electrons in energy levels of atoms. But, these models do not show the true structure of atoms.

Outer-Level Electrons and Bonding

Not all of the electrons in an atom make chemical bonds. Most atoms form bonds using only the electrons in the outermost energy level. An electron in the outermost energy level of an atom is a **valence electron** (VAY luhns ee LEK TRAHN). The models in **Figure 3** show the valence electrons for two atoms.

✓ **Reading Check**   Which electrons are used to form bonds?
(*See the Appendix for answers to Reading Checks.*)

chemical bonding the combining of atoms to form molecules or ionic compounds

chemical bond an interaction that holds atoms or ions together

valence electron an electron that is found in the outermost shell of an atom and that determines the atom's chemical properties

Figure 3   Counting Valence Electrons

Oxygen
Electron total: 8
First level: 2 electrons
Second level: 6 electrons

An oxygen atom has 6 valence electrons.

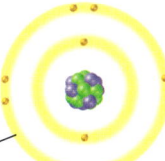

Sodium
Electron total: 11
First level: 2 electrons
Second level: 8 electrons
Third level: 1 electron

A sodium atom has 1 valence electron.

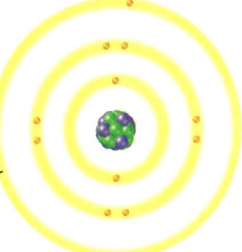

Is That a Fact!

An early theory of chemical bonding was developed by Swedish chemist Jöns Jacob Berzelius (1779–1848). Berzelius theorized that all elements had either a positive or a negative charge and that only positive and negative elements would bond with each other. His theory

was widely accepted. In many ways, it is a fairly accurate explanation of ionic bonding. Berzelius's theory fell short, however, by implying that molecules containing more than one atom of the same element could not exist because those atoms would repel each other.

Using the Periodic Table Have students refer to **Figure 4** and answer the following questions:

1. How many valence electrons are there in an atom of radium? of lead? of iodine? of neon? of cesium? (2, 4, 7, 8, 1)

2. Which elements from question 1 would be likely to bond with other atoms? (radium, lead, iodine, cesium)

3. Which element from question 1 would be least likely to bond with other atoms? (neon)

LS Visual/Logical — English Language Learners

Quiz — GENERAL

1. What is the joining of atoms to form new substances called? (chemical bonding)

2. Why do some atoms rarely bond? (They have a filled outermost energy level with 8 electrons, which makes them very nonreactive.)

Alternative Assessment — GENERAL

Concept Mapping Have students create a concept map using the following terms: *chemical bonding*, *chemical bond*, and *valence electron*. Students should include examples of elements from the periodic table to clarify their concept map. LS Visual

---

Figure 4 Determining the Number of Valence Electrons

Atoms of elements in **Groups 1 and 2** have the same number of valence electrons as their group number.

Atoms of elements in **Groups 13–18** have 10 fewer valence electrons than their group number. However, helium atoms have only 2 valence electrons.

Atoms of elements in **Groups 3–12** do not have a rule relating their valence electrons to their group number.

1	2	3	4	5	6	7	8	9	10	11	12	13	14	15	16	17	18
H																	He
Li	Be											B	C	N	O	F	Ne
Na	Mg											Al	Si	P	S	Cl	Ar
K	Ca	Sc	Ti	V	Cr	Mn	Fe	Co	Ni	Cu	Zn	Ga	Ge	As	Se	Br	Kr
Rb	Sr	Y	Zr	Nb	Mo	Tc	Ru	Rh	Pd	Ag	Cd	In	Sn	Sb	Te	I	Xe
Cs	Ba	La	Hf	Ta	W	Re	Os	Ir	Pt	Au	Hg	Tl	Pb	Bi	Po	At	Rn
Fr	Ra	Ac	Rf	Db	Sg	Bh	Hs	Mt	Dm	Uuu	Uub		Uuq				

CONNECTION TO Social Studies

WRITING SKILL **History of a Noble Gas** When Dmitri Mendeleev organized the first periodic table, he did not include the noble gases. The noble gases had not been discovered at that time. Research the history of the discovery of one of the noble gases. Write a paragraph in your **science journal** to summarize what you learned.

Valence Electrons and the Periodic Table

You can use a model to determine the number of valence electrons of an atom. But what would you do if you didn't have a model? You can use the periodic table to determine the number of valence electrons for atoms of some elements.

Elements are grouped based on similar properties. Within a group, or family, the atoms of each element have the same number of valence electrons. So, the group numbers can help you determine the number of valence electrons for some atoms, as shown in **Figure 4.**

To Bond or Not to Bond

Not all atoms bond in the same manner. In fact, some atoms rarely bond at all! The number of electrons in the outermost energy level of an atom determines whether an atom will form bonds.

Atoms of the noble gases (Group 18) do not usually form chemical bonds. Atoms of Group 18 elements (except helium) have 8 valence electrons. Having 8 valence electrons is a special condition. In fact, atoms that have 8 electrons in their outermost energy level do not usually form bonds. The outermost energy level of an atom is considered to be full if the energy level contains 8 electrons.

✓ **Reading Check** The atoms of which group in the periodic table rarely form chemical bonds?

INCLUSION Strategies

- Behavior Control Issues   • Gifted and Talented
- Attention Deficit Disorder

Friendly competition makes learning more fun for many students. Give students a chance to understand valence electrons by dividing the students into teams, randomly calling out different elements, and awarding points to the first team that holds up a card with the correct number of valence electrons.

LS Interpersonal

Answer to Reading Check

Atoms in Group 18 (the noble gases) rarely form chemical bonds.

Filling The Outermost Level

An atom that has fewer than 8 valence electrons is much more likely to form bonds than an atom that has 8 valence electrons is. Atoms bond by gaining, losing, or sharing electrons to have a filled outermost energy level. A filled outermost level contains 8 valence electrons. **Figure 5** describes how atoms can achieve a filled outermost energy level.

Is Two Electrons a Full Set?

Not all atoms need 8 valence electrons to have a filled outermost energy level. Helium atoms need only 2 valence electrons. The outermost energy level in a helium atom is the first energy level. The first energy level of any atom can hold only 2 electrons. So, the outermost energy level of a helium atom is full if the energy level has only 2 electrons. Atoms of hydrogen and lithium also form bonds by gaining, losing, or sharing electrons to achieve 2 electrons in the first energy level.

Figure 5 Filling Outermost Energy Levels

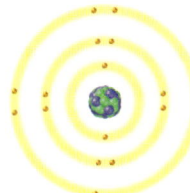

Sulfur
An atom of sulfur has 6 valence electrons. It can have 8 valence electrons by sharing 2 electrons with or gaining 2 electrons from other atoms.

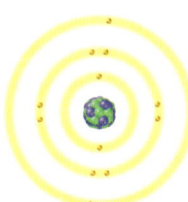

Magnesium
An atom of magnesium has 2 valence electrons. It can have a full outer level by losing 2 electrons. The second energy level becomes the outermost energy level and contains 8 electrons.

SECTION Review

Summary

- Chemical bonding is the joining of atoms to form new substances. A chemical bond is an interaction that holds two atoms together.

- A valence electron is an electron in the outermost energy level of an atom.

- Most atoms form bonds by gaining, losing, or sharing electrons until they have 8 valence electrons. Atoms of some elements need only 2 electrons to fill their outermost level.

Using Key Terms

1. Use the following terms in the same sentence: *chemical bond* and *valence electron*.

Understanding Key Ideas

2. Which of the following atoms do not usually form bonds?
   - **a.** calcium
   - **b.** neon
   - **c.** hydrogen
   - **d.** oxygen

3. Describe chemical bonding.

4. Explain how to use the valence electrons in an atom to predict if the atom will form bonds.

Critical Thinking

5. **Making Inferences** How can an atom that has 5 valence electrons achieve a full set of valence electrons?

6. **Applying Concepts** Identify the number of valence electrons in a barium atom.

Interpreting Graphics

7. Look at the model below. How many valence electrons are in a fluorine atom? Will fluorine atoms form bonds? Explain.

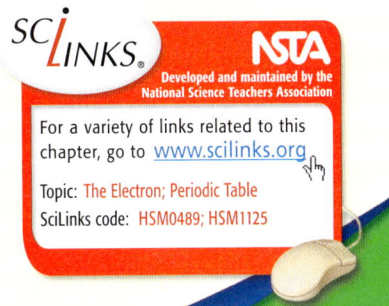

Fluorine

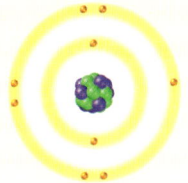

Developed and maintained by the National Science Teachers Association

For a variety of links related to this chapter, go to www.scilinks.org

Topic: The Electron; Periodic Table
SciLinks code: HSM0489; HSM1125

CHAPTER RESOURCES

Chapter Resource File

- Section Quiz **GENERAL**
- Section Review **GENERAL**
- Vocabulary and Section Summary **GENERAL**

Technology

Transparencies
- Determining the Number of Valence Electrons

Focus

Overview

This section introduces ionic bonds and describes how ions are formed. Students will also learn about the properties of ionic compounds.

 **Bellringer**

Salts are ionic compounds. Have students brainstorm uses for salts, things that contain salts, or words and phrases containing the term *salt*. (Sample answer: salt water, rubbing salt in a wound, salt on icy roads, salt in tears)

Motivate

Discussion ——— GENERAL

Penny Analogy Most convenience stores have a small container near the cash register with pennies in it. Discuss with students what purpose this container of pennies serves. (Many people leave a penny or two in it when they receive change. Others take a penny or two when they need to.) Discuss with students how this give-a-penny and take-a-penny strategy is similar to what atoms do when they form ionic bonds. **LS Logical**

READING WARM-UP

Objectives
- Explain how ionic bonds form.
- Describe how positive ions form.
- Describe how negative ions form.
- Explain why ionic compounds are neutral.

Terms to Learn
ionic bond
ion
crystal lattice

READING STRATEGY

Paired Summarizing Read this section silently. In pairs, take turns summarizing the material. Stop to discuss ideas that seem confusing.

ionic bond a bond that forms when electrons are transferred from one atom to another, which results in a positive ion and a negative ion

ion a charged particle that forms when an atom or group of atoms gains or loses one or more electrons

Ionic Bonds

Have you ever accidentally tasted sea water? If so, you probably didn't enjoy it. What makes sea water taste different from the water in your home?

Sea water tastes different because salt is dissolved in it. One of the salts in sea water is the same as the salt that you eat. The chemical bonds in salt are ionic (ie AHN ik) bonds.

Forming Ionic Bonds

An **ionic bond** is a bond that forms when electrons are transferred from one atom to another atom. During ionic bonding, one or more valence electrons are transferred from one atom to another. Like all chemical bonds, ionic bonds form so that the outermost energy levels of the atoms in the bonds are filled. **Figure 1** shows another substance that contains ionic bonds.

Charged Particles

An atom is neutral because the number of electrons in an atom equals the number of protons. So, the charges of the electrons and protons cancel each other. A transfer of electrons between atoms changes the number of electrons in each atom. But the number of protons stays the same in each atom. The negative charges and positive charges no longer cancel out, and the atoms become ions. **Ions** are charged particles that form when atoms gain or lose electrons. An atom normally cannot gain electrons without another atom nearby to lose electrons (or cannot lose electrons without a nearby atom to gain them). But it is easier to study the formation of ions one at a time.

✔ **Reading Check** Why are atoms neutral? (*See the Appendix for answers to Reading Checks.*)

Figure 1 Calcium carbonate in this snail's shell contains ionic bonds.

CHAPTER RESOURCES

Chapter Resource File

- Lesson Plan
- Directed Reading A **BASIC**
- Directed Reading B **SPECIAL NEEDS**

Technology

Transparencies
- Bellringer
- Forming Positive and Negative Ions

Answer to Reading Check
Atoms are neutral because the number of protons in an atom always equals the number of electrons in the atom.

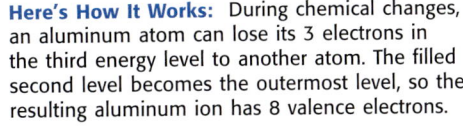
Figure 2 Forming Positive Ions

Here's How It Works: During chemical changes, a sodium atom can lose its 1 electron in the third energy level to another atom. The filled second level becomes the outermost level, so the resulting sodium ion has 8 valence electrons.

Here's How It Works: During chemical changes, an aluminum atom can lose its 3 electrons in the third energy level to another atom. The filled second level becomes the outermost level, so the resulting aluminum ion has 8 valence electrons.

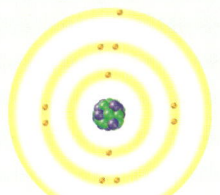

Sodium atom (Na)
11+	protons
11−	electrons
0	charge

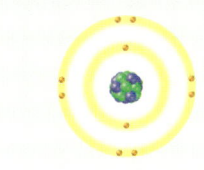

Sodium ion (Na$^+$)
11+	protons
10−	electrons
1+	charge

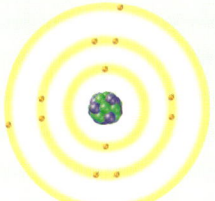

Aluminum atom (Al)
13+	protons
13−	electrons
0	charge

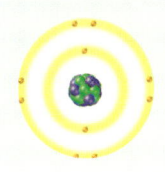

Aluminum ion (Al^{3+})
13+	protons
10−	electrons
3+	charge

Forming Positive Ions

Ionic bonds form during chemical changes when atoms pull electrons away from other atoms. The atoms that lose electrons form ions that have fewer electrons than protons. Because the positive charges outnumber the negative charges, these ions have a positive charge.

Metal Atoms and the Loss of Electrons

Atoms of most metals have few valence electrons. Metal atoms tend to lose these valence electrons and form positive ions. Look at the models in **Figure 2.** When a sodium atom loses its only valence electron to another atom, the sodium atom becomes a sodium ion. A sodium ion has 1 more proton than it has electrons. So, the sodium ion has a 1+ charge. The chemical symbol for this ion is written as Na$^+$. Notice that the charge is written to the upper right of the chemical symbol. **Figure 2** also shows a model for the formation of an aluminum ion.

The Energy Needed to Lose Electrons

Energy is needed to pull electrons away from atoms. Only a small amount of energy is needed to take electrons from metal atoms. In fact, the energy needed to remove electrons from atoms of elements in Groups 1 and 2 is so small that these elements react very easily. The energy needed to take electrons from metals comes from the formation of negative ions.

Studying Salt

Spread several grains of salt on a dark sheet of construction paper. Use a magnifying lens to examine the salt. Ask an adult at home to examine the salt. Discuss what you saw. Then, gently tap the salt with a small hammer. Examine the salt again. Describe your observations in your **science journal.**

SCIENCE HUMOR

Q: What do you call a bond that is fond of sarcasm?

A: an ironic bond

WEIRD SCIENCE

An ionic compound has properties different from those of the elements that form it. Table salt, or sodium chloride, is a good example. Elemental sodium is highly reactive—when it is placed in water, it bursts into flame! Elemental chlorine gas is toxic to humans. But when these two elements join, the resultant compound is nonreactive and harmless.

Answer to Math Practice
$(16+) + (18-) = 2-S^{2-}$; sulfide ion

Reteaching — BASIC

Never Transfer Protons Ions form by transferring electrons, never by transferring protons. To reinforce this concept, draw models of a sodium atom and a chlorine atom on the board that show the atom's electrons. Then, draw an arrow to show how the valence electron from the sodium is transferred to the chlorine atom. **LS Visual**

Quiz — GENERAL

1. How does an atom develop a charge? (by gaining or losing electrons)

2. What is a crystal lattice? (the regular pattern in which a crystal is arranged)

Alternative Assessment — GENERAL

Ion Model Have students build models of atoms with "moveable" electrons. For example, a stack of quarters can represent the nucleus of an atom and pennies around the stack can represent the electrons. Have students use their models to demonstrate the formation of positive and negative ions. **LS Kinesthetic**

Answer to Reading Check

Atoms in Group 17 give off the most energy when forming negative ions.

MATH PRACTICE

Calculating Charge

Calculating the charge of an ion is the same as adding integers (positive or negative whole numbers and 0) that have opposite signs. You write the number of protons as a positive integer and the number of electrons as a negative integer. Then, you add the integers. Calculate the charge of an ion that contains 16 protons and 18 electrons. Write the ion's symbol and name.

Forming Negative Ions

Some atoms gain electrons from other atoms during chemical changes. The ions that form have more electrons than protons. So, these ions have a negative charge.

Nonmetal Atoms Gain Electrons

The outermost energy level of nonmetal atoms is almost full. Only a few electrons are needed to fill the outer level of a nonmetal atom. So, atoms of nonmetals tend to gain electrons from other atoms. Look at the models in **Figure 3.** When an oxygen atom gains 2 electrons, it becomes an oxide ion that has a 2– charge. The symbol for the oxide ion is O^{2-}. Notice that the name of the negative ion formed from oxygen ends with *-ide*. This ending is used for the names of the negative ions formed when atoms gain electrons. **Figure 3** also shows a model of how a chloride ion is formed.

The Energy of Gaining Electrons

Energy is given off by most nonmetal atoms when they gain electrons. The more easily an atom gains an electron, the more energy the atom releases. Atoms of Group 17 elements give off the most energy when they gain an electron. These elements are very reactive. An ionic bond will form between a metal and a nonmetal if the nonmetal releases more energy than is needed to take electrons from the metal.

✓ **Reading Check** Atoms of which group on the periodic table give off the most energy when forming negative ions?

Figure 3 Forming Negative Ions

Here's How It Works: During chemical changes, an oxygen atom gains 2 electrons in the second energy level from another atom. An oxide ion that has 8 valence electrons is formed. Thus, its outermost energy level is filled.

Here's How It Works: During chemical changes, a chlorine atom gains 1 electron in the third energy level from another atom. A chloride ion that has 8 valence electrons is formed. Thus, its outermost energy level is filled.

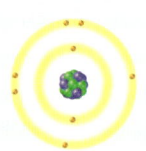
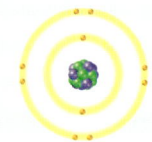
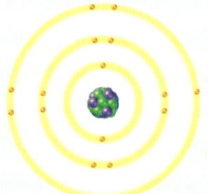
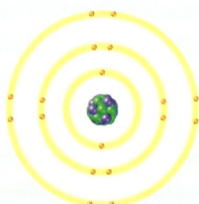

Oxygen atom (O)	
8+	protons
8–	electrons
0	charge

Oxide ion (O^{2-})	
8+	protons
10–	electrons
2–	charge

Chlorine atom (Cl)	
17+	protons
17–	electrons
0	charge

Chloride ion (Cl^-)	
17+	protons
18–	electrons
1–	charge

INCLUSION Strategies

- *Learning Disabled*   • *Developmentally Delayed*
- *Hearing Impaired*

Help students understand the concept of adding positive and negative charges. Ask six students to pair up. Have one student from each pair tape a minus sign to his or her shirt, and have the rest tape plus signs to their shirts. Explain that the group is neutral because the number of pluses is equal to the number of minuses. Then, ask one "negative" to sit down. Ask the partners to stand together. Explain that the group is no longer neutral because a "positive" has no partner. Tell students that this is what happens in atoms when an atom loses an electron—the atom is no longer neutral. Tell students that if the group in front of the room were an ion, it would be a positive ion because it has an extra, unmatched positive part. **English Language Learners**
LS Kinesthetic/Interpersonal

Ionic Compounds

When ionic bonds form, the number of electrons lost by the metal atoms equals the number gained by the nonmetal atoms. The ions that bond are charged, but the compound formed is neutral because the charges of the ions cancel each other. When ions bond, they form a repeating three-dimensional pattern called a **crystal lattice** (KRIS tuhl LAT is), like the one shown in **Figure 4.** The strong attraction between ions in a crystal lattice gives ionic compounds certain properties, which include brittleness, high melting points, and high boiling points.

crystal lattice the regular pattern in which a crystal is arranged

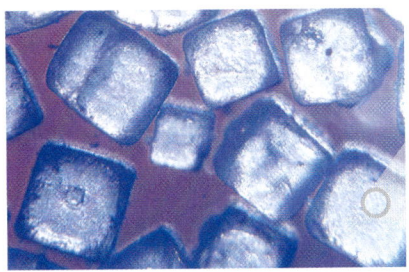

Figure 4 *This model of the crystal lattice of sodium chloride, or table salt, shows a three-dimensional view of the bonded ions. In the model, the sodium ions are pink and the chloride ions are green.*

SECTION Review

Summary

- An ionic bond is a bond that forms when electrons are transferred from one atom to another. During ionic bonding, the atoms become oppositely charged ions.
- Ionic bonding usually occurs between atoms of metals and atoms of nonmetals.
- Energy is needed to remove electrons from metal atoms. Energy is released when most nonmetal atoms gain electrons.

Using Key Terms

1. Use the following terms in the same sentence: *ion* and *ionic bond*.

2. In your own words, write a definition for the term *crystal lattice*.

Understanding Key Ideas

3. Which types of atoms usually become negative ions?
   a. metals
   b. nonmetals
   c. noble gases
   d. All of the above

4. How does an atom become a positive ion? a negative ion?

5. What are two properties of ionic compounds?

Math Skills

6. What is the charge of an ion that has 12 protons and 10 electrons? Write the ion's symbol.

Critical Thinking

7. **Applying Concepts** Which group of elements gains two valence electrons when the atoms form ionic bonds?

8. **Identifying Relationships** Explain why ionic compounds are neutral even though they are made up of charged particles.

9. **Making Comparisons** Compare the formation of positive ions with the formation of negative ions in terms of energy changes.

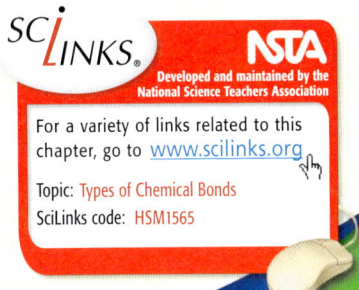

SCILINKS

NSTA
Developed and maintained by the National Science Teachers Association

For a variety of links related to this chapter, go to www.scilinks.org

Topic: Types of Chemical Bonds
SciLinks code: HSM1565

Focus

Overview

In this section, students will learn how covalent and metallic bonds are formed. Students will also learn how to draw electron-dot diagrams and will study the properties of metals.

 Bellringer

Give students one minute to brainstorm a list of things made of metal. Then, ask them to use their list to describe three properties of metals.

Motivate

ACTIVITY ————— GENERAL

Cereal-Dot Diagrams Use cereal pieces to represent electrons when making electron-dot diagrams. Have students write chemical symbols on index cards and place the correct number of cereal pieces around the symbol. Once students learn how to place the cereal pieces, have them make cereal-dot diagrams of water, H_2O, and ammonia, NH_3. Have students use cereal pieces of a different color for each atom to help them see where each electron originated. For example, a student may use green cereal pieces for oxygen electrons and red cereal pieces for the hydrogen electrons. **LS Kinesthetic**

READING WARM-UP

Objectives
- Explain how covalent bonds form.
- Describe molecules.
- Explain how metallic bonds form.
- Describe the properties of metals.

Terms to Learn
covalent bond
molecule
metallic bond

READING STRATEGY

Reading Organizer As you read this section, create an outline of the section. Use the headings from the section in your outline.

covalent bond a bond formed when atoms share one or more pairs of electrons

Covalent and Metallic Bonds

Imagine bending a wooden coat hanger and a wire coat hanger. The wire one would bend easily, but the wooden one would break. Why do these things behave differently?

One reason is that the bonds between the atoms of each object are different. The atoms of the wooden hanger are held together by covalent bonds (KOH VAY luhnt BAHNDZ). But the atoms of the wire hanger are held together by metallic bonds. Read on to learn about the difference between these kinds of chemical bonds.

Covalent Bonds

Most things around you, such as water, sugar, oxygen, and wood, are held together by covalent bonds. Substances that have covalent bonds tend to have low melting and boiling points and are brittle in the solid state. For example, oxygen has a low boiling point, which is why it is a gas at room temperature. And wood is brittle, so it breaks when bent.

A **covalent bond** forms when atoms share one or more pairs of electrons. When two atoms of nonmetals bond, a large amount of energy is needed for either atom to lose an electron. So, two nonmetals don't transfer electrons to fill the outermost energy levels of their atoms. Instead, two nonmetal atoms bond by sharing electrons with each another, as shown in the model in **Figure 1**.

✓ **Reading Check** What is a covalent bond? (*See the Appendix for answers to Reading Checks.*)

Figure 1 *By sharing electrons in a covalent bond, each hydrogen atom (the smallest atom) has a full outermost energy level containing two electrons.*

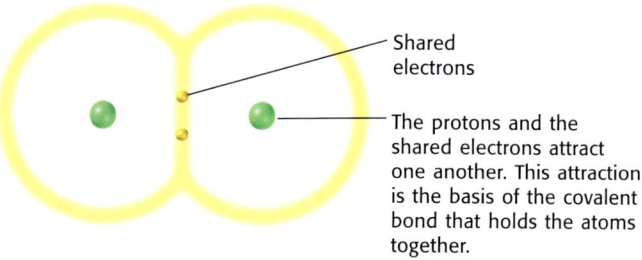

Shared electrons

The protons and the shared electrons attract one another. This attraction is the basis of the covalent bond that holds the atoms together.

CHAPTER RESOURCES

Chapter Resource File

- Lesson Plan
- Directed Reading A **BASIC**
- Directed Reading B **SPECIAL NEEDS**

Technology

Transparencies
- Bellringer
- Covalent Bond
- Covalent Bonds in a Water Molecule

Answer to Reading Check

A covalent bond is a bond that forms when atoms share one or more pairs of electrons.

Figure 2 — Covalent Bonds in a Water Molecule

The oxygen atom shares one of its electrons with each of the two hydrogen atoms. It now has an outer level filled with 8 electrons.

Each hydrogen atom shares its 1 electron with the oxygen atom. Each hydrogen atom now has an outer level filled with 2 electrons.

This electron-dot diagram for water shows only the outermost level of electrons for each atom. But you still see how the atoms share electrons.

Covalent Bonds and Molecules

Substances containing covalent bonds consist of individual particles called molecules (MAHL i KYOOLZ). A **molecule** usually consists of two or more atoms joined in a definite ratio. A hydrogen molecule is composed of two covalently bonded hydrogen atoms. However, most molecules are composed of atoms of two or more elements. The models in **Figure 2** show two ways to represent the covalent bonds in a water molecule.

One way to represent atoms and molecules is to use electron-dot diagrams. An electron-dot diagram is a model that shows only the valence electrons in an atom. Electron-dot diagrams can help you predict how atoms might bond. To draw an electron-dot diagram, write the symbol of the element and place one dot around the symbol for every valence electron in the atom, as shown in **Figure 3.** Place the first 4 dots alone on each side, and then pair up any remaining dots.

molecule the smallest unit of a substance that keeps all of the physical and chemical properties of that substance

Figure 3 — Using Electron–Dot Diagrams

Carbon atoms have 4 valence electrons. A carbon atom needs 4 more electrons to have a filled outermost energy level.

Oxygen atoms have 6 valence electrons. An oxygen atom needs only 2 more electrons to have a filled outermost energy level.

Krypton atoms have 8 valence electrons. Krypton is nonreactive. Krypton atoms do not need any more electrons.

This diagram represents a hydrogen molecule. The dots between the letters represent a pair of shared electrons.

SCIENCE HUMOR

Q: What is the one thing that atoms in molecules do not have to teach their children?

A: how to share with others

MISCONCEPTION ALERT

Electron Sharing In certain covalent compounds, atoms may have fewer or more than 8 electrons in their outer energy level. In boron trifluoride, BF_3, the boron atom has only 6 electrons in its outer energy level. In sulfur hexafluoride, SF_6, sulfur has 12 electrons in its outer energy level. For the purposes of this chapter, students can assume that sharing electrons to have 8 electrons in the outer energy level is the rule.

ACTIVITY — ADVANCED

Hydrogen Bonds In addition to the covalent bonds that link hydrogen atoms to an oxygen atom in a molecule of water, another type of bond is important in water. This bond, which is actually an intermolecular force, is called a *hydrogen bond*. Have students investigate hydrogen bonding and the properties they impart to water and other substances. Ask students to make a poster to illustrate what they learn. **LS Visual**

Making Models — GENERAL

Three-Dimensional Models
Demonstrate how to make three-dimensional models of hydrogen sulfide, H_2S, molecules using gumdrops and toothpicks. One color of gumdrop represents the sulfur atom, and another color represents the two hydrogen atoms. Use toothpicks to "bond" the hydrogen gumdrops to the sulfur gumdrop. Give students gumdrops and toothpicks, and have them make their own models of ammonia, NH_3, and methane, CH_4. **English Language Learners** **LS Kinesthetic**

Discussion — GENERAL

Double and Triple Bonds
Explain to students that atoms can form double or triple covalent bonds if they need more than one electron to complete their outermost energy level. An oxygen atom forms a double bond with another oxygen atom. Draw the electron-dot diagram for oxygen, O_2, on the chalkboard. Point out that each oxygen atom's outer energy level has 4 shared electrons (2 per bond) and 4 unshared electrons, so the outermost energy level has a total of 8 electrons. Ask students to make electron-dot diagrams for nitrogen, N_2, and carbon dioxide, CO_2. **LS Logical**

Valence Electrons Remind students that they can use the periodic table to find the number of valence electrons for atoms of elements in Groups 13–18. The atoms in these groups have 10 fewer valence electrons than their group number. Display a periodic table, and have students determine the number of valence electrons for silicon (6) and iodine (7). **LS Visual/Logical**

ACTIVITY — GENERAL

Drawing Diagrams Follow these guidelines to construct electron-dot diagrams for chlorine gas, Cl_2, and ammonia, NH_3:

1. Add up the valence electrons for all of the atoms that make up the molecule.

2. Use one pair of electrons to indicate the bond(s) shared by atoms.

3. Arrange the remaining electrons to form a stable molecule. Each atom (except hydrogen) needs 8 electrons to fill its outermost energy level. **English Language Learners**
LS Visual

Answer to Reading Check

There are two atoms in a diatomic molecule.

Figure 4 The water in this fishbowl is made up of many tiny water molecules. Each molecule is the smallest particle that has the chemical properties of water.

INTERNET ACTIVITY

For another activity related to this chapter, go to **go.hrw.com** and type in the keyword **HP5BNDW**.

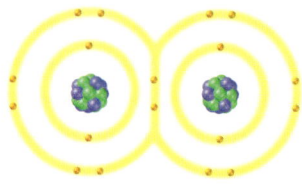

Figure 5 Two covalently bonded fluorine atoms have filled outermost energy levels. The two electrons shared by the atoms are counted as valence electrons for each atom.

Covalent Compounds and Molecules

An atom is the smallest particle into which an element can be divided and still be the same element. Likewise, a molecule is the smallest particle into which a covalently bonded compound can be divided and still be the same compound. Look at the three-dimensional models in **Figure 4.** They show how a sample of water is made up of many individual molecules of water. Imagine dividing water over and over. You would eventually end up with a single molecule of water. What would happen if you separated the hydrogen and oxygen atoms that make up a water molecule? Then, you would no longer have water.

The Simplest Molecules

Molecules are composed of at least two covalently bonded atoms. The simplest molecules are made up of two bonded atoms. Molecules made up of two atoms of the same element are called *diatomic molecules*. Elements that are found in nature as diatomic molecules are called *diatomic elements*. Hydrogen is a diatomic element. Oxygen, nitrogen, and the halogens fluorine, chlorine, bromine, and iodine are also diatomic elements. Look at **Figure 5.** The shared electrons are counted as valence electrons for each atom. So, both atoms of the molecule have filled outer energy levels.

✓ **Reading Check** How many atoms are in a diatomic molecule?

Is That a Fact!

Even though a molecule of water is bigger than the hydrogen and oxygen atoms it comprises, it is still an extremely tiny particle. There are about 2 million quadrillion (2 followed by 21 zeros) water molecules in a single drop of water!

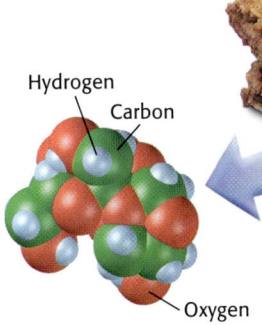

Hydrogen
Carbon
Oxygen

Figure 6 *A granola bar contains sucrose, or table sugar. A molecule of sucrose is composed of carbon atoms, hydrogen atoms, and oxygen atoms joined by covalent bonds.*

More-Complex Molecules

Diatomic molecules are the simplest molecules. They are also some of the most important molecules. You could not live without diatomic oxygen molecules. But other important molecules are much more complex. Soap, plastic bottles, and even proteins in your body are examples of complex molecules. Carbon atoms are the basis of many of these complex molecules. Each carbon atom needs to make four covalent bonds to have 8 valence electrons. These bonds can be with atoms of other elements or with other carbon atoms, as shown in the model in **Figure 6.**

Metallic Bonds

Look at the unusual metal sculptures shown in **Figure 7.** Some metal pieces have been flattened, while other metal pieces have been shaped into wires. How could the artist change the shape of the metal into all of these different forms without breaking the metal into pieces? Metal can be shaped because of the presence of a metallic bond, a special kind of chemical bond. A **metallic bond** is a bond formed by the attraction between positively charged metal ions and the electrons in the metal. Positively charged metal ions form when metal atoms lose electrons.

metallic bond a bond formed by the attraction between positively charged metal ions and the electrons around them

Figure 7 *The different shapes of metal in these sculptures are possible because of the bonds that hold the metal together.*

CONNECTION TO Biology

Proteins Proteins perform many functions throughout your body. A single protein can have thousands of covalently bonded atoms. Proteins are built from smaller molecules called *amino acids*. Make a poster showing how amino acids are joined to make proteins.

ACTIVITY

CHAPTER RESOURCES

Technology

Transparencies
• **LINK TO LIFE SCIENCE** Making of a Protein A and B

Reteaching — BASIC

Malleability and Ductility To help students understand the difference between malleability and ductility, show students metal objects of different shapes. Ask students to identify whether malleability or ductility was more important to molding the metal into each shape. **LS Visual**

Quiz — GENERAL

1. What is the smallest particle of a covalently bonded compound? (a molecule)

2. List three common materials that contain covalent bonds (Sample answer: water, soap, and plastic bottles)

3. Why can metals conduct electric current? (Metals can conduct electric current because the electrons in metallic bonds are free to move around.)

Alternative Assessment — GENERAL

Bonding Charades Organize students in small groups. Have them develop a charade that depicts either covalent or metallic bonding. Students may want to use a few props, such as balls to represent electrons. Have each group present a charade. Then, have the remainder of the class determine the type of bond being portrayed. **LS Kinesthetic**

Answer to Reading Check

Ductility is the ability to be drawn into wires.

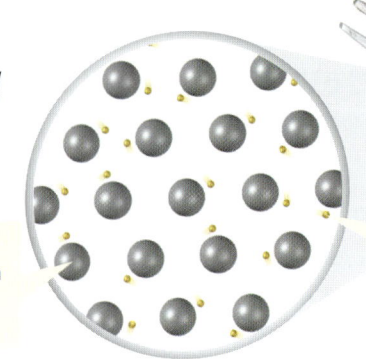

Figure 8 *Moving electrons are attracted to the metal ions, and the attraction forms metallic bonds.*

The positive metal ions are in fixed positions in the metal.

Negative electrons are free to move.

Bending with Bonds

1. Straighten out a **wire paper clip.** Record your observations.

2. Bend a **piece of chalk.** Record your observations.

3. Chalk is composed of calcium carbonate, a compound containing ionic bonds. What kind of bond is present in the paper clip?

4. Explain why you could change the shape of the paper clip but could not bend the chalk without breaking it.

Movement of Electrons Throughout a Metal

Bonding in metals is a result of the metal atoms being so close to one another that their outermost energy levels overlap. This overlapping allows valence electrons to move throughout the metal, as shown in **Figure 8.** You can think of a metal as being made up of positive metal ions that have enough valence electrons "swimming" around to keep the ions together. The electrons also cancel the positive charge of the ions. Metallic bonds extend throughout the metal in all directions.

Properties of Metals

Metallic bonding is what gives metals their particular properties. These properties include electrical conductivity, malleability, and ductility.

Conducting Electric Current

Metallic bonding allows metals to conduct electric current. For example, when you turn on a lamp, electrons move within the copper wire that connects the lamp to the outlet. The electrons that move are the valence electrons in the copper atoms. These electrons are free to move because the electrons are not connected to any one atom.

Reshaping Metals

Because the electrons swim freely around the metal ions, the atoms in metals can be rearranged. As a result, metals can be reshaped. The properties of *ductility* (the ability to be drawn into wires) and *malleability* (the ability to be hammered into sheets) describe a metal's ability to be reshaped. For example, copper is made into wires for use in electrical cords. Aluminum can be pounded into thin sheets and made into aluminum foil.

✓ **Reading Check** What is ductility?

MATERIALS

FOR EACH STUDENT
- chalk, piece
- paper clip, wire

Safety Caution: Remind students to review all safety cautions and icons before beginning this lab activity.

Answers

3. metallic bonds

4. The metallic bonds give the paper clip the ability to bend without breaking because the electrons move within the metal. The ionic bonds in the piece of chalk cause the chalk to be brittle.

Bending Without Breaking

When a piece of metal is bent, some of the metal ions are forced closer together. You might expect the metal to break because all of the metal ions are positively charged. Positively charged ions repel one another. However, positive ions in a metal are always surrounded by and attracted to the electrons in the metal—even if the metal ions move. The electrons constantly move around and between the metal ions. The moving electrons maintain the metallic bonds no matter how the shape of the metal changes. So, metal objects can be bent without being broken, as shown in **Figure 9.**

Figure 9 *Metal can be reshaped without breaking because metallic bonds occur in many directions.*

SECTION Review

Summary

- In covalent bonding, two atoms share electrons. A covalent bond forms when atoms share one or more pairs of electrons.
- Covalently bonded atoms form a particle called a *molecule*. A molecule is the smallest particle of a compound that has the chemical properties of the compound.
- In metallic bonding, the valence electrons move throughout the metal. A bond formed by the attraction between positive metal ions and the electrons in the metal is a metallic bond.
- Properties of metals include conductivity, ductility, and malleability.

Using Key Terms

1. Use each of the following terms in a separate sentence: *covalent bond* and *metallic bond*.

2. In your own words, write a definition for the term *molecule*.

Understanding Key Ideas

3. Between which of the following atoms is a covalent bond most likely to occur?
   a. calcium and lithium
   b. sodium and fluorine
   c. nitrogen and oxygen
   d. helium and argon

4. What happens to the electrons in covalent bonding?

5. How many dots does an electron-dot diagram of a sulfur atom have?

6. List three properties of metals that are a result of metallic bonds.

7. Describe how the valence electrons in a metal move.

8. Explain the difference between ductility and malleability. Give an example of when each property is useful.

Critical Thinking

9. **Identifying Relationships** How do the metallic bonds in a staple allow it to function properly?

10. **Applying Concepts** Draw an electron-dot diagram for ammonia (a nitrogen atom covalently bonded to three hydrogen atoms).

Interpreting Graphics

11. This electron-dot diagram is not complete. Which atom needs to form another bond? Explain.

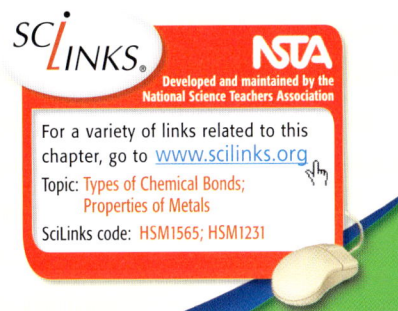

For a variety of links related to this chapter, go to www.scilinks.org
Topic: Types of Chemical Bonds; Properties of Metals
SciLinks code: HSM1565; HSM1231

Answers to Section Review

1. Sample answer: A carbon monoxide molecule contains a covalent bond. Metals can be reshaped because of the nature of the metallic bond.

2. Sample answer: A molecule is the smallest piece of a substance joined by covalent bonds.

3. c

4. The electrons are shared between two atoms.

5. 6

6. Three properties that are a result of metallic bonds are ductility, malleability, and electrical conductivity.

7. The valence electrons in a metal "swim" around the metal ions to hold the ions together.

8. Sample answer: Ductility is the ability to be drawn into wires. Malleability is the ability to be pounded into thin sheets. Copper can be drawn into long wires. Long wires are useful to connect a television to an outlet that is on the other side of the room. Steel can be pounded into thin sheets. The sheets are useful for building car bodies.

9. Sample answer: A staple must bend to hold papers together. The metallic bonds in the staple allow it to bend without breaking.

10. Students should draw an N surrounded by 8 dots: 2 above, 2 on the right, 2 on the left, and 2 below. Three Hs should be drawn, one on each of three sides of the N.

11. The carbon atom needs to form another bond. The electron dot diagram shows that carbon has only 7 valence electrons. It needs one more electron to have a filled outer energy level (8 valence electrons).

Covalent Marshmallows

Teacher's Notes

Time Required
One 45-minute class period

Lab Ratings

EASY ——————————→ HARD

Teacher Prep 🧪🧪
Student Set-Up 🧪
Concept Level 🧪🧪
Clean Up 🧪

MATERIALS

Materials listed are for one to two students. Colored marshmallows are available in some grocery stores. To create different colored marshmallows, "paint" the marshmallows lightly with diluted food coloring. To discourage students from eating the marshmallows, dust the marshmallows lightly with alum, a bitter spice that can be purchased at a grocery store. An alternative method of coloring the marshmallows is to spray them lightly with hair spray, and sprinkle them with different colors of glitter. Be sure students do not eat the marshmallows.

Safety Caution
Remind students to review all safety cautions and icons before beginning this lab activity.

Model-Making Lab

OBJECTIVES

Build a three-dimensional model of a water molecule.

Draw an electron-dot diagram of a water molecule.

MATERIALS

- marshmallows (two of one color, one of another color)
- toothpicks

SAFETY

Covalent Marshmallows

A hydrogen atom has 1 electron in its outer energy level, but 2 electrons are required to fill its outer level. An oxygen atom has 6 electrons in its outer level, but 8 electrons are required to fill its outer level. To fill their outer energy levels, two atoms of hydrogen and one atom of oxygen can share electrons, as shown below. Such a sharing of electrons to fill the outer level of atoms is called *covalent bonding*. When hydrogen and oxygen bond in this manner, a molecule of water is formed. In this lab, you will build a three-dimensional model of water to better understand the covalent bonds formed in a water molecule.

A Model of a Water Molecule

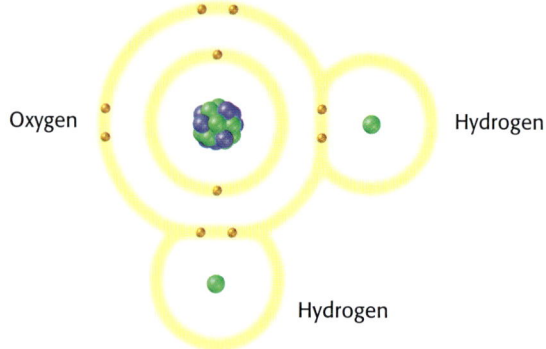

Procedure

1 Using the marshmallows and toothpicks, create a model of a water molecule. Use the diagram above for guidance in building your model.

2 Draw a sketch of your model. Be sure to label the hydrogen and oxygen atoms on your sketch.

3 Draw an electron-dot diagram of the water molecule.

Procedure Notes
To extend this activity, you may use additional colors to create marshmallow models of various molecules. In addition, different-sized marshmallows can be used to represent the relative sizes of atoms of different elements.

CHAPTER RESOURCES

Chapter Resource File
- Datasheet for Chapter Lab
- Lab Notes and Answers

Technology

Classroom Videos
- Lab Video

Analyze the Results

1 **Classifying** What do the marshmallows represent? What do the toothpicks represent?

2 **Evaluating Models** Why are the marshmallows different colors?

3 **Analyzing Results** Compare your model with the diagram on the previous page. How might your model be improved to more accurately represent a water molecule?

Draw Conclusions

4 **Making Predictions** Hydrogen in nature can covalently bond to form hydrogen molecules, H₂. How could you use the marshmallows and toothpicks to model this bond?

5 **Applying Conclusions** Draw an electron-dot diagram of a hydrogen molecule.

6 **Drawing Conclusions** Which do you think would be more difficult to create—a model of an ionic bond or a model of a covalent bond? Explain your answer.

Applying Your Data

Create a model of a carbon dioxide molecule, which consists of two oxygen atoms and one carbon atom. The structure is similar to the structure of water, although the three atoms bond in a straight line instead of at angles. The bond between each oxygen atom and the carbon atom in a carbon dioxide molecule is a *double bond,* so use two connections. Do the double bonds in carbon dioxide appear stronger or weaker than the single bonds in water? Explain your answer.

Analyze the Results

1. The marshmallows represent atoms. The toothpicks represent the pairs of electrons that create the covalent bonds.

2. Marshmallows are different colors to represent atoms of different elements.

3. Accept all reasonable answers. Sample answer: I could use marshmallows of different sizes to show the difference in size between the atoms. Then, I could make sure the atoms form an angle and are not in a straight line.

Draw Conclusions

4. A hydrogen molecule could be modeled by using a toothpick to connect the two hydrogen marshmallows together.

5. An electron-dot diagram of the hydrogen molecule should look like the following: H : H.

6. Accept all reasonable answers. Sample answer: A model of an ionic bond would be more difficult to create because it involves the transfer of electrons. You would have to break off a little piece of one marshmallow and glue it to another marshmallow. But then there is no way to hold the marshmallows together.

Applying Your Data

The double bonds appear stronger than single bonds because there is more "attraction" (more toothpicks) holding atoms together. It is more difficult to separate two shared pairs of electrons (break two toothpicks) than to separate one shared pair of electrons (break one toothpick).

CHAPTER RESOURCES

Workbooks

 Long-Term Projects & Research Ideas
• The Wonders of Water **ADVANCED**

CLASSROOM TESTED & APPROVED

Rebecca Ferguson
North Ridge Middle School
North Richland Hills, Texas

Chapter Review

Assignment Guide

Section	Questions
1	1, 5, 7
2	2, 6, 8, 10–11, 13, 15
3	3–4, 9, 12, 18–19, 21–23
2 and 3	14, 17, 20
1, 2, and 3	16

ANSWERS

Using Key Terms

1. chemical bond
2. ion
3. covalent bond
4. metallic bond
5. valence electron
6. crystal lattice

Understanding Key Ideas

7. b
8. c
9. c
10. a
11. b
12. Answers may include a low melting point, a low boiling point, and brittleness in the solid state. (Students should list at least two properties.)

USING KEY TERMS

Complete each of the following sentences by choosing the correct term from the word bank.

crystal lattice   ionic bond
molecule   chemical bond
chemical bonding   metallic bond
valence electron   ion
covalent bond

1 An interaction that holds two atoms together is a(n) ___.

2 A charged particle that forms when an atom transfers electrons is a(n) ___.

3 A bond formed when atoms share electrons is a(n) ___.

4 Electrons free to move throughout a material are associated with a(n) ___.

5 An electron in the outermost energy level of an atom is a(n) ___.

6 Ionic compounds are bonded in a three-dimensional pattern called a(n) ___.

UNDERSTANDING KEY IDEAS

Multiple Choice

7 Which element has a full outermost energy level containing only two electrons?

a. fluorine, F   c. hydrogen, H
b. helium, He   d. oxygen, O

8 Which of the following describes what happens when an atom becomes an ion with a 2– charge?

a. The atom gains 2 protons.
b. The atom loses 2 protons.
c. The atom gains 2 electrons.
d. The atom loses 2 electrons.

9 The properties of ductility and malleability are associated with which type of bonds?

a. ionic   c. metallic
b. covalent   d. All of the above

10 What type of element tends to lose electrons when it forms bonds?

a. metal   c. nonmetal
b. metalloid   d. noble gas

11 Which pair of atoms can form an ionic bond?

a. sodium, Na, and potassium, K
b. potassium, K, and fluorine, F
c. fluorine, F, and chlorine, Cl
d. sodium, Na, and neon, Ne

Short Answer

12 List two properties of covalent compounds.

13 Explain why an iron ion is attracted to a sulfide ion but not to a zinc ion.

14 Compare the three types of bonds based on what happens to the valence electrons of the atoms.

13. Metal atoms tend to lose electrons and form positive ions. Both iron and zinc are metals, and both form ions that are positively charged. Ions with the same charge repel one another, so an iron ion is not attracted to a zinc ion. Nonmetal atoms tend to gain electrons and form negative ions. Sulfur is a nonmetal, so a sulfide ion is negatively charged. The positively charged iron ion is attracted to the sulfide ion.

14. Ionic bonds involve the transfer of valence electrons between atoms. Covalent bonds involve the sharing of valence electrons between atoms. Metallic bonds involve the movement of valence electrons between many atoms within a metal.

Math Skills

15 For each atom below, write the number of electrons it must gain or lose to have 8 valence electrons. Then, calculate the charge of the ion that would form.

a. calcium, Ca

b. phosphorus, P

c. bromine, Br

d. sulfur, S

CRITICAL THINKING

16 **Concept Mapping** Use the following terms to create a concept map: *chemical bonds, ionic bonds, covalent bonds, metallic bonds, molecule,* and *ions.*

17 **Identifying Relationships** Predict the type of bond each of the following pairs of atoms would form:

a. zinc, Zn, and zinc, Zn

b. oxygen, O, and nitrogen, N

c. phosphorus, P, and oxygen, O

d. magnesium, Mg, and chlorine, Cl

18 **Applying Concepts** Draw electron-dot diagrams for each of the following atoms, and state how many bonds it will have to make to fill its outer energy level.

a. sulfur, S

b. nitrogen, N

c. neon, Ne

d. iodine, I

e. silicon, Si

19 **Predicting Consequences** Using your knowledge of valence electrons, explain the main reason so many different molecules are made from carbon atoms.

20 **Making Inferences** Does the substance being hit in the photo below contain ionic or metallic bonds? Explain your answer.

INTERPRETING GRAPHICS

Use the picture of a wooden pencil below to answer the questions that follow.

21 In which part of the pencil are metallic bonds found?

22 List three materials in the pencil that are composed of molecules that have covalent bonds.

23 Identify two differences between the properties of the material that has metallic bonds and the materials that have covalent bonds.

15. a. lose 2 electrons; 2+

b. gain 3 electrons; 3–

c. gain 1 electron; 1–

d. gain 2 electrons; 2–

Critical Thinking

16. An answer to this exercise can be found at the end of this book.

17. a. metallic

b. covalent

c. covalent

d. ionic

18. a. 6 dots; 2 bonds

b. 5 dots; 3 bonds

c. 8 dots; no bonds

d. 7 dots; 1 bond

e. 4 dots; 4 bonds

19. Carbon atoms have 4 valence electrons. Each carbon atom must make 4 bonds to fill its outermost energy level with 8 electrons. Because each carbon atom can bond with up to 4 atoms (including other carbon atoms), carbon forms the basis of many different compounds.

20. It contains ionic bonds because the substance is breaking into smaller pieces as the hammer hits it. The substance is brittle, so the bonds are more likely to be ionic.

Interpreting Graphics

21. the metal band near the eraser

22. graphite, wood, and rubber (eraser)

23. Sample answer: The metallically bonded material is shiny, and the covalently bonded materials are not shiny. The metal can be bent without breaking, but the wood or graphite will break if bent.

Standardized Test Preparation

Teacher's Note

To provide practice under more realistic testing conditions, give students 20 minutes to answer all of the questions in this Standardized Test Preparation.

MISCONCEPTION ALERT

Answers to the standardized test preparation can help you identify student misconceptions and misunderstandings.

READING

Passage 1

1. B
2. F
3. C

TEST DOCTOR

Question 2: Students may be tempted to select answer choice I because the *Voyager* aircraft flew around the world without refueling. Explain to students that this statement means that *Voyager* did not need additional fuel during its trip, but it was filled with fuel before the flight started.

READING

Read each of the passages below. Then, answer the questions that follow each passage.

Passage 1 In 1987, pilots Richard Rutan and Jeana Yeager flew the *Voyager* aircraft around the world without refueling. The record-breaking trip lasted a little more than nine days. To carry enough fuel for the trip, the plane had to be as lightweight as possible. Using fewer bolts than the number of bolts usually used to attach parts would make the airplane lighter. But without bolts, what would hold the parts together? The designers decided to use glue!

They could not use regular glue. They used superglue. When superglue is applied, it combines with water from the air to form chemical bonds. So, the materials stick together as if they were one material. Superglue is so strong that the weight of a two-ton elephant cannot separate two metal plates glued together with just a few drops!

1. Who are Richard Rutan and Jeana Yeager?
   - A the designers of the *Voyager* aircraft
   - B the pilots of the *Voyager* aircraft
   - C the inventors of superglue
   - D chemists that study superglue

2. In the passage, what does *aircraft* mean?
   - F an airplane
   - G a helicopter
   - H a hot-air balloon
   - I an airplane that doesn't need fuel

3. The author probably wrote this passage to
   - A encourage people to fly airplanes.
   - B tell airplane designers how to make airplanes that need less fuel.
   - C explain why superglue was a good substitute for bolts in the *Voyager* aircraft.
   - D explain why people should buy superglue instead of regular glue.

Passage 2 One of the first contact lenses was developed by a Hungarian physician named Joseph Dallos in 1929. He came up with a way to make a mold of the human eye. He used these molds to make a glass lens that followed the shape of the eye. Unfortunately, the glass lenses he made were not very comfortable.

Many years later, in an effort to solve the comfort problem of contact lenses, Czechoslovakian chemists Otto Wichterle and Drahoslav Lim invented a water-absorbing plastic gel. The lenses made from this gel were soft and <u>pliable</u>, and they allowed air to pass through the lens to the eye. These characteristics made the lenses more comfortable to wear than glass lenses.

1. In the passage, what does *pliable* mean?
   - A able to be bent
   - B very stiff
   - C spongelike
   - D similar to glass

2. Which of the following statements is a fact from the passage?
   - F The first contact lenses were plastic.
   - G Two Hungarian physicians developed a way of making molds of human eyes.
   - H Glass contact lenses were not comfortable.
   - I Joseph Dallos was a chemist.

3. What is a possible reason that glass contact lenses were not comfortable?
   - A Glass contact lenses allow air to pass through the lens to the eye.
   - B Glass contact lenses did not follow the shape of the human eye.
   - C Glass contact lenses absorb water.
   - D Glass contact lenses are very hard.

Passage 2

1. A
2. H
3. D

TEST DOCTOR

Question 3: The reason why glass contact lenses were not comfortable is not directly stated in the passage. However, students should notice that the lenses made of plastic gel are described as both soft and comfortable. Therefore, they can infer that hard, glass lenses are uncomfortable.

INTERPRETING GRAPHICS

The graph below shows chemicals used by the science department at Harding Middle School. Use the graph below to answer the questions that follow.

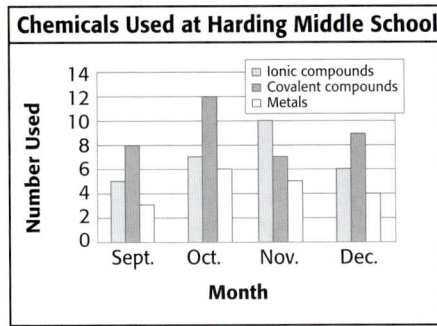

Chemicals Used at Harding Middle School

1. In which month were the most ionic compounds used?

   A September

   B October

   C November

   D December

2. Which type of chemical was used the least number of times?

   F ionic compounds

   G covalent compounds

   H metals

   I both ionic compounds and metals

3. How many covalent compounds were used during all four months?

   A 16

   B 25

   C 28

   D 36

4. In which month were the most compounds used?

   F September

   G October

   H November

   I December

MATH

Read each question below, and choose the best answer.

1. Protons have a charge of 1+ and electrons have a charge of 1−. A magnesium ion has 12 protons and 10 electrons. What is the charge of the ion?

   A 2+

   B 2−

   C 10−

   D 12+

2. Fructose is the chemical name for a sugar found in some fruits. The chemical formula for fructose is $C_6H_{12}O_6$. The C is the symbol for carbon, the O is the symbol for oxygen, and the H is the symbol for hydrogen. The numbers after each letter tell you how many atoms of each element are in one molecule of fructose. What percentage of the atoms in fructose are carbon atoms?

   F 0.25%

   G 6%

   H 25%

   I 33%

3. The density of an object is found by dividing its mass by its volume. Katie has a piece of silver metal that has a mass of 5.4 g and a volume of 2.0 cm³. What is the density of Katie's metal?

   A 0.37 cm³/g

   B 2.7 g/cm³

   C 7.4 g/cm³

   D 10.8 g•cm³

4. Ms. Mazza is a chemistry teacher. During class, her students ask her four to six questions every 10 min. What is a reasonable estimate of the number of questions asked during a 45 min class period?

   F 12 questions

   G 15 questions

   H 23 questions

   I 40 questions

Standardized Test Preparation

INTERPRETING GRAPHICS

1. C
2. H
3. D
4. G

 TEST DOCTOR

Question 1: Students may select answer choice B because the tallest bar is in the month of October. Remind students that they need to look at the key to the graph to know which bars to compare. The correct answer is C.

MATH

1. A
2. H
3. B
4. H

 **TEST DOCTOR**

Question 2: To find the answer to this question, students need to find the total number of atoms in a molecule of fructose by adding the subscripts on the chemical formula (6 + 12 + 6 = 24). Then, students should divide the number of carbon atoms by the total number of atoms and multiply by 100% to find the percentage of carbon (6 ÷ 24 × 100% = 25%).

CHAPTER RESOURCES

Chapter Resource File

▪ **Standardized Test Preparation** GENERAL

State Resources

For specific resources for your state, visit **go.hrw.com** and type in the keyword **HSMSTR**.

Science, Technology, and Society

Discussion — GENERAL

Lead a discussion comparing the use of superglue, stitches, and bandages to cover or close wounds. Ask students to name some advantages and disadvantages of using superglue to close wounds. (Sample answer: Bandages often fall off when they get wet, but superglue will stay on in water. Superglue bandages are more expensive than regular bandages.)

Weird Science

Background

A van der Waals force is a type of intermolecular force (a force between molecules). Van der Waals forces are attractions resulting from the uneven distribution of electrons and the creation of temporary dipoles. The positive end of a dipole on one molecule attracts the negative end of a dipole on another molecule, and the molecules are briefly held together. Van der Waals forces are compared to ionic bonds in the text because ionic bonds are also attractions between opposite charges. However, be sure your students understand that van der Waals forces are much weaker than ionic bonds.

Science in Action

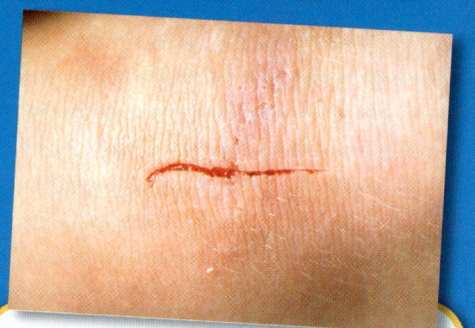

Science, Technology, and Society

Superglue Bandages and Stitches

If you aren't careful when using superglue, you may accidentally learn that superglue quickly bonds skin together! This property of superglue led to the development of new kinds of superglue that can be used as alternatives for bandages and stitches. Using superglue to close wounds has several advantages over using bandages and stitches. For example, superglue bandages can cover cuts on parts of the body that are difficult to cover with regular bandages. And superglue stitches are less painful than regular stitches. Finally, wounds closed with superglue are easier to care for than wounds covered by bandages or closed with stitches.

Math ACTIVITY

A wound can be closed 3 times faster with glue than it can be with stitches. If it takes a doctor 27 min to close a wound by using stitches, how long would it take to close the same wound by using glue?

Weird Science

How Geckos Stick to Walls

Geckos are known for their ability to climb up smooth surfaces. Recently, scientists found the secret to the gecko's sticky talent. Geckos have millions of microscopic hairs on the bottom of their feet. Each hair splits into as many as 1,000 tinier hairs called *hairlets*. At the end of each hairlet is a small pad. As the gecko walks, each pad forms a van der Waals force with the surface on which the gecko is walking. A van der Waals force is an attraction similar to an ionic bond, but the van der Waals force is much weaker than an ionic bond and lasts for only an instant. But because there are so many pads on a gecko's foot, the van der Waals forces are strong enough to keep the gecko from falling.

Language Arts ACTIVITY

WRITING SKILL Imagine that you could stick to walls as well as a gecko can. Write a five-paragraph short story describing what you would do with your wall-climbing ability.

Answer to Math Activity

27 min ÷ 3 = 9 min

Answer to Language Arts Activity

Accept all reasonable responses. All stories should describe how students would use their wall-climbing ability. For example, students may discuss becoming a superhero and using their wall-climbing ability to fight crime.

Roberta Jordan

Analytical Chemist Have you ever looked at something and wondered what chemicals it contained? That's what analytical chemists do for a living. They use tests to find the chemical makeup of a sample. Roberta Jordan is an analytical chemist at the Idaho National Engineering and Environmental Laboratory in Idaho Falls, Idaho.

Jordan's work focuses on the study of radioactive waste generated by nuclear power plants and nuclear-powered submarines. Jordan works with engineers to develop safe ways to store the radioactive waste. She tells the engineers which chemicals need to be studied and which techniques to use to study those chemicals.

Jordan enjoys her job because she is always learning new techniques. "One of the things necessary to be a good chemist is you have to be creative. You have to be able to think above and beyond the normal ways of doing things to come up with new ideas, new experiments," she explains. Jordan believes that a person interested in a career in chemistry has many opportunities. "There are a lot of things out there that need to be discovered," says Jordan.

Social Studies ACTIVITY

Many elements in the periodic table were discovered by analytical chemists. Pick an element from the periodic table, and research its history. Make a poster about the discovery of that element.

To learn more about these Science in Action topics, visit **go.hrw.com** and type in the keyword **HP5BNDF.**

Current Science

Check out Current Science® articles related to this chapter by visiting **go.hrw.com.** Just type in the keyword **HP5CS13.**

Careers

Background

Originally, analytical chemists worked using tests in which they reacted the unknown substance with other substances. These methods often involved changing the original sample. But in the last 50 years, the field of instrumental analysis (in which instruments are used to analyze substances) has grown. Many of these instruments, such as infrared spectrophotometers and nuclear magnetic resonance spectrometers, are able to analyze a substance without changing it chemically.

Answer to Social Studies Activity

Encourage students to focus their research on the main group elements or the transition metals in periods 4–6. The discovery of these elements is better documented, and the elements often have interesting histories.

Accept all reasonable responses. All posters should include the name of the element researched, the name(s) of the person who discovered it, and the year it was discovered.

2 Chemical Reactions
Chapter Planning Guide

Compression guide:
To shorten instruction because of time limitations, omit the Chapter Lab.

OBJECTIVES	LABS, DEMONSTRATIONS, AND ACTIVITIES	TECHNOLOGY RESOURCES
PACING • 90 min pp. 26–31 **Chapter Opener**	**SE** Start-up Activity, p. 27 GENERAL	**OSP** Parent Letter ■ GENERAL **CD** Student Edition CD-ROM **CD** Guided Reading Audio CD ■ **TR** Chapter Starter Transparency* **VID** Brain Food Video Quiz
Section 1 Forming New Substances • Describe how chemical reactions produce new substances that have different chemical and physical properties. • Identify four signs that indicate that a chemical reaction might be taking place. • Explain what happens to chemical bonds during a chemical reaction.	**TE** Activity Performing a Chemical Reaction, p. 29 ◆ BASIC **SE** Quick Lab Reaction Ready, p. 31 GENERAL **CRF** Datasheet for Quick Lab* **SE** Model-Making Labs Finding a Balance, p. 110 GENERAL **CRF** Datasheet for LabBook*	**CRF** Lesson Plans* **TR** Bellringer Transparency* **TR** Reaction of Hydrogen and Chlorine* **CRF** SciLinks Activity* GENERAL
PACING • 45 min pp. 32–37 **Section 2 Chemical Formulas and Equations** • Interpret and write simple chemical formulas. • Write and balance simple chemical equations. • Explain how a balanced equation shows the law of conservation of mass.	**TE** Activity Names of Compounds at Home, p. 33 ADVANCED **TE** Connection Activity Math, p. 35 GENERAL **TE** Activity Reactions in the Atmosphere, p. 35 ADVANCED **SE** Connection to Language Arts Diatomic Molecules, p. 36 GENERAL **SE** Quick Labs Conservation of Mass, p. 37 GENERAL **CRF** Datasheet for Quick Lab*	**CRF** Lesson Plans* **TR** Bellringer Transparency* **TR** Writing Chemical Formulas and Equations* **TR** Balancing a Chemical Equation*
PACING • 45 min pp. 38–41 **Section 3 Types of Chemical Reactions** • Describe four types of chemical reactions. • Classify a chemical equation as one of four types of chemical reactions.	**TE** Demonstration Single-Displacement Reaction, p. 39 GENERAL **SE** Quick Lab Identifying Reactions, p. 40 GENERAL **SE** Skills Practice Lab Putting Elements Together, p. 112 GENERAL **LB** Inquiry Labs Curses, Foiled Again!* BASIC **LB** Labs You Can Eat How to Fluff a Muffin* GENERAL **SE** Science in Action Math, Social Studies, and Language Arts, Activities, pp. 54–55 GENERAL	**CRF** Lesson Plans* **TR** Bellringer Transparency* **TR** Models of Reactions* **SE** Internet Activity, p. 40 GENERAL
PACING • 90 min pp. 42–47 **Section 4 Energy and Rates of Chemical Reactions** • Compare exothermic and endothermic reactions. • Explain activation energy. • Interpret an energy diagram. • Describe the five factors that affect the rate of a reaction.	**SE** Quick Lab Endo Alert, p. 43 GENERAL **SE** Connection to Social Studies The Strike-Anywhere Match, p. 44 GENERAL **SE** Quick Lab Which Is Quicker?, p. 45 GENERAL **TE** Activity Factors Affecting Rates, p. 45 BASIC **SE** Connection to Biology Enzymes and Inhibitors, p. 46 GENERAL **SE** Skills Practice Lab Speed Control, p. 48 GENERAL **CRF** Datasheet for Chapter Lab* **SE** Skills Practice Lab Cata-what? Catalyst!, p. 111 GENERAL **CRF** Datasheet for LabBook* **LB** Whiz-Bang Demonstrations Fire and Ice* GENERAL **LB** Long-Term Projects & Research Ideas Fruitful Chemistry* ADVANCED	**CRF** Lesson Plans* **TR** Bellringer Transparency* **TR** Energy Diagrams* **TR** LINK TO LIFE SCIENCE Photosynthesis* **VID** Lab Videos for Physical Science

PACING • 90 min

CHAPTER REVIEW, ASSESSMENT, AND STANDARDIZED TEST PREPARATION

CRF Vocabulary Activity* GENERAL
SE Chapter Review, pp. 50–51 GENERAL
CRF Chapter Review* ■ GENERAL
CRF Chapter Tests A* ■ GENERAL, B* ADVANCED, C* SPECIAL NEEDS
SE Standardized Test Preparation, pp. 52–53 GENERAL
CRF Standardized Test Preparation* GENERAL
CRF Performance-Based Assessment* GENERAL
OSP Test Generator GENERAL
CRF Test Item Listing* GENERAL

Online and Technology Resources

Visit **go.hrw.com** for a variety of free resources related to this textbook. Enter the keyword **HP5REA.**

Students can access interactive problem-solving help and active visual concept development with the *Holt Science and Technology* Online Edition available at **www.hrw.com.**

Guided Reading Audio CD

A direct reading of each chapter using instructional visuals as guideposts. For auditory learners, reluctant readers, and Spanish-speaking students. Available in English and Spanish.

SKILLS DEVELOPMENT RESOURCES	SECTION REVIEW AND ASSESSMENT	STANDARDS CORRELATIONS
SE Pre-Reading Activity, p. 26 `GENERAL` **OSP** Science Puzzlers, Twisters & Teasers `GENERAL`		National Science Education Standards SAI 2
CRF Directed Reading A* ■ `BASIC`, B* `SPECIAL NEEDS` **CRF** Vocabulary and Section Summary* ■ `GENERAL` **SE** Reading Strategy Reading Organizer, p. 28 `GENERAL`	**SE** Reading Checks, pp. 29, 30 `GENERAL` **TE** Homework, p. 29 `GENERAL` **TE** Reteaching, p. 30 `BASIC` **TE** Quiz, p. 30 `GENERAL` **TE** Alternative Assessment, p. 30 `GENERAL` **SE** Section Review,* p. 31 ■ `GENERAL` **CRF** Section Quiz* ■ `GENERAL`	UCP 3; PS 1b, 3a, 3e; *LabBook:* PS 3e
CRF Directed Reading A* ■ `BASIC`, B* `SPECIAL NEEDS` **CRF** Vocabulary and Section Summary* ■ `GENERAL` **SE** Reading Strategy Discussion, p. 32 `GENERAL` **TE** Inclusion Strategies, p. 34 **SE** Math Practice Counting Atoms, p. 35 `GENERAL` **MS** Math Skills for Science Balancing Chemical Equations* `GENERAL`	**SE** Reading Checks, pp. 33, 34, 36 `GENERAL` **TE** Homework, p. 33 `GENERAL` **TE** Reteaching, p. 36 `BASIC` **TE** Quiz, p. 36 `GENERAL` **TE** Alternative Assessment, p. 36 `GENERAL` **SE** Section Review,* p. 37 ■ `GENERAL` **CRF** Section Quiz* ■ `GENERAL`	UCP 3; SAI 1; PS 1b
CRF Directed Reading A* ■ `BASIC`, B* `SPECIAL NEEDS` **CRF** Vocabulary and Section Summary* ■ `GENERAL` **SE** Reading Strategy Mnemonics, p. 38 `GENERAL` **CRF** Reinforcement Worksheet Fabulous Food Reactions* `BASIC`	**SE** Reading Checks, pp. 38, 39, 40 `GENERAL` **TE** Reteaching, p. 40 `BASIC` **TE** Quiz, p. 40 `GENERAL` **TE** Alternative Assessment, p. 40 `GENERAL` **SE** Section Review,* p. 41 ■ `GENERAL` **CRF** Section Quiz* ■ `GENERAL`	UCP 3; PS 1b; *LabBook:* UCP 3; SAI 1, 2; PS 1b, 3e
CRF Directed Reading A* ■ `BASIC`, B* `SPECIAL NEEDS` **CRF** Vocabulary and Section Summary* ■ `GENERAL` **SE** Reading Strategy Paired Summarizing, p. 42 `GENERAL` **TE** Inclusion Strategies, p. 46 **CRF** Reinforcement Worksheet Activation Energy* `BASIC` **CRF** Critical Thinking Shedding Light on Landfills* `ADVANCED`	**SE** Reading Checks, pp. 43, 44, 46 `GENERAL` **TE** Reteaching, p. 46 `BASIC` **TE** Quiz, p. 46 `GENERAL` **TE** Alternative Assessment, p. 46 `GENERAL` **SE** Section Review,* p. 47 ■ `GENERAL` **CRF** Section Quiz* ■ `GENERAL`	UCP 3; SAI 1; PS 1b, 3a, 3e; *Chapter Lab:* SAI 1, 2; *LabBook:* UCP 3; SAI 1; PS 3e

One-Stop Planner® CD-ROM

This convenient CD-ROM includes:
- Lab Materials QuickList Software
- Holt Calendar Planner
- Customizable Lesson Plans
- Printable Worksheets
- ExamView® Test Generator

cnnstudentnews.com

Find the latest news, lesson plans, and activities related to important scientific events.

NSTA

www.scilinks.org

Maintained by the **National Science Teachers Association.** See Chapter Enrichment pages for a complete list of topics.

Current Science®

Check out *Current Science* articles and activities by visiting the HRW Web site at **go.hrw.com.** Just type in the keyword **HP5CS14T.**

Classroom Videos
- **Lab Videos** demonstrate the chapter lab.
- **Brain Food Video Quizzes** help students review the chapter material.

Visual Resources

CHAPTER STARTER TRANSPARENCY

BELLRINGER TRANSPARENCIES

TEACHING TRANSPARENCIES

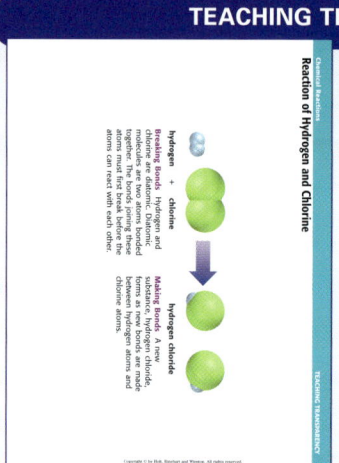

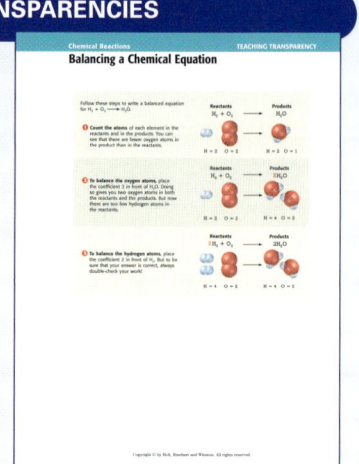

TEACHING TRANSPARENCIES

CONCEPT MAPPING TRANSPARENCY

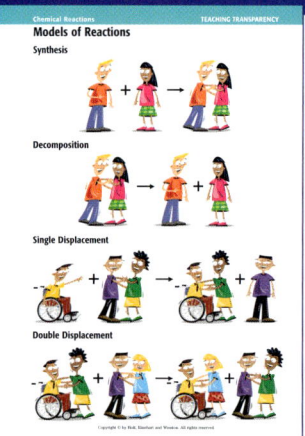

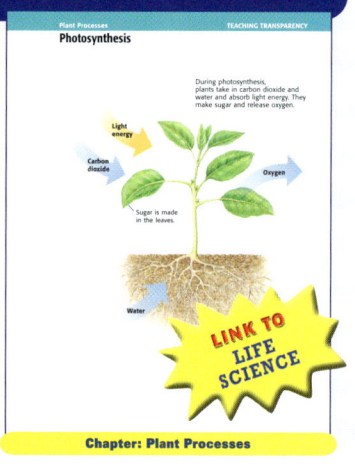

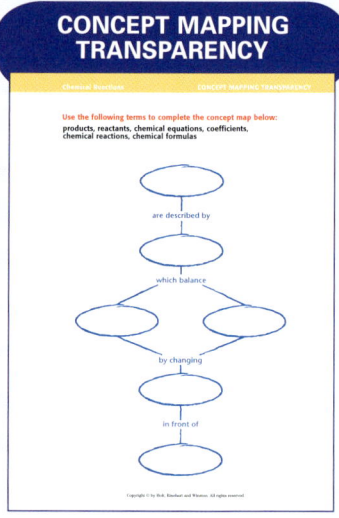

Chapter: Plant Processes

Planning Resources

LESSON PLANS

PARENT LETTER

TEST ITEM LISTING

One-Stop Planner® CD-ROM

This CD-ROM includes all of the resources shown here and the following time-saving tools:

- *Lab Materials QuickList Software*
- *Customizable lesson plans*
- *Holt Calendar Planner*
- *The powerful ExamView® Test Generator*

Meeting Individual Needs

DIRECTED READING A
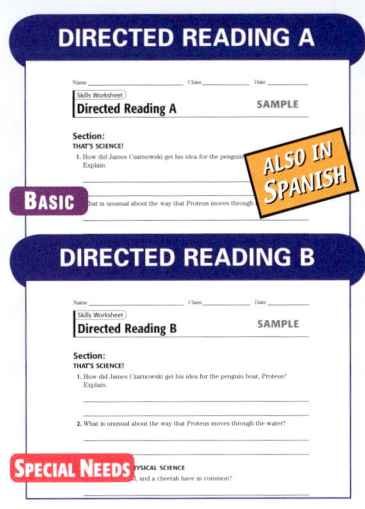
Skills Worksheet
Directed Reading A — SAMPLE
Section: THAT'S SCIENCE!
1. How did James Czarnowski get his idea for the penguin...
Explain.
2. What is unusual about the way that Proteus moves through...
BASIC — ALSO IN SPANISH

DIRECTED READING B
Skills Worksheet
Directed Reading B — SAMPLE
Section: THAT'S SCIENCE!
1. How did James Czarnowski get his idea for the penguin boat, Proteus? Explain.
2. What is unusual about the way that Proteus moves through the water?
SPECIAL NEEDS — PHYSICAL SCIENCE
...and a cheetah have in common?

VOCABULARY ACTIVITY
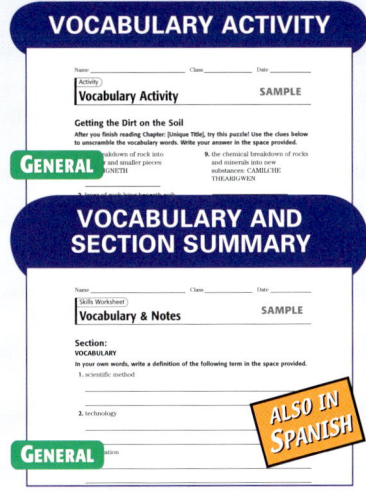
Activity
Vocabulary Activity — SAMPLE
Getting the Dirt on the Soil
After you finish reading Chapter: [Unique Title], try this puzzle! Use the clues below to unscramble the vocabulary words. Write your answer in the space provided.
...breakdown of rock into ...and smaller pieces ...NGTH
...the chemical breakdown of rocks and minerals into new substances CAMILCHE THEAIRGWEN
9. ...
GENERAL

VOCABULARY AND SECTION SUMMARY
Skills Worksheet
Vocabulary & Notes — SAMPLE
Section: VOCABULARY
In your own words, write a definition of the following term in the space provided.
1. scientific method
2. technology
GENERAL — ALSO IN SPANISH

REINFORCEMENT
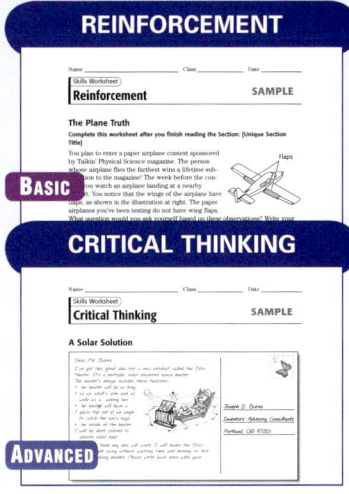
Skills Worksheet
Reinforcement — SAMPLE
The Plane Truth
Complete this worksheet after you finish reading the Section: [Unique Section Title]
You plan to enter a paper airplane contest sponsored by Talkin' Physical Science magazine. The person whose airplane flies the farthest wins a lifetime subscription to the magazine! The week before the contest, you watch an airplane landing at a nearby airport. You notice that the wings of the airplane have flaps, as shown in the illustration at right. The paper airplanes you've been testing do not have wing flaps. What question would you ask yourself based on this observation? Write your...
BASIC

CRITICAL THINKING
Skills Worksheet
Critical Thinking — SAMPLE
A Solar Solution
Dear Mr. Burns...
Joseph D. Burns
Inventors' Advisory Consultants
Portland, OR 97201
ADVANCED

SCILINKS ACTIVITY
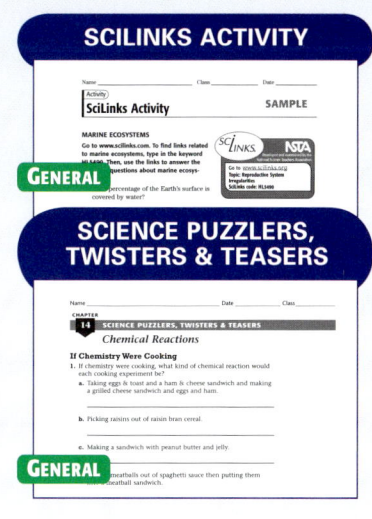
Activity
SciLinks Activity — SAMPLE
MARINE ECOSYSTEMS
Go to www.scilinks.com. To find links related to marine ecosystems, type in the keyword HL5400. Then, use the links to answer the questions about marine ecosys-
...percentage of the Earth's surface is covered by water?
GENERAL

SCIENCE PUZZLERS, TWISTERS & TEASERS
CHAPTER 14 SCIENCE PUZZLERS, TWISTERS & TEASERS
Chemical Reactions
If Chemistry Were Cooking
1. If chemistry were cooking, what kind of chemical reaction would each cooking experiment be?
a. Taking eggs & toast and a ham & cheese sandwich and making a grilled cheese sandwich and eggs and ham.
b. Picking raisins out of raisin bran cereal.
c. Making a sandwich with peanut butter and jelly.
...meatballs out of spaghetti sauce then putting them ...meatball sandwich.
GENERAL

Labs and Activities

LONG-TERM PROJECTS & RESEARCH IDEAS
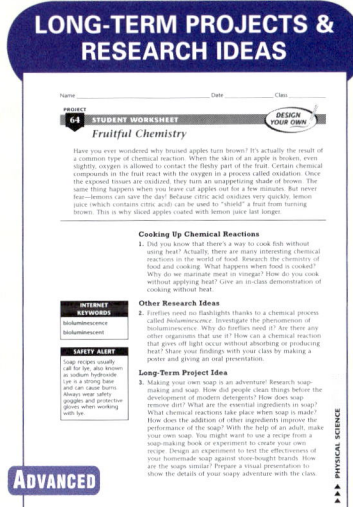
PROJECT 64 STUDENT WORKSHEET — DESIGN YOUR OWN
Fruitful Chemistry
Have you ever wondered why bruised apples turn brown? It's actually the result of a common type of chemical reaction. When the skin of an apple is broken, even slightly, oxygen is allowed to contact the fleshy part of the fruit. Certain chemical compounds in the fruit react with the oxygen in a process called oxidation. Once the exposed tissues are oxidized, they turn an unappetizing shade of brown. The same thing happens when you leave cut apples out for a few minutes. But never fear—lemons can save the day! Because citric acid oxidizes very quickly, lemon juice (which contains citric acid) can be used to "shield" a fruit from turning brown. This is why sliced apples coated with lemon juice last longer.
Cooking Up Chemical Reactions
1. Did you know that there's a way to cook fish without using heat? Actually, there are many interesting chemical reactions in the world of food. Research the chemistry of food and cooking. What happens when food is cooked? Why do we marinate meat or vinegar? How do you cook without applying heat? Give an in-class demonstration of cooking without heat.
INTERNET KEYWORDS
bioluminescence
bioluminescent
Other Research Ideas
2. Fireflies need no flashlights thanks to a chemical process called bioluminescence. Investigate the phenomenon of bioluminescence. Why do fireflies need it? Are there any other organisms that use it? How can a chemical reaction that gives off light occur without absorbing or producing heat? Share your findings with your class by making a poster and giving an oral presentation.
SAFETY ALERT
Soap recipes usually call for lye, also known as sodium hydroxide Lye is a strong base and can cause burns. Always wear safety goggles and protective gloves when working with lye.
Long-Term Project Idea
3. Making your own soap is an adventure! Research soap-making and soap. How do people clean things before the development of modern detergents? How does soap remove dirt? What are the essential ingredients in soap? What chemical reactions take place when soap is made? How does the addition of other ingredients improve the performance of the soap? With the help of an adult, make your own soap. You might want to use a recipe from a soap-making book or experiment to create your own recipe. Design an experiment to test the effectiveness of your homemade soap against store-bought brands. How are the soaps similar? Prepare a visual presentation to show the details of your soapy adventure with the class.
ADVANCED — PHYSICAL SCIENCE

WHIZ-BANG DEMONSTRATIONS
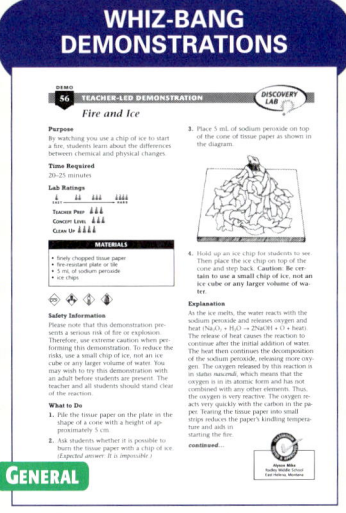
DEMO 56 TEACHER-LED DEMONSTRATION — DISCOVERY LAB
Fire and Ice
Purpose
By watching you use a chip of ice to start a fire, students learn about the differences between chemical and physical changes.
Time Required
20–25 minutes
Lab Ratings
TEACHER PREP ▲▲
CONCEPT LEVEL ▲▲▲
CLEAN UP ▲▲▲
MATERIALS
• finely chopped tissue paper
• fire-resistant plate or tile
• 5 mL of sodium peroxide
• ice chips
Safety Information
Please note that this demonstration presents a serious risk of fire or explosion. Therefore, use extreme caution when performing this demonstration. To reduce the risks, use a small chip of ice, not an ice cube or any larger volume of water. You may wish to try this demonstration with an adult before students are present. The teacher and all students should stand clear of the reaction.
What to Do
1. Pile the tissue paper on the plate in the shape of a cone with a height of approximately 5 cm.
2. Ask students whether it is possible to burn the tissue paper with a chip of ice. (Expected answer: It is impossible.)
3. Place 5 mL of sodium peroxide on top of the cone of tissue paper as shown in the diagram.
4. Hold up an ice chip for students to see. Then place the ice chip on top of the cone and step back. **Caution:** Be certain to use a small chip of ice, not an ice cube or any larger volume of water.
Explanation
As the ice melts, the water reacts with the sodium peroxide and releases oxygen and heat ($Na_2O_2 + H_2O \rightarrow 2NaOH + O + heat$). The release of heat causes the reaction to continue after the initial addition of water. The heat then continues the decomposition of the sodium peroxide, releasing more oxygen. The oxygen released by this reaction is in status nascendi, which means that the oxygen is in its atomic form and has not combined with any other elements. Thus the oxygen is very reactive. The oxygen reacts very quickly with the carbon in the paper. Tearing the tissue paper into small pieces increases the paper's kindling temperature and aids in starting the fire.
continued...
Alyssa Mike
Radley Middle School
East Helena, Montana
GENERAL

INQUIRY LABS
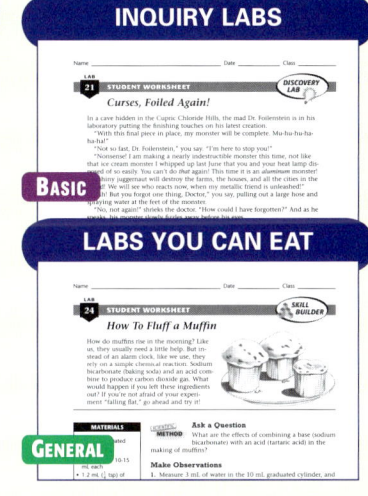
LAB 21 STUDENT WORKSHEET — DISCOVERY LAB
Curses, Foiled Again!
In a cave hidden in the Cupric Chloride Hills, the mad Dr. Foilenstein is in his laboratory putting the finishing touches on his latest creation.
"With this foil piece to play, my monster will be complete. Mu-ha-ha-ha-ha-ha!"
"Not so fast, Dr. Foilenstein," you say. "I'm here to stop you!"
"Nonsense! I am making a nearly indestructible monster this time, not like that ice cream monster I whipped up last June that you and your heat lamp disposed of so easily. You can't do that again! This time it is an aluminum monster! When my juggernaut will destroy the farms, the houses, and all the cities in the land! We will see who reacts now, when my metallic friend is unleashed!"
"But you forgot one thing, Doctor," you say, pulling out a large hose and applying water at the feet of the monster.
"No, not again!" shrieks the doctor. "How could I have forgotten?" And as he speaks, his monster slowly fizzles away before his eyes.
BASIC

LABS YOU CAN EAT
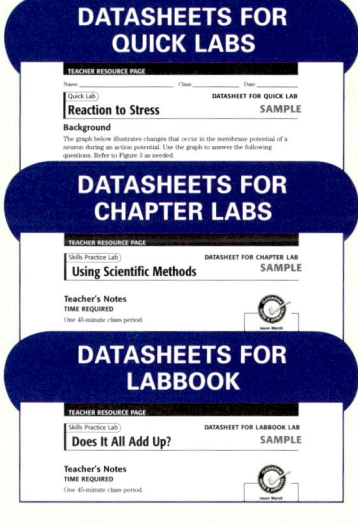
LAB 24 STUDENT WORKSHEET — SKILL BUILDER
How To Fluff a Muffin
How do muffins rise in the morning? Like us, they usually need a little help. But instead of an alarm clock, like we use, they rely on a simple chemical reaction. Sodium bicarbonate (baking soda) and an acid combine to produce carbon dioxide gas. What would happen if you left the ingredients out? If you're not afraid of your experiment "falling flat," go ahead and try it!
MATERIALS
...ate
...60-15
...mL
• 1.2 mL (⅛ tsp) of
Ask a Question
What are the effects of combining a base (sodium bicarbonate) and an acid (tartaric acid) in the making of muffins?
Make Observations
1. Measure 3 mL of water in the 10 mL graduated cylinder, and
GENERAL

DATASHEETS FOR QUICK LABS
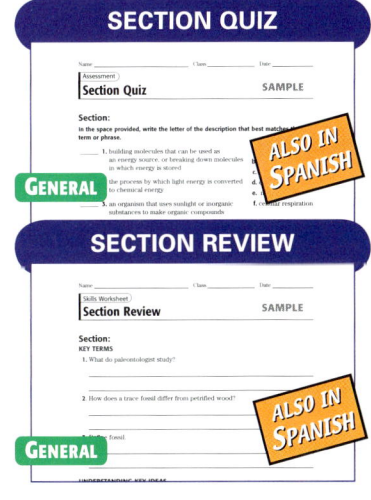
TEACHER RESOURCE PAGE
Quick Lab — DATASHEET FOR QUICK LAB
Reaction to Stress — SAMPLE
Background
The graph below illustrates changes that occur in the membrane potential of a neuron during an action potential. Use the graph to answer the following questions. Refer to Figure 3 as needed.

DATASHEETS FOR CHAPTER LABS
TEACHER RESOURCE PAGE
Skills Practice Lab — DATASHEET FOR CHAPTER LAB
Using Scientific Methods — SAMPLE
Teacher's Notes
TIME REQUIRED
One 45-minute class period.

DATASHEETS FOR LABBOOK
TEACHER RESOURCE PAGE
Skills Practice Lab — DATASHEET FOR LABBOOK LAB
Does It All Add Up? — SAMPLE
Teacher's Notes
TIME REQUIRED
One 45-minute class period.

Review and Assessments

SECTION QUIZ
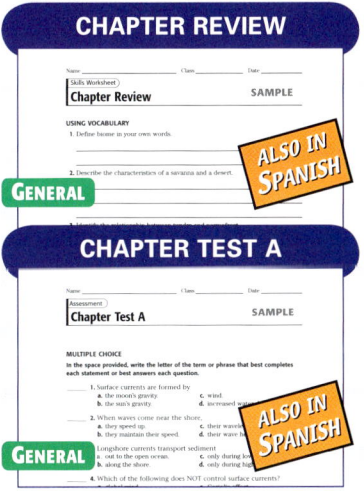
Assessment
Section Quiz — SAMPLE
Section:
In the space provided, write the letter of the description that best matches the term or phrase.
_____ 1. building molecules that can be used as an energy source, or breaking down molecules in which energy is scored
_____ 2. the process by which light energy is converted to chemical energy
_____ 3. an organism that uses sunlight or inorganic substances to make complex compounds
a. ...
b. ...
c. ...
d. ...
e. ...
f. cellular respiration
GENERAL — ALSO IN SPANISH

SECTION REVIEW
Skills Worksheet
Section Review — SAMPLE
Section: KEY TERMS
1. What do paleontologists study?
2. How does a trace fossil differ from petrified wood?
3. Define fossil.
UNDERSTANDING KEY IDEAS
GENERAL

CHAPTER REVIEW
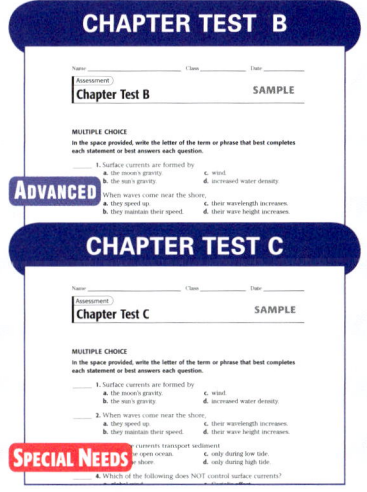
Skills Worksheet
Chapter Review — SAMPLE
USING VOCABULARY
1. Define biome in your own words.
2. Describe the characteristics of a savanna and a desert.
GENERAL — ALSO IN SPANISH

CHAPTER TEST A
Assessment
Chapter Test A — SAMPLE
MULTIPLE CHOICE
In the space provided, write the letter of the term or phrase that best completes each statement or best answers each question.
_____ 1. Surface currents are formed by
a. the moon's gravity.   c. wind.
b. the sun's gravity.   d. increased water...
_____ 2. When waves come near the shore,
a. they speed up.   c. their wavelength...
b. they maintain their speed. d. their wave...
...Longshore currents transport sediment
a. out to the open ocean.   c. only during low...
b. along the shore.   d. only during high...
4. Which of the following does NOT control surface currents?
GENERAL — ALSO IN SPANISH

CHAPTER TEST B
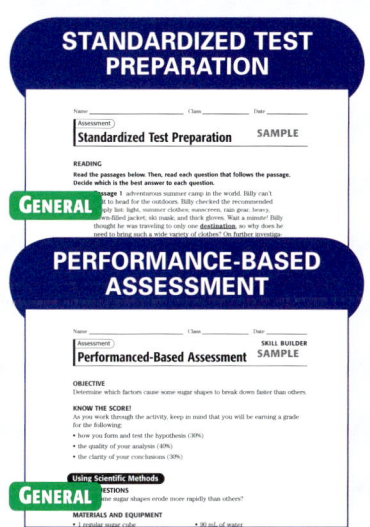
Assessment
Chapter Test B — SAMPLE
MULTIPLE CHOICE
In the space provided, write the letter of the term or phrase that completes each statement or best answers each question.
_____ 1. Surface currents are formed by
a. the moon's gravity.   c. wind.
b. the sun's gravity.   d. increased water density.
_____ 2. When waves come near the shore,
a. they speed up.   c. their wavelength increases.
b. they maintain their speed. d. their wave height increases.
ADVANCED

CHAPTER TEST C
Assessment
Chapter Test C — SAMPLE
MULTIPLE CHOICE
In the space provided, write the letter of the term or phrase that best completes each statement or best answers each question.
_____ 1. Surface currents are formed by
a. the moon's gravity.   c. wind.
b. the sun's gravity.   d. increased water density.
_____ 2. When waves come near the shore,
a. they speed up.   c. their wavelength increases.
b. they maintain their speed. d. their wave height increases.
...currents transport sediment
...to the open ocean.   c. only during low tide.
...along the shore.   d. only during high tide.
4. Which of the following does NOT control surface currents?
SPECIAL NEEDS

STANDARDIZED TEST PREPARATION
Assessment
Standardized Test Preparation — SAMPLE
READING
Read the passages below. Then, read each question that follows the passage. Decide which is the best answer to each question.
Passage 1 adventurous summer camp in the world. Billy can't wait to head for the outdoors. Billy checked the recommended ...supply list: light, summer clothes, sunscreen, rain gear, heavy, ...down-filled jacket, ski mask, and thick gloves. Wait a minute! Billy ...thought he was traveling to only one destination, so why does he ...need to bring such a wide variety of clothes? On further investiga...
GENERAL

PERFORMANCE-BASED ASSESSMENT
Assessment
Performanced-Based Assessment — SKILL BUILDER SAMPLE
OBJECTIVE
Determine which factors cause some sugar shapes to break down faster than others.
KNOW THE SCORE!
As you work through the activity, keep in mind that you will be earning a grade for the following:
• how you form and test the hypothesis (30%)
• the quality of your analysis (40%)
• the clarity of your conclusions (30%)
Using Scientific Methods
QUESTIONS
...some sugar shapes erode more rapidly than others?
MATERIALS AND EQUIPMENT
• 1 regular sugar cube   • 30 mL of water
GENERAL

This Chapter Enrichment provides relevant and interesting information to expand and enhance your presentation of the chapter material.

Section 1

Forming New Substances

Chemical Symbols

- To be able to discuss the nature of chemical reactions, scientists identify elements with one- or two-letter symbols. In this way, the language of chemical reactions can be understood universally.

Conserved Quantities

- In any chemical or physical change, the total amount of mass and energy is unchanged in the reaction. Both the law of conservation of mass and the law of conservation of energy apply to the chemical reactions discussed in this chapter.

Section 2

Chemical Formulas and Equations

Chemical Formulas

- Chemical formulas describe compounds and elements. A chemical formula describes one formula unit of a compound or element. If the substance is molecular, the formula unit represents one molecule. For example, the chemical formula for the ionic compound calcium chloride is $CaCl_2$, because each formula unit consists of one calcium ion and two chloride ions. The chemical formula for the covalent compound water is H_2O, because each molecule of water consists of two hydrogen atoms and one oxygen atom.

Chemical Equations

- Chemical formulas are used together to form chemical equations that describe a chemical reaction. A chemical equation states which elements or compounds are used up and which are formed, and it shows the relative amounts of each.

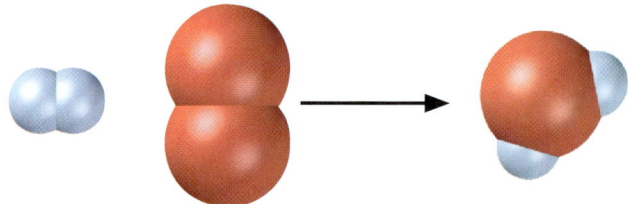

Is That a Fact!

- ◆ The reaction between hydrogen gas and oxygen gas to form water can be started with a small flame. However, water cannot be changed back into hydrogen and oxygen merely by cooling. The reverse chemical reaction can be accomplished only if some type of energy, such as electrical energy, is used to break the bonds between the hydrogen and oxygen atoms in the water molecules.

Section 3

Types of Chemical Reactions

Synthesis Reactions

- The formation of one product from two or more reactants is a synthesis reaction. For example, the formation of magnesium oxide from magnesium and oxygen (in early flashbulbs) and the formation of ammonia from nitrogen and hydrogen are synthesis reactions.

Decomposition Reactions

- A decomposition reaction is one in which a single compound produces two or more simpler substances. For example, the breakdown of water molecules into hydrogen and oxygen molecules is a decomposition reaction in which energy is used to break the bonds in the water molecules.

Single-Displacement Reactions

- Both metals and nonmetals undergo single-replacement reactions; for example, zinc, a metal, will react with hydrochloric acid to form zinc chloride and hydrogen gas. Chlorine, a nonmetal, will replace the bromine in sodium bromide to form sodium chloride and bromine.

Double-Displacement Reactions

- In a double-displacement reaction, two compounds exchange their ions. When a person takes milk of magnesia (magnesium hydroxide) to neutralize stomach acid, a double-replacement reaction occurs. The two products formed are magnesium chloride and water.

Is That a Fact!

- The reaction between baking soda, $NaHCO_3$, and tartaric acid, $H_2C_4H_4O_6$, in baking powder is a double-replacement reaction, followed by a decomposition reaction that produces carbon dioxide, CO_2. The bubbles of CO_2 make some doughs rise.

Section 4

Energy and Rates of Chemical Reactions

Chemical Kinetics

- For a chemical reaction to occur, reactant molecules must collide with enough energy and in the proper orientation to allow bonds to break and new bonds to form.

Endothermic Reactions

- Endothermic reactions are those that absorb energy. Photosynthesis in plants is an example of an endothermic reaction. Light energy from the sun drives the formation of glucose from carbon dioxide and water.

Exothermic Reactions

- Exothermic reactions are those in which energy is given off. For example, when hydrogen and oxygen react to form water, light and thermal energy are given off.

- In the body, exothermic reactions take place when food molecules are broken down and absorbed by cells in a series of reactions.

Catalysts

- *Catalysts* are substances that significantly increase the rate of a reaction. *Enzymes* are catalysts in living systems. These large protein molecules speed up many of the reactions in our body. For instance, the enzyme *amylase,* found in saliva, breaks down starch.

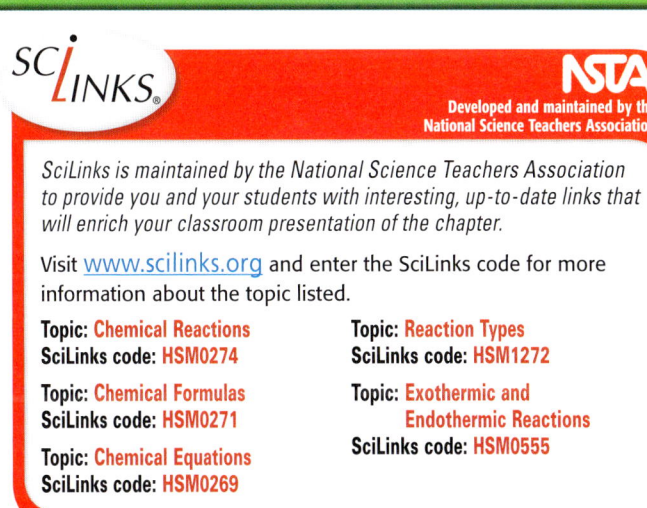

SCILINKS.

NSTA
Developed and maintained by the
National Science Teachers Association

SciLinks is maintained by the National Science Teachers Association to provide you and your students with interesting, up-to-date links that will enrich your classroom presentation of the chapter.

Visit www.scilinks.org and enter the SciLinks code for more information about the topic listed.

Topic: **Chemical Reactions**
SciLinks code: **HSM0274**

Topic: **Chemical Formulas**
SciLinks code: **HSM0271**

Topic: **Chemical Equations**
SciLinks code: **HSM0269**

Topic: **Reaction Types**
SciLinks code: **HSM1272**

Topic: **Exothermic and Endothermic Reactions**
SciLinks code: **HSM0555**

Overview

Tell students that this chapter will help them learn about chemical reactions. The chapter describes what is involved in a chemical reaction, how chemical reactions are expressed, what the different kinds of chemical reaction are, and how energy and rates are involved in chemical reactions.

Assessing Prior Knowledge

Students should be familiar with the following topics:

- matter
- energy
- atoms
- elements
- chemical bonding

Identifying Misconceptions

The conservation of atoms in a chemical reaction will probably have to be reinforced repeatedly, as students may believe that chemical reactions change matter intrinsically. Once the conservation of atoms in a chemical reaction is established, many other concepts pertaining to reactions, such as formulas, equations, and types of reactions, should follow quite naturally. Students may have initial difficulty understanding that some reactions require energy input in order to go forward.

2

Chemical Reactions

About the PHOTO

Dazzling fireworks and the Statue of Liberty are great examples of chemical reactions. Chemical reactions cause fireworks to soar, explode, and light up the sky. And the Statue of Liberty has its distinctive green color because of the reaction between the statue's copper and chemicals in the air.

PRE-READING ACTIVITY

FOLDNOTES **Four-Corner Fold**
Before you read the chapter, create the FoldNote entitled "Four-Corner Fold" described in the **Study Skills** section of the Appendix. Label the flaps of the four-corner fold with "Chemical formulas," "Chemical equations," "Types of chemical reactions," and "Rates of chemical reactions." Write what you know about each topic under the appropriate flap. As you read the chapter, add other information that you learn.

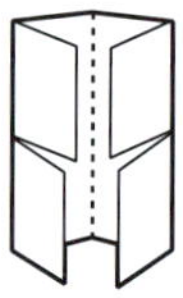

Standards Correlations

National Science Education Standards

The following codes indicate the National Science Education Standards that correlate to this chapter. The full text of the standards is at the front of the book.

Chapter Opener
SAI 2

Section 1 Forming New Substances
UCP 3; PS 1b, 3a, 3e; *LabBook:* PS 3e

Section 2 Chemical Formulas and Equations
UCP 3; SAI 1; PS 1b

Section 3 Types of Chemical Reactions
UCP 3; PS 1b; *LabBook:* UCP 3; SAI 1, 2; PS 1b, 3e

Section 4 Energy and Rates of Chemical Reactions
UCP 3; SAI 1; PS 1b, 3a, 3e; *LabBook:* UCP 3; SAI 1; PS 3e

Chapter Lab
SAI 1, 2

START-UP ACTIVITY

MATERIALS

FOR EACH STUDENT
- marshmallow models of hydrogen peroxide (3)

FOR EACH GROUP
- marshmallow model of water (1)
- marshmallow model of oxygen (1)

Safety Caution: Remind students to wear safety goggles while performing this activity. Students should not eat any of the marshmallows.

Answers
1. two
2. two; one
3. six; three

START-UP ACTIVITY

A Model Formula

Chemicals react in very precise ways. In this activity, you will model a chemical reaction and will predict how chemicals react.

Procedure

1. You will receive **several marshmallow models.** The models are marshmallows attached by **toothpicks.** Each of these models is a Model A.

2. Your teacher will show you an example of Model B and Model C. Take apart one or more Model As to make copies of Model B and Model C.

3. If you have marshmallows left over, use them to make more Model Bs and Model Cs. If you need more parts to complete a Model B or Model C, take apart another Model A.

4. Repeat step 3 until you have no parts left over.

Analysis

1. How many Model As did you use to make copies of Model B and Model C?

2. How many Model Bs did you make? How many Model Cs did you make?

3. Suppose you needed to make six Model Bs. How many Model As would you need? How many Model Cs could you make with the leftover marshmallows?

Chapter Review
PS 1b, 3a, 3e

Science in Action
SPSP 3, 5; ST 2

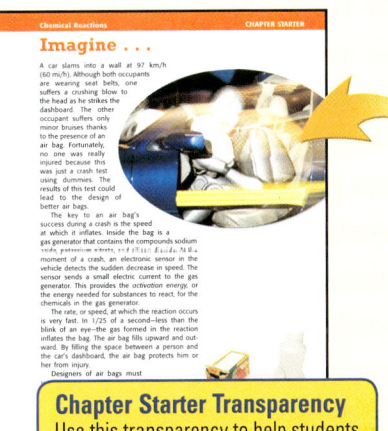

Chemical Reactions   **CHAPTER STARTER**

Imagine . . .

A car slams into a wall at 97 km/h (60 mi/h). Although both occupants are wearing seat belts, one suffers a crushing blow to the head as he strikes the dashboard. The other occupant suffers only minor bruises thanks to the presence of an air bag. Fortunately, no one was really injured because this was just a crash test using dummies. The results of this test could lead to the design of better air bags.

The key to an air bag's success during a crash is the speed at which it inflates. Inside the bag is a gas generator that contains the compounds sodium  At the moment of a crash, an electronic sensor in the vehicle detects the sudden decrease in speed. The sensor sends a small electric current to the gas generator. This provides the activation energy, or the energy needed for substances to react, for the chemicals in the gas generator.

The rate, or speed, at which the reaction occurs is very fast. In 1/25 of a second—less than the blink of an eye—the gas formed in the reaction inflates the bag. The air bag fills upward and outward. By filling the space between a person and the car's dashboard, the air bag protects him or  her from injury.

Designers of air bags must

Chapter Starter Transparency
Use this transparency to help students begin thinking about chemical reactions.

CHAPTER RESOURCES

Technology

Transparencies
- Chapter Starter Transparency

READING SKILLS

Student Edition on CD-ROM

Guided Reading Audio CD
- English or Spanish

Classroom Videos
- Brain Food Video Quiz

Workbooks

Science Puzzlers, Twisters & Teasers
- Chemical Reactions **GENERAL**

Focus

Overview

This section discusses the nature of chemical reactions. Students will learn some of the signs that a chemical reaction has taken place. They will also learn that bonds are broken and new bonds are formed during a chemical reaction.

 Bellringer

Ask students the following question: "What do baking bread, launching the space shuttle, and digesting food have in common?" (They all involve chemical reactions.)

Have them write their answer in their **science journal.**

Motivate

Discussion —————— GENERAL

Chemical Reactions at School
Ask students to think about the chemical reactions that occur in school every day. Have them consider the reactions in the meals cooking in the school cafeteria and the reactions in the batteries that provide energy to run equipment. List their responses on the board, and discuss the signs to look for in each chemical reaction. **LS Logical**

READING WARM-UP

Objectives

● Describe how chemical reactions produce new substances that have different chemical and physical properties.

● Identify four signs that indicate that a chemical reaction might be taking place.

● Explain what happens to chemical bonds during a chemical reaction.

Terms to Learn

chemical reaction
precipitate

READING STRATEGY

Reading Organizer As you read this section, create an outline of the section. Use the headings from the section in your outline.

chemical reaction the process by which one or more substances change to produce one or more different substances

---

Forming New Substances

Each fall, a beautiful change takes place when leaves turn colors. You see bright reds, oranges, and yellows that had been hidden by green all year. What causes this change?

To answer this question, you need to know what causes leaves to be green. Leaves are green because they contain a green substance, or *pigment*. This pigment is called *chlorophyll* (KLAWR uh FIL). During the spring and summer, the leaves have a large amount of chlorophyll in them. But in the fall, when temperatures drop and there are fewer hours of sunlight, chlorophyll breaks down to form new substances that have no color. The green chlorophyll is no longer present to hide the other pigments. You can now see the red, orange, and yellow colors that were present all along.

Chemical Reactions

A chemical change takes place when chlorophyll breaks down into new substances. This change is an example of a chemical reaction. A **chemical reaction** is a process in which one or more substances change to make one or more new substances. The chemical and physical properties of the new substances differ from those of the original substances. Some results of chemical reactions are shown in **Figure 1.**

| **Figure 1** | **Results of Chemical Reactions** |

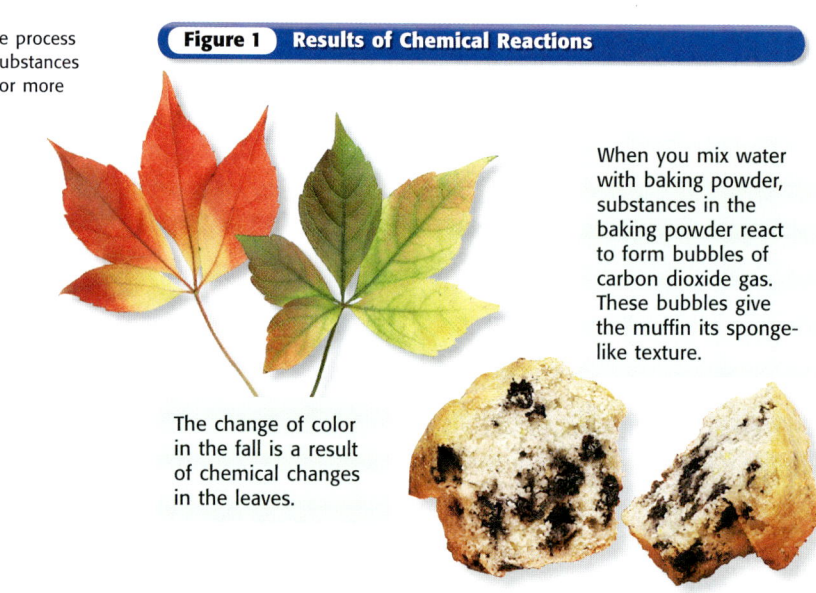

When you mix water with baking powder, substances in the baking powder react to form bubbles of carbon dioxide gas. These bubbles give the muffin its sponge-like texture.

The change of color in the fall is a result of chemical changes in the leaves.

CHAPTER RESOURCES

Chapter Resource File

- Lesson Plan
- Directed Reading A **BASIC**
- Directed Reading B **SPECIAL NEEDS**

Technology

Transparencies
- Bellringer

Is That a Fact!

Carbon dioxide has a number of well-known uses. When dissolved under pressure, it produces the effervescence in carbonated beverages. Because carbon dioxide gas does not combust and is denser than air, it is used in fire extinguishers to smother flames. Dry ice, or solid carbon dioxide, is valuable for its cooling effect, which is almost twice that of water ice. Carbon dioxide also changes directly from a solid to a gas, bypassing the liquid state.

Signs of Chemical Reactions

How can you tell when a chemical reaction is taking place? **Figure 2** shows some signs that tell you that a reaction may be taking place. In some chemical reactions, gas bubbles form. Other reactions form solid precipitates (pree SIP uh TAYTS). A **precipitate** is a solid substance that is formed in a solution. During other chemical reactions, energy is given off. This energy may be in the form of light, thermal energy, or electricity. Reactions often have more than one of these signs. And the more of these signs that you see, the more likely that a chemical reaction is taking place.

precipitate a solid that is produced as a result of a chemical reaction in solution

✔ **Reading Check** What is a precipitate? (*See the Appendix for answers to Reading Checks.*)

Figure 2 Some Signs of Chemical Reactions

Gas Formation
The chemical reaction in the beaker has formed a brown gas, nitrogen dioxide. This gas is formed when a strip of copper is placed into nitric acid.

Solid Formation
Here you see potassium chromate solution being added to a silver nitrate solution. The dark red solid is a precipitate of silver chromate.

Energy Change
Energy is released during some chemical reactions. The fire in this photo gives off light energy and thermal energy. During some other chemical reactions, energy is taken in.

Color Change
Don't spill chlorine bleach on your jeans! The bleach reacts with the blue dye on the fabric and causes the color of the material to change.

Answer to Reading Check
A precipitate is a solid substance that is formed in a solution.

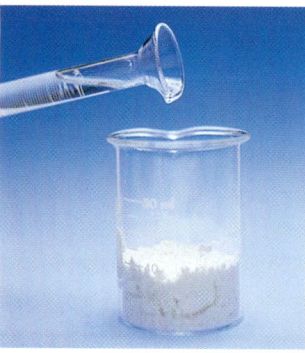

Figure 3 The top photo shows the starting substances: table sugar and sulfuric acid, a clear liquid. The substances formed in this chemical reaction are very different from the starting substances.

A Change of Properties

Even though the signs we look for to see if a reaction is taking place are good signals of chemical reactions, they do not guarantee that a reaction is happening. For example, gas can be given off when a liquid boils. But this example is a physical change, not a chemical reaction.

So, how can you be sure that a chemical reaction is occurring? The most important sign is the formation of new substances that have different properties. Look at **Figure 3.** The starting materials in this reaction are sugar and sulfuric acid. Several things tell you that a chemical reaction is taking place. Bubbles form, a gas is given off, and the beaker becomes very hot. But most important, new substances form. And the properties of these substances are very different from those of the starting substances.

Bonds: Holding Molecules Together

A *chemical bond* is a force that holds two atoms together in a molecule. For a chemical reaction to take place, the original bonds must break and new bonds must form.

Breaking and Making Bonds

How do new substances form in a chemical reaction? First, chemical bonds in the starting substances must break. Molecules are always moving. If the molecules bump into each other with enough energy, the chemical bonds in the molecules break. The atoms then rearrange, and new bonds form to make the new substances. **Figure 4** shows how bonds break and form in the reaction between hydrogen and chlorine.

✔ **Reading Check** What happens to the bonds of substances during a chemical reaction?

Figure 4 Reaction of Hydrogen and Chlorine

hydrogen + chlorine   hydrogen chloride

Breaking Bonds Hydrogen and chlorine are diatomic. Diatomic molecules are two atoms bonded together. The bonds joining these atoms must first break before the atoms can react with each other.

Making Bonds A new substance, hydrogen chloride, forms as new bonds are made between hydrogen atoms and chlorine atoms.

Answer to Reading Check

In a chemical reaction, the chemical bonds in the starting substances break, and then new bonds form to make new substances.

New Bonds, New Substances

What happens when hydrogen and chlorine are combined? A chlorine gas molecule is a diatomic (DIE uh TAHM ik) molecule. That is, a chlorine molecule is made of two atoms of chlorine. Chlorine gas has a greenish yellow color. Hydrogen gas is also a diatomic molecule. Hydrogen gas is a flammable, colorless gas. When chlorine gas and hydrogen gas react, the bond between the hydrogen atoms breaks. And the bond between the chlorine atoms also breaks. A new bond forms between each hydrogen and chlorine atom. A new substance, hydrogen chloride, is formed. Hydrogen chloride is a nonflammable, colorless gas. Its properties differ from the properties of both of the starting substances.

Let's look at another example. Sodium is a metal that reacts violently in water. Chlorine gas is poisonous. When chlorine gas and sodium react, the result is a familiar compound—table salt. Sodium chloride, or table salt, is a harmless substance that almost everyone uses. The salt's properties are very different from sodium's or chlorine's. Salt is a new substance.

Quick Lab

Reaction Ready

1. Place a **piece of chalk** in a **plastic cup.**
2. Add **5 mL of vinegar** to the cup. Record your observations.
3. What evidence of a chemical reaction do you see?
4. What type of new substance was formed?

Quick Lab

MATERIALS

FOR EACH PAIR OF STUDENTS
• chalk
• cup, clear plastic
• vinegar, 5 mL

Safety Caution: Remind students to wear safety goggles, gloves, and aprons.

Answers

3. fizzing, bubbles form, gas given off
4. a white, solid precipitate

Answers to Section Review

1. Sample answer: In one class of chemical reaction, two liquids are mixed, and a solid precipate forms.
2. a
3. no
4. no; This is a physical change: the steam that is being formed is just another form of water.
5. Charcoal burning in a grill is a chemical change because new substances are formed in the process.
6. Formation of gas and light are clues that a chemical reaction is taking place.
7. Bonds in the starting substances are being broken.

SECTION Review

Summary

● A chemical reaction is a process by which substances change to produce new substances with new chemical and physical properties.

● Signs that indicate a chemical reaction has taken place are a color change, formation of a gas or a solid, and release of energy.

● During a reaction, bonds are broken, atoms are rearranged, and new bonds are formed.

Using Key Terms

1. Use the following terms in the same sentence: *chemical reaction* and *precipitate*.

Understanding Key Ideas

2. Most chemical reactions
   a. have starting substances that collide with each other.
   b. do not break bonds.
   c. do not rearrange atoms.
   d. cannot be seen.

3. If the chemical properties of a substance have not changed, has a chemical reaction occurred?

Critical Thinking

4. **Analyzing Processes** Steam is escaping from a teapot. Is this a chemical reaction? Explain.

5. **Applying Concepts** Explain why charcoal burning in the grill is a chemical change.

Interpreting Graphics

Use the photo below to answer the questions that follow.

6. What evidence of a chemical reaction is shown in the photo?

7. What is happening to the bonds of the starting substances?

SCiLINKS

Developed and maintained by the
National Science Teachers Association

For a variety of links related to this chapter, go to www.scilinks.org

Topic: Chemical Reactions
SciLinks code: HSM0247

CHAPTER RESOURCES

Chapter Resource File
• Section Quiz **GENERAL**
• Section Review **GENERAL**
• Vocabulary and Section Summary **GENERAL**
• SciLinks Activity **GENERAL**
• Datasheet for Quick Lab

Technology

Transparencies
• Reaction of Hydrogen and Chlorine

Focus

Overview

In this section, students will learn how to write chemical formulas and how to balance chemical equations. This section also explains how the law of conservation of mass is maintained in a balanced equation.

 Bellringer

Write the symbols for several common elements on the board. Have students list the symbols, and then have them try to remember the names of the matching elements. When they have finished, have them check their answers with the periodic table in their book.

Motivate

Discussion ———— GENERAL

Rearranging Atoms Revisit the teaching transparency "Reaction of Hydrogen and Chlorine" to review the nature of a chemical reaction. Help students trace where each atom in the chemical reaction ends up. Use this as a springboard into the idea of balanced chemical equations.

LS Visual

Chemical Formulas and Equations

How many words can you make using the 26 letters of the alphabet? Many thousands? Now, think of how many sentences you can make with all of those words.

Letters are used to form words. In the same way, chemical symbols are put together to make chemical formulas that describe substances. Chemical formulas can be placed together to describe a chemical reaction, just like words can be put together to make a sentence.

Chemical Formulas

All substances are formed from about 100 elements. Each element has its own chemical symbol. A **chemical formula** is a shorthand way to use chemical symbols and numbers to represent a substance. A chemical formula shows how many atoms of each kind are present in a molecule.

As shown in **Figure 1,** the chemical formula for water is H_2O. This formula tells you that one water molecule is made of two atoms of hydrogen and one atom of oxygen. The small 2 in the formula is a subscript. A *subscript* is a number written below and to the right of a chemical symbol in a formula. Sometimes, a symbol, such as O for oxygen in water's formula, has no subscript. If there is no subscript, only one atom of that element is present. Look at **Figure 1** for more examples of chemical formulas.

Figure 1 **Chemical Formulas of Different Substances**

Water	Oxygen	Glucose
H_2O	O_2	$C_6H_{12}O_6$

Water molecules are made up of 3 atoms—2 atoms of hydrogen bonded to 1 atom of oxgen.

Oxygen is a diatomic molecule. Each molecule has 2 atoms of oxygen bonded together.

Glucose molecules have 6 atoms of carbon, 12 atoms of hydrogen, and 6 atoms of oxygen.

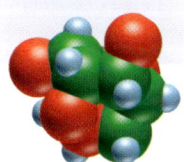

CHAPTER RESOURCES

Chapter Resource File

- Lesson Plan
- Directed Reading A **BASIC**
- Directed Reading B **SPECIAL NEEDS**

Technology

 Transparencies
- Bellringer

Is That a Fact!

When hydrogen chloride gas is dissolved in water, it is known as *hydrochloric acid*. There is a small amount of concentrated hydrochloric acid in your stomach, where it is necessary for the digestion of food.

Carbon dioxide

$$CO_2$$

The *absence of a prefix* indicates one carbon atom.

The prefix *di-* indicates two oxygen atoms.

Dinitrogen monoxide

$$N_2O$$

The prefix *di-* indicates two nitrogen atoms.

The prefix *mono-* indicates one oxygen atom.

Figure 2 *The formulas of these covalent compounds can be written by using the prefixes in the names of the compounds.*

Writing Formulas for Covalent Compounds

If you know the name of the covalent compound, you can often write the chemical formula for that compound. Covalent compounds are usually composed of two nonmetals. The names of many covalent compounds use prefixes. Each prefix represents a number, as shown in **Table 1.** The prefixes tell you how many atoms of each element are in a formula. **Figure 2** shows you how to write a chemical formula from the name of a covalent compound.

Table 1	Prefixes Used in Chemical Names		
mono-	1	hexa-	6
di-	2	hepta-	7
tri-	3	octa-	8
tetra-	4	nona-	9
penta-	5	deca-	10

Writing Formulas for Ionic Compounds

If the name of a compound contains the name of a metal and the name of a nonmetal, the compound is ionic. To write the formula for an ionic compound, make sure the compound's charge is 0. In other words, the formula must have subscripts that cause the charges of the ions to cancel out. **Figure 3** shows you how to write a chemical formula from the name of an ionic compound.

✓ Reading Check What kinds of elements make up an ionic compound? (*See the Appendix for answers to Reading Checks.*)

Sodium chloride

$$NaCl$$

A sodium ion has a **1+ charge.**

A chloride ion has a **1− charge.**

One sodium ion and one chloride ion have an overall **charge of (1+) + (1−) = 0**

Magnesium chloride

$$MgCl_2$$

A magnesium ion has a **2+ charge.**

A chloride ion has a **1− charge.**

One magnesium ion and two chloride ions have an overall **charge of (2+) + 2(1−) = 0.**

Figure 3 *The formula of an ionic compound is written by using enough of each ion so that the overall charge is 0.*

S C I E N C E
HUMOR

Q: Name the compound $Ba(Na)_2$.

A: banana

Answer to Reading Check

Ionic compounds are made up of a metal and a nonmetal.

Section 2 • Chemical Formulas and Equations **33**

Teach

Homework ——— GENERAL

Formulas of Ionic and Covalent Compounds Write the chemical names given below on the board. Have students identify each compound as ionic or covalent. Then, have them use the table of prefixes on this page and a periodic table to write the formula for each compound.

• sulfur trioxide (covalent; SO_3)

• calcium fluoride (ionic; CaF_2)

• phosphorus pentachloride (covalent; PCl_5)

• dinitrogen trioxide (covalent; N_2O_3)

• lithium oxide (ionic; Li_2O)

LS Logical

ACTIVITY ——— ADVANCED

Names of Compounds at Home Have students write the names of compounds in the list of ingredients of products found around their homes. Encourage students to identify the compounds as ionic or covalent and to attempt to write the chemical formulas for each of them. Many of the compounds can be found in chemistry references.

LS Logical

Cultural Awareness GENERAL

Greek Roots The prefixes in the names of covalent compounds have their origins in the Greek language. Each prefix is a Greek numeric representation.

Discussion — GENERAL

Yields/Equals Have students discuss how the "yields" sign in a chemical equation is like an "equals" sign in a mathematical equation. Also, discuss how the two signs are different. **LS Logical**

INCLUSION Strategies

- **Hearing Impaired**
- **Developmentally Delayed**
- **Learning Disabled**

Help students understand that combining the same atoms into different groupings creates different substances. Trace 30 quarters onto a piece of paper. Label each circle as an element. Include the following numbers of elements: 12 hydrogen, 6 carbon, 9 oxygen, 3 nitrogen. Make a copy for each student and have students cut out the circles. Ask students to rearrange their "atoms" to assemble each of the following substances (actual structures will be considered unimportant): oxygen (O_2), glucose ($C_6H_{12}O_6$), carbon dioxide (CO_2), vitamin C ($C_6H_8O_6$), nitroglycerin ($C_3H_5N_3O_9$), rubbing alcohol (C_3H_7OH), and water (H_2O). **LS Kinesthetic**

Answer to Reading Check

Reactants are the starting substances in a chemical reaction, and products are the substances that are formed.

Figure 4 Like chemical symbols, the symbols on this musical score are understood around the world!

chemical equation a representation of a chemical reaction that uses symbols to show the relationship between the reactants and the products

reactant a substance or molecule that participates in a chemical reaction

product the substance that forms in a chemical reaction

Chemical Equations

Think about a piece of music, such as the one in **Figure 4.** Someone writing music must tell the musician what notes to play, how long to play each note, and how each note should be played. Words aren't used to describe the musical piece. Instead, musical symbols are used. The symbols can be understood by anyone who can read music.

Describing Reactions by Using Equations

In the same way that composers use musical symbols, chemists around the world use chemical symbols and chemical formulas. Instead of changing words and sentences into other languages to describe reactions, chemists use chemical equations. A **chemical equation** uses chemical symbols and formulas as a shortcut to describe a chemical reaction. A chemical equation is short and is understood by anyone who understands chemical formulas.

From Reactants to Products

When carbon burns, it reacts with oxygen to form carbon dioxide. **Figure 5** shows how a chemist would use an equation to describe this reaction. The starting materials in a chemical reaction are **reactants** (ree AK tuhnts). The substances formed from a reaction are **products.** In this example, carbon and oxygen are reactants. Carbon dioxide is the product.

✔ **Reading Check** What is the difference between reactants and products in a chemical reaction?

Figure 5 | The Parts of a Chemical Equation

Charcoal is used to cook food on a barbecue grill. When carbon in charcoal reacts with oxygen in the air, the primary product is carbon dioxide, as shown by the chemical equation.

The formulas of the **reactants** are written before the arrow.

The formulas of the **products** are written after the arrow.

$$C + O_2 \longrightarrow CO_2$$

A **plus sign** separates the formulas of two or more reactants or products from one another.

The **arrow,** also called the *yields sign,* separates the formulas of the reactants from the formulas of the products.

SCIENCE HUMOR

Q: What does a doctor do with an injured chemist?

A: helium

Q: And what does a doctor do with a sick chemist?

A: curium

Q: And if the doctor can't cure him?

A: barium

MISCONCEPTION ALERT

Triatomic Molecules Some students may assume that triatomic molecules such as H_2O, CO_2, and N_2O consist of only two particles bonded together. For example, students may think that H_2O consists of a molecule of hydrogen, H_2, bonded to an atom of oxygen. Be sure to point out that in such molecules, there are three particles bonded together. In the case of H_2O, there are two hydrogen atoms bonded to one oxygen atom.

Figure 6 Examples of Similar Symbols and Formulas

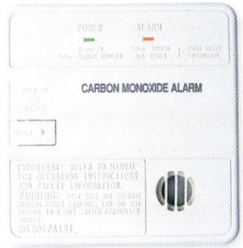

CO₂

The chemical formula for the compound **carbon dioxide** is CO_2. Carbon dioxide is a colorless, odorless gas that you exhale.

CO

The chemical formula for the compound **carbon monoxide** is CO. Carbon monoxide is a colorless, odorless, and poisonous gas.

Co

The chemical symbol for the element **cobalt** is Co. Cobalt is a hard, bluish gray metal.

The Importance of Accuracy

The symbol or formula for each substance in the equation must be written correctly. For a compound, use the correct chemical formula. For an element, use the proper chemical symbol. An equation that has the wrong chemical symbol or formula will not correctly describe the reaction. In fact, even a simple mistake can make a huge difference. **Figure 6** shows how formulas and symbols can be mistaken.

The Reason Equations Must Be Balanced

Atoms are never lost or gained in a chemical reaction. They are just rearranged. Every atom in the reactants becomes part of the products. When writing a chemical equation, make sure the number of atoms of each element in the reactants equals the number of atoms of those elements in the products. This is called balancing the equation.

Balancing equations comes from the work of a French chemist, Antoine Lavoisier (lah vwah ZYAY). In the 1700s, Lavoisier found that the total mass of the reactants was always the same as the total mass of the products. Lavoisier's work led to the **law of conservation of mass.** This law states that mass is neither created nor destroyed in ordinary chemical and physical changes. This law means that a chemical equation must show the same numbers and kinds of atoms on both sides of the arrow.

MATH PRACTICE

Counting Atoms

Some chemical formulas contain parentheses. When counting atoms, multiply everything inside the parentheses by the subscript. For example, $Ca(NO_3)_2$ has one calcium atom, two (2×1) nitrogen atoms, and six (2×3) oxygen atoms. Find the number of atoms of each element in the formulas $Mg(OH)_2$ and $Al_2(SO_4)_3$.

law of conservation of mass the law that states that mass cannot be created or destroyed in ordinary chemical and physical changes

Is That a Fact!

Hydrogen peroxide is a compound that can be made by combining barium peroxide and phosphoric acid. In a 3% solution, hydrogen peroxide is an effective antiseptic and germicide. In a 30% solution, however, it is caustic and highly toxic.

CHAPTER RESOURCES

Technology

Transparencies
• Writing Chemical Formulas and Equations

Reteaching — **BASIC**

Writing Chemical Formulas

Remind students of the importance of writing chemical formulas correctly. As an example, write the formulas for water, H_2O, and hydrogen peroxide, H_2O_2, or for oxygen, O_2, and ozone, O_3, on the board. Discuss with students the properties of each substance and the dangers of mistaking one for the other.

English Language Learners

LS Logical

Quiz — **GENERAL**

1. What is the difference between a reactant and a product? (reactant—a starting material in a chemical reaction; product—a substance formed in a chemical reaction)

2. What is the difference between the formula CO and the symbol Co? (CO is the compound carbon monoxide, and Co is the element cobalt.)

Alternative Assessment — **GENERAL**

Concept Mapping Write the following equation on the board:

$$C_6H_{12}O_6 + O_2 \rightarrow CO_2 + H_2O$$

Have students create a concept map that both shows all the components of this equation (the reactants and products) and balances the equation. **LS** Logical

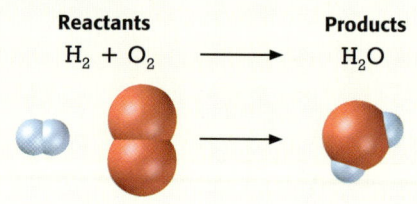

CONNECTION TO Language Arts

WRITING SKILL **Diatomic Molecules** Seven of the chemical elements exist as diatomic molecules. Do research to find out which seven elements these are. Write a short report that describes each diatomic molecule. Be sure to include the formula for each molecule.

How to Balance an Equation

To balance an equation, you must use coefficients (KOH uh FISH uhnts). A *coefficient* is a number that is placed in front of a chemical symbol or formula. For example, 2CO represents two carbon monoxide molecules. The number *2* is the coefficent.

For an equation to be balanced, all atoms must be counted. So, you must multiply the subscript of each element in a formula by the formula's coefficient. For example, $2H_2O$ contains a total of four hydrogen atoms and two oxygen atoms. Only coefficients—not subscripts—are changed when balancing equations. Changing the subscripts in the formula of a compound would change the compound. **Figure 7** shows you how to use coefficients to balance an equation.

✓ **Reading Check** If you see $4O_2$ in an equation, what is the coefficient?

Figure 7 **Balancing a Chemical Equation**

Follow these steps to write a balanced equation for $H_2 + O_2 \longrightarrow H_2O$.

❶ Count the atoms of each element in the reactants and in the products. You can see that there are fewer oxygen atoms in the product than in the reactants.

Reactants	Products
$H_2 + O_2$	H_2O
H = 2 O = 2	H = 2 O = 1

❷ To balance the oxygen atoms, place the coefficient 2 in front of H_2O. Doing so gives you two oxygen atoms in both the reactants and the products. But now there are too few hydrogen atoms in the reactants.

Reactants	Products
$H_2 + O_2$	$2H_2O$
H = 2 O = 2	H = 4 O = 2

❸ To balance the hydrogen atoms, place the coefficient 2 in front of H_2. But to be sure that your answer is correct, always double-check your work!

Reactants	Products
$2H_2 + O_2$	$2H_2O$
H = 4 O = 2	H = 4 O = 2

MATERIALS

FOR EACH STUDENT
- bag, large, strong, sealable plastic
- baking soda, 5 g (1 tsp)
- balance
- film canister, plastic, with lid
- vinegar, 5 mL (1 tsp)

Safety Caution: Remind students to use goggles, aprons, and gloves and NOT to squeeze the bag. The bag must be sealed completely.

Answer

7. The mass should be the same before and after the reaction. (Students may note an apparent loss of mass because the bag is buoyed up by the air that surrounds it. As the bag fills with CO_2, the overall density of the system decreases.)

Conservation of Mass

1. Place **5 g of baking soda** into a **sealable plastic bag.**
2. Place **5 mL of vinegar** into a **plastic film canister.** Put the lid on the canister.
3. Place the canister into the bag. Squeeze the air out of the bag. Seal the bag tightly.
4. Use a **balance** to measure the mass of the bag and its contents. Record the mass.
5. Keeping the bag closed, open the canister in the bag. Mix the vinegar with the baking soda. Record your observations.
6. When the reaction has stopped, measure the mass of the bag and its contents. Record the mass.
7. Compare the mass of the materials before the reaction and the mass of the materials after the reaction. Explain your observations.

SECTION Review

Summary

- A chemical formula uses symbols and subscripts to describe the makeup of a compound.
- Chemical formulas can often be written from the names of covalent and ionic compounds.
- A chemical equation uses chemical formulas, chemical symbols, and coefficients to describe a reaction.
- Balancing an equation requires that the same numbers and kinds of atoms be on each side of the equation.
- A balanced equation illustrates the law of conservation of mass: mass is neither created nor destroyed during ordinary physical and chemical changes.

Using Key Terms

The statements below are false. For each statement, replace the underlined word to make a true statement.

1. A chemical <u>formula</u> describes a chemical reaction.
2. The substances formed from a chemical reaction are <u>reactants</u>.

Understanding Key Ideas

3. The correct chemical formula for carbon tetrachloride is
   a. CCl_3.
   b. C_3Cl.
   c. CCl.
   d. CCl_4.
4. Calcium oxide is used to make soil less acidic. Its formula is
   a. Ca_2O_2.
   b. CaO.
   c. CaO_2.
   d. Ca_2O.
5. Balance the following equations by adding the correct coefficients.
   a. $Na + Cl_2 \longrightarrow NaCl$
   b. $Mg + N_2 \longrightarrow Mg_3N_2$
6. How does a balanced chemical equation illustrate that mass is never lost or gained in a chemical reaction?

7. What is the difference between a subscript and a coefficient?

Math Skills

8. Calculate the number of atoms of each element represented in each of the following: $2Na_3PO_4$, $4Al_2(SO_4)_3$, and $6PCl_5$.

Critical Thinking

9. **Analyzing Methods** Describe how to write a formula for a covalent compound. Give an example of a covalent compound.
10. **Applying Concepts** Explain why the subscript in a formula of a chemical compound cannot be changed when balancing an equation.

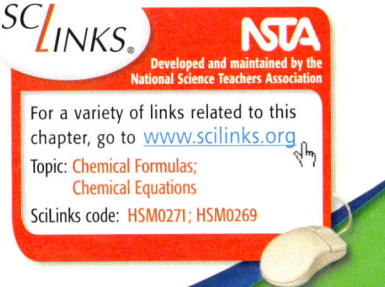

For a variety of links related to this chapter, go to www.scilinks.org
Topic: Chemical Formulas;
    Chemical Equations
SciLinks code: HSM0271; HSM0269

Answers to Section Review

1. equation
2. products
3. d
4. b
5. a. $2Na + Cl_2 \rightarrow 2NaCl$
   b. $3Mg + N_2 \rightarrow Mg_3N_2$
6. There is the same number of atoms of each element on each side of the equation.
7. A subscript shows the number of atoms of a particular element that are present in a certain compound. A coefficient is a number that is placed in front of a chemical symbol or formula in order to balance a chemical equation.
8. $2Na_3PO_4$:
   $2 \times 3 = 6$ atoms Na
   $2 \times 1 = 2$ atoms P
   $2 \times 4 = 8$ atoms O
   $4Al_2(SO_4)_3$:
   $4 \times 2 = 8$ atoms Al
   $4 \times 1 \times 3 = 12$ atoms S
   $4 \times 4 \times 3 = 48$ atoms O
   $6PCl_5$:
   $6 \times 1 = 6$ atoms P
   $6 \times 5 = 30$ atoms Cl
9. To write a formula for a covalent compound, use the prefixes in the name of the compound to find how many atoms of each element are present. Use subscripts to indicate how many atoms of each element are present.
10. The subscript in a formula cannot be changed when balancing an equation because the identity of the substance would be changed.

Answer to Reading Check

The coefficient is 4.

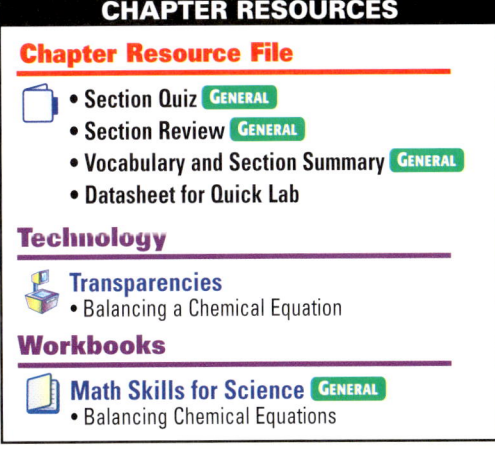

CHAPTER RESOURCES

Chapter Resource File

- Section Quiz **GENERAL**
- Section Review **GENERAL**
- Vocabulary and Section Summary **GENERAL**
- Datasheet for Quick Lab

Technology

Transparencies
- Balancing a Chemical Equation

Workbooks

Math Skills for Science **GENERAL**
- Balancing Chemical Equations

Focus

Overview

This section describes four types of chemical reactions. Students will learn how to determine the type of reaction that is represented by a chemical equation.

 **Bellringer**

Have students answer the following questions in their **science journal:** "Are the products of a reaction always more complex than the reactants? Could products be simpler than the reactants? Explain."

Motivate

Discussion ——— GENERAL

Decomposition and Synthesis

Ask students to define the words *decompose* and *synthesize*. Challenge them to explain what happens during decomposition and synthesis reactions. **LS Verbal/Logical**

READING WARM-UP

Objectives

- Describe four types of chemical reactions.
- Classify a chemical equation as one of four types of chemical reactions.

Terms to Learn

synthesis reaction
decomposition reaction
single-displacement reaction
double-displacement reaction

READING STRATEGY

Mnemonics As you read this section, create a mnemonic device to help you remember the four types of chemical reactions.

synthesis reaction a reaction in which two or more substances combine to form a new compound

---

Types of Chemical Reactions

There are thousands of known chemical reactions. Can you imagine having to memorize even 50 of them?

Remembering all of them would be impossible! But fortunately, there is help. In the same way that the elements are divided into groups based on their properties, reactions can be classified based on what occurs during the reaction.

Most reactions can be placed into one of four categories: synthesis (SIN thuh sis), decomposition, single-displacement, and double-displacement. Each type of reaction has a pattern that shows how reactants become products. One way to remember what happens in each type of reaction is to imagine people at a dance. As you learn about each type of reaction, study the models of students at a dance. The models will help you recognize each type of reaction.

Synthesis Reactions

A **synthesis reaction** is a reaction in which two or more substances combine to form one new compound. For example, a synthesis reaction takes place when sodium reacts with chlorine. This synthesis reaction produces sodium chloride, which you know as table salt. A synthesis reaction would be modeled by two people pairing up to form a dancing couple, as shown in **Figure 1.**

✓ **Reading Check** What is a synthesis reaction? (*See the Appendix for answers to Reading Checks.*)

$$2Na + Cl_2 \longrightarrow 2NaCl$$

Figure 1 *Sodium reacts with chlorine to form sodium chloride in this synthesis reaction.*

CHAPTER RESOURCES

Chapter Resource File

- Lesson Plan
- Directed Reading A BASIC
- Directed Reading B SPECIAL NEEDS

Technology

Transparencies
- Bellringer
- Models of Reactions

Answer to Reading Check

A synthesis reaction is a reaction in which two or more substances combine to form one new compound.

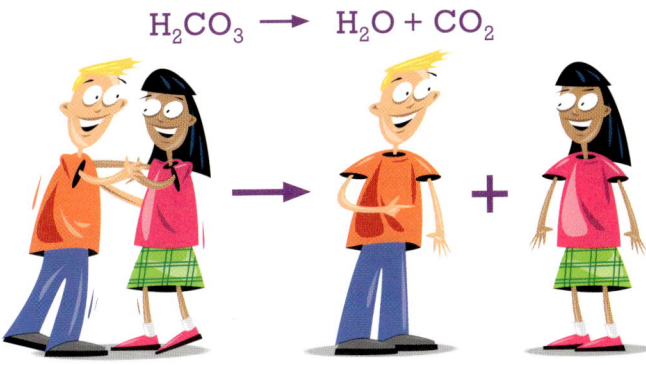

$$H_2CO_3 \longrightarrow H_2O + CO_2$$

Figure 2 In this decomposition reaction, carbonic acid, H_2CO_3, decomposes to form water and carbon dioxide.

Decomposition Reactions

A **decomposition reaction** is a reaction in which a single compound breaks down to form two or more simpler substances. Decomposition is the reverse of synthesis. The dance model for a decomposition reaction would be a couple that finishes a dance and separates, as shown in **Figure 2**.

✓ **Reading Check** How is a decomposition reaction different from a synthesis reaction?

decomposition reaction a reaction in which a single compound breaks down to form two or more simpler substances

single-displacement reaction a reaction in which one element or radical takes the place of another element or radical in a compound

Single-Displacement Reactions

Sometimes, an element replaces another element that is a part of a compound. This type of reaction is called a **single-displacement reaction.** The products of single-displacement reactions are a new compound and a different element. The dance model for a single-displacement reaction would show a person cutting in on a couple who is dancing. A new couple is formed. And a different person is left alone, as shown in **Figure 3.**

Figure 3 Zinc replaces the hydrogen in hydrochloric acid to form zinc chloride and hydrogen gas in this single-displacement reaction.

$$Zn + 2HCl \longrightarrow ZnCl_2 + H_2$$

CONNECTION to Environmental Science ——— GENERAL

Some synthesis reactions have two compounds or an element and a compound as reactants forming a compound. For example, the formation of sulfuric acid, a major component of acid precipitation, involves the following synthesis reactions:

$$S + O_2 \rightarrow SO_2$$

$$2SO_2 + O_2 \rightarrow 2SO_3$$

$$SO_3 + H_2O \rightarrow H_2SO_4$$

Reteaching — BASIC

Role-Playing Organize the class into groups of four, and have them simulate each type of reaction in a manner similar to the "dancing partners" models (they would not necessarily need to dance, however!). **LS Interpersonal**

Quiz — GENERAL

1. Two reactants exchange ions to form two new compounds. What kind of reaction is this? (double-displacement)

2. One element takes the place of another element in a compound. What kind of reaction is this? (single-displacement)

3. The following is the reaction for the explosion of nitroglycerin:
$C_3H_5(NO_3)_3 \rightarrow 12CO_2 + 10H_2O + 6N_2 + O_2$

What kind of reaction is this? (decomposition)

Alternative Assessment — ADVANCED

Other Models This section uses dance partners as an analogy to explain the different kinds of chemical reactions. Ask each student to come up with another analogy to describe chemical reactions. (For example, objects or food could be used.) **LS Logical**

Answer to Reading Check

In a single-displacement reaction, one element can replace another element if the replacing element is more reactive than the starting element.

Figure 4 **Reactivity of Elements**

$Cu + 2AgNO_3 \rightarrow 2Ag + Cu(NO_3)_2$
Copper is more reactive than silver.

$Ag + Cu(NO_3)_2 \rightarrow$ **no reaction**
Silver is less reactive than copper.

Reactivity of Elements

In a single-displacement reaction, a more reactive element can displace a less reactive element in a compound. For example, **Figure 4** shows that copper is more reactive than silver. Copper (Cu) can replace the silver (Ag) ion in the compound silver nitrate. But the opposite reaction does not occur, because silver is less reactive than copper.

The elements in Group 1 of the periodic table are the most reactive metals. Very few nonmetals are involved in single-displacement reactions. In fact, only Group 17 nonmetals participate in single-displacement reactions.

✔ **Reading Check** Why can one element sometimes replace another element in a single-displacement reaction?

INTERNET ACTIVITY

For another activity related to this chapter, go to **go.hrw.com** and type in the keyword **HP5REAW**.

QUICK Lab

Identifying Reactions

1. Study each of the following equations:

$4Na + O_2 \rightarrow 2Na_2O$ $P_4 + 5O_2 \rightarrow 2P_2O_5$

$2Ag_3N \rightarrow 6Ag + N_2$ $Zn + 2HCl \rightarrow ZnCl_2 + H_2$

2. Build models of each of these reactions using **colored clay.** Choose a different color of clay to represent each kind of atom.

3. Identify each type of reaction as a synthesis, decomposition, or single-displacement reaction.

MATERIALS

FOR EACH GROUP OR STUDENT
• clay, 3 colors

Answer

3. $4Na + O_2 \rightarrow 2Na_2O$: synthesis

$2Ag_3N \rightarrow 6Ag + N_2$: decomposition

$P_4 + 5O_2 \rightarrow 2P_2O_5$: synthesis

$Zn + 2HCl \rightarrow ZnCl_2 + H_2$: single-displacement

Double-Displacement Reactions

A **double-displacement reaction** is a reaction in which ions from two compounds exchange places. One of the products of this type of reaction is often a gas or a precipitate. A dance model of a double-displacement reaction would be two couples dancing and then trading partners, as shown in **Figure 5**.

double-displacement reaction
a reaction in which a gas, a solid precipitate, or a molecular compound forms from the exchange of ions between two compounds

$$NaCl + AgF \longrightarrow NaF + AgCl$$

Figure 5 A double-displacement reaction occurs when sodium chloride reacts with silver fluoride to form sodium fluoride and silver chloride (a precipitate).

SECTION Review

Summary

- A synthesis reaction is a reaction in which two or more substances combine to form a compound.

- A decomposition reaction is a reaction in which a compound breaks down to form two or more simpler substances.

- A single-displacement reaction is a reaction in which an element takes the place of another element that is part of a compound.

- A double-displacement reaction is a reaction in which ions in two compounds exchange places.

Using Key Terms

1. In your own words, write a definition for each of the following terms: *synthesis reaction* and *decomposition reaction*.

Understanding Key Ideas

2. What type of reaction does the following equation represent?

   $$FeS + 2HCl \longrightarrow FeCl_2 + H_2S$$

   a. synthesis reaction
   b. double-displacement reaction
   c. single-displacement reaction
   d. decomposition reaction

3. Describe the difference between single- and double-displacement reactions.

Math Skills

4. Write the balanced equation in which potassium iodide, KI, reacts with chlorine to form potassium chloride, KCl, and iodine.

Critical Thinking

5. **Analyzing Processes** The first reaction below is a single-displacement reaction that could occur in a laboratory. Explain why the second single-displacement reaction could not occur.

   $$CuCl_2 + Fe \longrightarrow FeCl_2 + Cu$$
   $$CaS + Al \longrightarrow \text{no reaction}$$

6. **Making Inferences** When two white compounds are mixed in a solution, a yellow solid forms. What kind of reaction has taken place? Explain your answer.

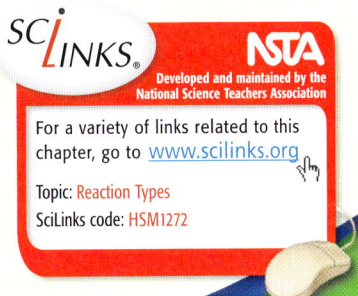

SCILINKS®

Developed and maintained by the National Science Teachers Association

For a variety of links related to this chapter, go to www.scilinks.org

Topic: Reaction Types
SciLinks code: HSM1272

Answers to Section Review

1. Sample answer: A synthesis reaction is a reaction in which atoms or molecules are joined together to form a larger molecule. A decomposition reaction is a reaction in which a compound is split apart to form simpler compounds.

2. b

3. In a single-displacement reaction, one kind of atom moves from one compound to another. In a double-displacement reaction, two different kinds of atoms exchange places between compounds.

4. $2KI + Cl_2 \rightarrow 2KCl + I_2$

5. Calcium is more reactive than aluminum, so no reaction will take place.

6. The reaction is a double-displacement reaction; You started with two solid compounds, each of which must have at least two types of atoms, and ended up with another solid compound, so two different types of atoms must have switched places.

MISCONCEPTION ALERT

Other Types of Reactions
The four types of reactions discussed in this section are not the only types of reactions that substances can undergo. For example, combustion (reaction with oxygen) is an important type of reaction not discussed here.

CHAPTER RESOURCES

Chapter Resource File

- Section Quiz **GENERAL**
- Section Review **GENERAL**
- Vocabulary and Section Summary **GENERAL**
- Reinforcement Worksheet **BASIC**
- Datasheet for Quick Lab

Focus

Overview

This section compares endothermic and exothermic reactions and examines the factors that affect reaction rates. Students will also learn how to interpret energy diagrams.

 Bellringer

Pose the following to students at the beginning of class: "Now that you know a little about chemical reactions, think about the many chemical reactions that take place around you every day. Describe your favorite chemical reaction. How do you think energy is involved in the reaction?"

Motivate

Discussion ——— GENERAL

Exothermic/Endothermic Write the words *exothermic* and *endothermic* on the board. If *thermic* refers to heat, ask students to infer what the two words might mean. Next, display an ordinary houseplant and a match. Ask students to guess which object represents an exothermic reaction and which represents an endothermic reaction. Encourage students to discuss their reasoning.

LS Logical

READING WARM-UP

Objectives

- Compare exothermic and endothermic reactions.
- Explain activation energy.
- Interpret an energy diagram.
- Describe five factors that affect the rate of a reaction.

Terms to Learn

exothermic reaction
endothermic reaction
law of conservation of energy
activation energy
inhibitor
catalyst

READING STRATEGY

Paired Summarizing Read this section silently. In pairs, take turns summarizing the material. Stop to discuss ideas that seem confusing.

Energy and Rates of Chemical Reactions

What is the difference between eating a meal and running a mile? You could say that a meal gives you energy, while running "uses up" energy.

Chemical reactions can be described in the same way. Some reactions release energy, and other reactions absorb energy.

Reactions and Energy

Chemical energy is part of all chemical reactions. Energy is needed to break chemical bonds in the reactants. As new bonds form in the products, energy is released. By comparing the chemical energy of the reactants with the chemical energy of the products, you can decide if energy is released or absorbed in the overall reaction.

Exothermic Reactions

A chemical reaction in which energy is released is called an **exothermic reaction.** *Exo* means "go out" or "exit." *Thermic* means "heat" or "energy." Exothermic reactions can give off energy in several forms, as shown in **Figure 1.** The energy released in an exothermic reaction is often written as a product in a chemical equation, as in this equation:

$$2Na + Cl_2 \rightarrow 2NaCl + energy$$

Figure 1 **Types of Energy Released in Exothermic Reactions**

Light energy is released in the exothermic reaction that is taking place in these light sticks.

Electrical energy is released in the exothermic reaction that will take place in this battery.

Light and thermal energy are released in the exothermic reaction taking place in this campfire.

CHAPTER RESOURCES

Chapter Resource File

- **Lesson Plan**
- **Directed Reading A** BASIC
- **Directed Reading B** SPECIAL NEEDS

Technology

Transparencies
- Bellringer
- **LINK TO LIFE SCIENCE** Photosynthesis

SCIENCE HUMOR

There once was a chemist named Rexo,

Who combined things in reactions quite "exo."

He took a swift fall,

And combusted it all,

From the tips of his toes to his necks-o!

Endothermic Reactions

A chemical reaction in which energy is taken in is called an **endothermic reaction.** *Endo* means "go in." The energy that is taken in during an endothermic reaction is often written as a reactant in a chemical equation. Energy as a reactant is shown in the following equation:

$$2H_2O + energy \rightarrow 2H_2 + O_2$$

An example of an endothermic process is photosynthesis. In photosynthesis, plants use light energy from the sun to produce glucose. Glucose is a simple sugar that is used for nutrition. The equation that describes photosynthesis is the following:

$$6CO_2 + 6H_2O + energy \rightarrow C_6H_{12}O_6 + 6O_2$$

exothermic reaction a chemical reaction in which heat is released to the surroundings

endothermic reaction a chemical reaction that requires heat

law of conservation of energy the law that states that energy cannot be created or destroyed but can be changed from one form to another

✓ Reading Check What is an endothermic reaction? (*See the Appendix for answers to Reading Checks.*)

The Law of Conservation of Energy

Neither mass nor energy can be created or destroyed in chemical reactions. The **law of conservation of energy** states that energy cannot be created or destroyed. However, energy can change forms. And energy can be transferred from one object to another in the same way that a baton is transferred from one runner to another runner, as shown in **Figure 2.**

The energy released in exothermic reactions was first stored in the chemical bonds in the reactants. And the energy taken in during endothermic reactions is stored in the products. If you could measure all the energy in a reaction, you would find that the total amount of energy (of all types) is the same before and after the reaction.

Figure 2 *Energy can be transferred from one object to another object in the same way that a baton is transferred from one runner to another runner in a relay race.*

Endo Alert

1. Fill a **plastic cup** half full with **calcium chloride solution.**
2. Measure the temperature of the solution by using a **thermometer.**
3. Carefully add **1 tsp of baking soda.**
4. Record your observations.
5. When the reaction has stopped, record the temperature of the solution.
6. What evidence that an endothermic reaction took place did you observe?

Is That a Fact!

Endothermic changes can be very useful if you sprain your ankle. Many manufacturers make "instant cold packs" that can be used to treat the swelling of a sprain or bruise. The packs are plastic bags that contain water and another chemical (such as ammonium nitrate) separated by a plastic barrier. When the barrier is broken, the water dissolves the ammonium nitrate, which is an endothermic change. This change causes the pack to get very cold without refrigeration.

MISCONCEPTION
ALERT

Endothermic Reactions and Energy
Students may mistakenly think that endothermic reactions do not produce any energy. Explain that a certain amount of activation energy is required for any reaction to occur. However, in endothermic reactions, the activation energy required is greater than the energy produced in the reaction, so overall, energy is absorbed.

Teach

CONNECTION to Life Science — GENERAL

Photosynthesis One of the most important endothermic reactions is the one carried on continuously by plants and some aquatic organisms—*photosynthesis.* The photosynthetic capability of terrestrial plants is well known. In addition, in the uppermost layer of the ocean (the top 100 m or so), tiny geometrically shaped organisms called *phytoplankton* also make food using photosynthesis. Many fish and other sea animals depend on phytoplankton as their food source. Use the teaching transparency "Photosynthesis" to help students understand this important chemical reaction.

Answer to Reading Check

An endothermic reaction is a chemical reaction in which energy is taken in.

Quick Lab

MATERIALS

FOR EACH STUDENT
- baking soda, 1 tsp
- calcium chloride solution, 5% (100 mL)
- cup, plastic
- thermometer

Safety Cautions: Remind students to review all safety cautions and icons before beginning this lab activity. Caution students to wear goggles and an apron when doing this activity. Students should not put the tablets in their mouth.

Answer

4. The temperature decreased.

MREs Some packaged meals used by the military, campers, hunters, and others are called *Meals Ready to Eat,* or MREs. MREs come in plastic containers and are fully cooked. The meals can be eaten cold but can also be heated by a flameless ration heater that uses an exothermic reaction. In about 12 min, the reaction in the ration heater releases enough energy to warm the MRE to 38°C.

Matches Humans have been using fire since before recorded history. However, until the early 1800s, no fire had ever been started with a match. Matches that light from friction were invented in the 1820s by a British chemist named John Walker. Walker's matches were coated with phosphorus at one end. They caught fire when the phosphorus ignited because of the thermal energy produced by the friction of rubbing the match on a rough surface. Many matches today are safety matches. They light only if rubbed against the striking surface of their package, because the red phosphorus necessary for the reaction is on that surface, not in the match itself.

Figure 3 *Chemical reactions need energy to get started in the same way that a bowling ball needs a push to get rolling.*

activation energy the minimum amount of energy required to start a chemical reaction

**CONNECTION TO
Social Studies**

WRITING SKILL **The Strike-Anywhere Match**
Research the invention of the strike-anywhere match. Find out who invented it, who patented it, and when the match was introduced to the public. In your **science journal,** write a short report about what you learn from your research.

Rates of Reactions

A reaction takes place only if the particles of reactants collide. But there must be enough energy to break the bonds that hold particles together in a molecule. The speed at which new particles form is called the *rate of a reaction.*

Activation Energy

Before the bowling ball in **Figure 3** can roll down the alley, the bowler must first put in some energy to start the ball rolling. A chemical reaction must also get a boost of energy before the reaction can start. This boost of energy is called activation energy. **Activation energy** is the smallest amount of energy that molecules need to react.

Another example of activation energy is striking a match. Before a match can be used to light a campfire, the match has to be lit! A strike-anywhere match has all the reactants it needs to burn. The chemicals on a match react and burn. But, the chemicals will not light by themselves. You must strike the match against a surface. The heat produced by this friction provides the activation energy needed to start the reaction.

✔ **Reading Check** What is activation energy?

Sources of Activation Energy

Friction is one source of activation energy. In the match example, friction provides the energy needed to break the bonds in the reactants and allow new bonds to form. An electric spark in a car's engine is another source of activation energy. This spark begins the burning of gasoline. Light can also be a source of activation energy for a reaction. **Figure 4** shows how activation energy relates to exothermic reactions and endothermic reactions.

Is That a Fact!

Diesel engines have no spark plugs to provide activation energy to ignite the fuel in their cylinders. Although this may seem to be a flaw in the design of the engine, spark plugs are actually not necessary in diesel engines. Air in the cylinders of a diesel engine is compressed so much that its temperature is very high. When the fuel is squirted into the cylinder, the fuel ignites instantly because of the high temperature of the compressed air.

Answer to Reading Check
Activation energy is the energy that is needed to start a chemical reaction.

Figure 4 Energy Diagrams

Exothermic Reaction Once an exothermic reaction starts, it can continue. The energy given off as the product forms continues to supply the activation energy needed for the substances to react.

Endothermic Reaction An endothermic reaction continues to absorb energy. Energy must be used to provide the activation energy needed for the substances to react.

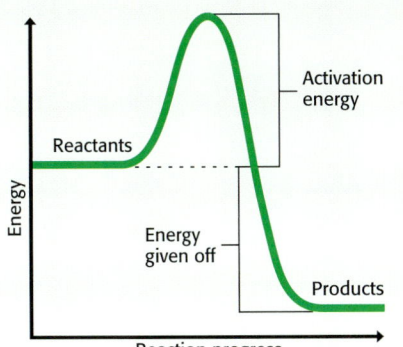

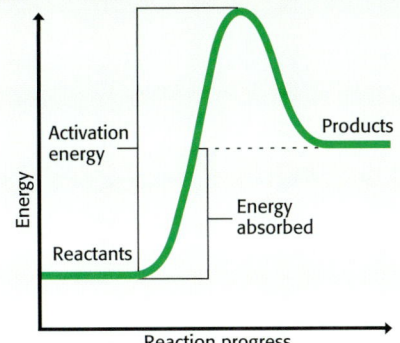

Factors Affecting Rates of Reactions

The rate of a reaction is a measure of how fast the reaction takes place. Recall that the rate of a reaction depends on how fast new particles form. There are four factors that affect the rate of a reaction. These factors are: temperature, concentration, surface area, and the presence of an inhibitor or catalyst.

Temperature

A higher temperature causes a faster rate of reaction, as shown in **Figure 5.** At high temperatures, particles of reactants move quickly. The rapid movement causes the particles to collide often and with a lot of energy. So, many particles have the activation energy to react. And many reactants can change into products in a short time.

Figure 5 *The light stick on the right glows brighter than the one on the left because the one on the right is warmer. The higher temperature causes the rate of the reaction to increase.*

Which Is Quicker?

1. Fill a **clear plastic cup** with **250 mL of warm water.** Fill a **second clear plastic cup** with **250 mL of cold water.**
2. Place **one-quarter of an effervescent tablet** in each of the two cups of water at the same time. Using a **stopwatch,** time each reaction.
3. Observe each reaction, and record your observations.
4. In which cup did the reaction occur at a faster rate?

Is That a Fact!

Cold temperatures can make diesel engines difficult to start. When it is cold, there is not enough thermal energy to ignite the fuel and start the reaction. Combustion chambers in diesel engines now have small heaters called *glow plugs* that have solved this problem. Before the engine starts, an electric current causes the glow plugs to warm up the combustion chambers to a temperature that will ignite the fuel.

Concept Mapping Have students create a concept map that includes the terms *chemical reaction, reaction energy, reaction rate, endothermic reaction, exothermic reaction, temperature, surface area, catalyst,* and *inhibitor.* **LS Visual**

Quiz ——————— GENERAL

1. Why does grinding a solid into a powder increase reaction rate? (The powdered form exposes more particles of the reactant, allowing more collisions with another reactant.)

2. What is the difference between a reactant and a catalyst? (Reactants are changed into products during a reaction. A catalyst increases the speed of a reaction without being used up.)

Alternative Assessment ——— GENERAL

Instruction Manual Have students create an instruction manual entitled *How to Change Chemical Reaction Rates.* Encourage them to be creative in describing the factors that affect reaction rate as discussed in this section. **LS Verbal**

Answer to Reading Check

A high concentration of reactants allows the particles to run into each other more often, so the reaction proceeds at a faster rate.

Figure 6 Concentration of Solutions

▼ When the amount of copper sulfate crystals dissolved in water is **small**, the concentration of the copper sulfate solution is **low**.

▼ When the amount of copper sulfate crystals dissolved in water is **large**, the concentration of the copper sulfate solution is **high**.

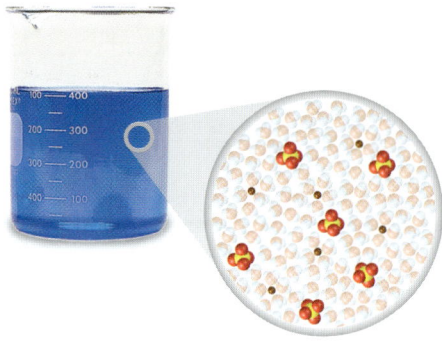

CONNECTION TO Biology

Enzymes and Inhibitors
Enzymes are proteins that speed up reactions in your body. Sometimes, chemicals called *inhibitors* stop the action of enzymes. Research how inhibitors are beneficial in reactions in the human body. Make a poster or a model that explains what you have learned, and present it to your class.

inhibitor a substance that slows down or stops a chemical reaction

catalyst a substance that changes the rate of a chemical reaction without being used up or changed very much

Concentration

In general, a high concentration of reactants causes a fast rate of a reaction. *Concentration* is a measure of the amount of one substance dissolved in another substance, as shown in **Figure 6.** When the concentration is high, there are many reactant particles in a given volume. So, there is a small distance between particles. The particles run into each other often. Thus, the particles react faster.

 **Reading Check** How does a high concentration of reactants increase the rate of a reaction?

Surface Area

Surface area is the amount of exposed surface of a substance. Increasing the surface area of solid reactants increases the rate of a reaction. Grinding a solid into a powder makes a larger surface area. Greater surface area exposes more particles of the reactant to other reactant particles. This exposure to other particles causes the particles of the reactants to collide with each other more often. So, the rate of the reaction is increased.

Inhibitors

An **inhibitor** is a substance that slows down or stops a chemical reaction. Slowing down or stopping a reaction may sometimes be useful. For example, preservatives are added to foods to slow down the growth of bacteria and fungi. The preservatives prevent bacteria and fungi from producing substances that can spoil food. Some antibiotics are examples of inhibitors. For example, penicillin prevents certain kinds of bacteria from making a cell wall. So, the bacteria die.

INCLUSION Strategies

- *Attention Deficit Disorder*
- *Behavior Control Issues*
- *Learning Disabled*

Clarify the meaning of the words *catalyst* and *inhibitor.* Blindfold one person. Tell the person to walk across the classroom and back within two minutes. After the student is blindfolded, clear a path from one side of the room to the other. Organize the rest of the students into two teams. One team is the catalyst and must help the blindfolded person get across the room safely. The other team is the inhibitor and must try to confuse the blindfolded person so he or she doesn't get across the room in time. Teams must not touch the blindfolded person or place anything in his or her path, and only two members of a team may speak at one time. Conclude the activity when the person makes it across the room or after 2 min, whichever comes first. **LS Interpersonal**

English Language Learners

Catalysts

Some chemical reactions would be too slow to be useful without a catalyst (KAT uh LIST). A **catalyst** is a substance that speeds up a reaction without being permanently changed. Because it is not changed, a catalyst is not a reactant. A catalyst lowers the activation energy of a reaction, which allows the reaction to happen more quickly. Catalysts called *enzymes* speed up most reactions in your body. Catalysts are even found in cars, as seen in **Figure 7.** The catalytic converter decreases air pollution. It does this by increasing the rate of reactions that involve the harmful products given off by cars.

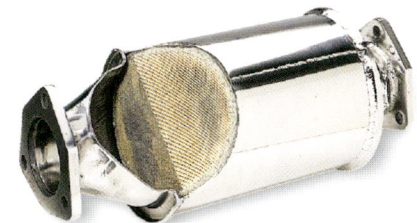

Figure 7 *This catalytic converter contains platinum and palladium. These two catalysts increase the rate of reactions that make the car's exhaust less harmful.*

SECTION Review

Summary

- Energy is given off in exothermic reactions.
- Energy is absorbed in an endothermic reaction.
- The law of conservation of energy states that energy is neither created nor destroyed.
- Activation energy is the energy needed for a reaction to occur.
- The rate of a chemical reaction is affected by temperature, concentration, surface area, and the presence of an inhibitor or catalyst.

Using Key Terms

The statements below are false. For each statement, replace the underlined term to make a true statement.

1. An <u>exothermic</u> reaction absorbs energy.

2. The rate of a reaction can be <u>increased</u> by adding an inhibitor.

Understanding Key Ideas

3. Which of the following will not increase the rate of a reaction?
   a. adding a catalyst
   b. increasing the temperature of the reaction
   c. decreasing the concentration of reactants
   d. grinding a solid into powder

4. How does the concentration of a solution affect the rate of reaction?

Critical Thinking

5. **Making Comparisons** Compare exothermic and endothermic reactions.

6. **Applying Concepts** Explain how chewing your food thoroughly can help your body digest food.

Interpreting Graphics

Use the diagram below to answer the questions that follow.

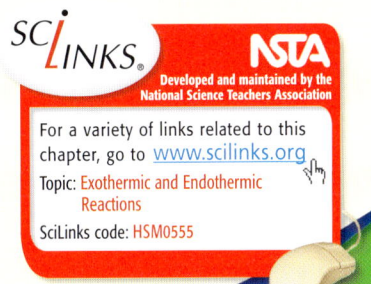

7. Does this energy diagram show an exothermic or an endothermic reaction? How can you tell?

8. A catalyst lowers the amount of activation energy needed to get a reaction started. What do you think the diagram would look like if a catalyst were added?

SCILINKS

NSTA
Developed and maintained by the National Science Teachers Association

For a variety of links related to this chapter, go to www.scilinks.org
Topic: Exothermic and Endothermic Reactions
SciLinks code: HSM0555

CHAPTER RESOURCES

Chapter Resource File

- Section Quiz **GENERAL**
- Section Review **GENERAL**
- Vocabulary and Section Summary **GENERAL**
- Reinforcement Worksheet **BASIC**
- Critical Thinking **ADVANCED**
- Datasheet for Quick Lab

Answers to Section Review

1. endothermic
2. decreased
3. c
4. A higher concentration increases the reaction rate.
5. An exothermic reaction gives off energy. An endothermic reaction absorbs energy.
6. Sample answer: Chewing food grinds it into smaller particles, which allows more surface area of the food to come into contact with digestive juices.
7. an exothermic reaction; You can tell that energy has been given off because the system is at a lower energy after the reaction has taken place.
8. Sample answer: The hump in the middle would not be as tall.

Speed Control

Teacher's Notes

Time Required

One to two 45-minute class periods

Lab Ratings

EASY —————→ HARD

Teacher Prep 🧪🧪🧪
Student Set-Up 🧪🧪
Concept Level 🧪🧪
Clean Up 🧪🧪

MATERIALS

Materials listed are for groups of 2–3 students. For one of the acid solutions, use hydrochloric acid with a concentration between 0.5 M and 1.0 M. For the other acid solution, use 0.1 M hydrochloric acid. When making a solution of acid, it is important always to add the acid to the water. If aluminum strips are not available, substitute strips cut from aluminum cans or aluminum foil.

Safety Caution

Only hydrochloric acid of concentrations 1.0 M or less should be used. Students should not handle more-concentrated hydrochloric acid. In case of an acid spill, dilute the spill first with water. Then, wearing disposable plastic gloves, mop up the spill with wet cloths designated for spill cleanup.

Preparation Notes

The folded aluminum may actually react *faster* because folding may break open the oxide that coats the aluminum. Sandpaper the metal first, or have students sandpaper each strip before they begin the lab.

OBJECTIVES

Describe how the surface area of a solid affects the rate of a reaction.

Explain how concentration of reactants will speed up or slow down a reaction.

MATERIALS

- funnels (2)
- graduated cylinders, 10 mL (2)
- hydrochloric acid, concentrated
- hydrochloric acid, dilute
- strips of aluminum, about 5 cm x 1 cm each (6)
- scissors
- test-tube rack
- test tubes, 30 mL (6)

SAFETY

Speed Control

The reaction rate (how fast a chemical reaction happens) is an important factor to control. Sometimes, you want a reaction to take place rapidly, such as when you are removing tarnish from a metal surface. Other times, you want a reaction to happen very slowly, such as when you are depending on a battery as a source of electrical energy.

In this lab, you will discover how changing the surface area and concentration of the reactants affects reaction rate. In this lab, you can estimate the rate of reaction by observing how fast bubbles form.

Part A: Surface Area

Ask a Question

1 How does changing the surface area of a metal affect reaction rate?

Form a Hypothesis

2 Write a statement that answers the question above. Explain your reasoning.

Test the Hypothesis

3 Use three identical strips of aluminum. Put one strip into a test tube. Place the test tube in the test-tube rack. **Caution:** The strips of metal may have sharp edges.

CHAPTER RESOURCES

Chapter Resource File

 • Datasheet for LabBook
• Lab Notes and Answers

Technology

 Classroom Videos
• Lab Video

• Finding a Balance
• Cata-what? Catalyst!
• Putting Elements Together

4 Carefully fold a second strip in half and then in half again. Use a textbook or other large object to flatten the folded strip as much as possible. Place the strip in a second test tube in the test-tube rack.

5 Use scissors to cut a third strip of aluminum into the smallest possible pieces. Place all of the pieces into a third test tube, and place the test tube in the test-tube rack.

6 Use a funnel and a graduated cylinder to pour 10 mL of concentrated hydrochloric acid into each of the three test tubes. **Caution:** Hydrochloric acid is corrosive. If any acid should spill on you, immediately flush the area with water and notify your teacher.

7 Observe the rate of bubble formation in each test tube. Record your observations.

Analyze the Results

1 **Organizing Data** Which form of aluminum had the greatest surface area? the smallest surface area?

2 **Analyzing Data** The amount of aluminum and the amount of acid were the same in all three test tubes. Which form of the aluminum seemed to react the fastest? Which form reacted the slowest? Explain your answers.

3 **Analyzing Results** Do your results support the hypothesis you made? Explain.

Draw Conclusions

4 **Making Predictions** Would powdered aluminum react faster or slower than the forms of aluminum you used? Explain your answer.

Part B: Concentration

Ask a Question

1 How does changing the concentration of acid affect the reaction rate?

Form a Hypothesis

2 Write a statement that answers the question above. Explain your reasoning.

Test the Hypothesis

3 Place one of the three remaining aluminum strips in each of the three clean test tubes. (Note: Do not alter the strips.) Place the test tubes in the test-tube rack.

4 Using the second funnel and graduated cylinder, pour 10 mL of water into one of the test tubes. Pour 10 mL of dilute acid into the second test tube. Pour 10 mL of concentrated acid into the third test tube.

5 Observe the rate of bubble formation in the three test tubes. Record your observations.

Analyze the Results

1 **Explaining Events** In this set of test tubes, the strips of aluminum were the same, but the concentration of the acid was different. Was there a difference between the test tube that contained water and the test tubes that contained acid? Which test tube formed bubbles the fastest? Explain.

2 **Analyzing Results** Do your results support the hypothesis you made? Explain.

Draw Conclusions

3 **Applying Conclusions** Why should spilled hydrochloric acid be diluted with water before it is wiped up?

Disposal Information

For disposal, neutralize all hydrochloric acid with 0.1 M NaOH as required until the pH is between 6 and 8, and pour down the drain. Aluminum strips that cannot be reused can be placed in the trash.

CLASSROOM TESTED & APPROVED

Tracy Jahn
Berkshire Junior-Senior High School
Canaan, New York

Part A: Form a Hypothesis

2 Sample answer: Increasing the surface area of a metal will increase the reaction rate.

Part A: Analyze the Results

1 The strip of aluminum cut into pieces had the greatest surface area; The folded strip of aluminum had the smallest surface area.

2 The cut-up strip reacted fastest; The folded strip reacted slowest; The cut-up strip had more surface area with which to react with the acid.

3 Answers may vary, depending on hypotheses, but students should understand that increased surface area leads to a faster reaction.

Part A: Draw Conclusions

4 Powdered aluminum would react faster than the other forms of aluminum because it has more surface area.

Part B: Form a Hypothesis

2 Sample answer: Increasing the concentration of acid will increase the reaction rate.

Part B: Analyze the Results

1 The test tube with water showed no signs of reaction; The test tubes with acid were bubbling; The first acid used caused more bubbling because it has the greatest concentration of acid.

2 Answers may vary, depending on hypotheses, but students should state that a higher acid concentration produces a faster reaction.

Part B: Draw Conclusions

3 Diluting acid with water will decrease the acid's concentration and slow down any reaction between the acid and a surface.

Assignment Guide

Section	Questions
1	11, 13
2	4–6, 14–17, 19–20
3	3, 7, 10, 21
4	1–2, 8–9, 12, 18, 22–23

ANSWERS

Using Key Terms

1. inhibitor
2. exothermic reaction
3. synthesis reaction
4. subscript

Understanding Key Ideas

5. c
6. d
7. d
8. c
9. a
10. **a.** synthesis
    b. single-displacement
    c. double-displacement
11. Chemical bonds are broken in a chemical reaction.
12. raise the temperature, increase the concentration of a reactant, increase the surface area of a reactant, and add a catalyst
13. gas formation, solid formation, color change, and energy change

Chapter Review

USING KEY TERMS

Complete each of the following sentences by choosing the correct term from the word bank.

subscript	exothermic reaction
inhibitor	synthesis reaction
coefficient	reactant

1. Adding a(n) ___ will slow down a chemical reaction.

2. A chemical reaction that gives off heat is called a(n) ___.

3. A chemical reaction that forms one compound from two or more substances is called a(n) ___.

4. The 2 in the formula Ag_2S is a (an) ___.

UNDERSTANDING KEY IDEAS

Multiple Choice

5. Balancing a chemical equation so that the same number of atoms of each element is found in both the reactants and the products is an example of
   a. activation energy.
   b. the law of conservation of energy.
   c. the law of conservation of mass.
   d. a double-displacement reaction.

6. Which of the following is the correct chemical formula for dinitrogen tetroxide?
   a. N_4O_2
   b. NO_2
   c. N_2O_5
   d. N_2O_4

7. In which type of reaction do ions in two compounds switch places?
   a. a synthesis reaction
   b. a decomposition reaction
   c. a single-displacement reaction
   d. a double-displacement reaction

8. Which of the following actions is an example of the use of activation energy?
   a. plugging in an iron
   b. playing basketball
   c. holding a lit match to paper
   d. eating

9. Enzymes in your body act as catalysts. Thus, the role of enzymes is
   a. to increase the rate of chemical reactions.
   b. to decrease the rate of chemical reactions.
   c. to help you breathe.
   d. to inhibit chemical reactions.

Short Answer

10. Name the type of reaction that each of the following equations represents.
    a. $2Cu + O_2 \rightarrow 2CuO$
    b. $2Na + MgSO_4 \rightarrow Na_2SO_4 + Mg$
    c. $Ba(CN)_2 + H_2SO_4 \rightarrow BaSO_4 + 2HCN$

11. Describe what happens to chemical bonds during a chemical reaction.

12. Name four ways that you can change the rate of a chemical reaction.

13. Describe four clues that signal that a chemical reaction is taking place.

Math Skills

14 Write balanced equations for the following:

  a. $Fe + O_2 \rightarrow Fe_2O_3$

  b. $Al + CuSO_4 \rightarrow Al_2(SO_4)_3 + Cu$

  c. $Mg(OH)_2 + HCl \rightarrow MgCl_2 + H_2O$

15 Calculate the number of atoms of each element shown in the formulas below:

  a. $CaSO_4$

  b. $4NaOCl$

  c. $Fe(NO_3)_2$

  d. $2Al_2(CO_3)_3$

CRITICAL THINKING

16 **Concept Mapping** Use the following terms to create a concept map: *products, chemical reaction, chemical equation, chemical formulas, reactants, coefficients,* and *subscripts.*

17 **Evaluating Assumptions** Your friend is very worried by rumors that he has heard about a substance called *dihydrogen monoxide* in the city's water system. What could you say to your friend to calm his fears? (Hint: Write the formula of the substance.)

18 **Analyzing Ideas** As long as proper safety precautions have been taken, why can explosives be transported long distances without exploding?

19 **Applying Concepts** You measured the mass of a steel pipe before leaving it outdoors. One month later, the pipe had rusted, and its mass had increased. Does this change violate the law of conservation of mass? Explain your answer.

20 **Applying Concepts** Acetic acid, a compound found in vinegar, reacts with baking soda to produce carbon dioxide, water, and sodium acetate. Without writing an equation, identify the reactants and the products of this reaction.

INTERPRETING GRAPHICS

Use the photo below to answer the questions that follow.

21 What evidence in the photo supports the claim that a chemical reaction is taking place?

22 Is this reaction an exothermic or endothermic reaction? Explain your answer.

23 Draw and label an energy diagram of both an exothermic and endothermic reaction. Identify the diagram that describes the reaction shown in the photo above.

 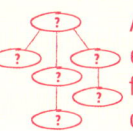
CHAPTER RESOURCES

Chapter Resource File

• Chapter Review **GENERAL**
• Chapter Test A **GENERAL**
• Chapter Test B **ADVANCED**
• Chapter Test C **SPECIAL NEEDS**
• Vocabulary Activity **GENERAL**

Workbooks

Study Guide
• Assessment resources are also available in Spanish.

Standardized Test Preparation

Teacher's Note

To provide practice under more realistic testing conditions, give students 20 minutes to answer all of the questions in this Standardized Test Preparation.

MISCONCEPTION ALERT

Answers to the standardized test preparation can help you identify student misconceptions and misunderstandings.

READING

Passage 1

1. D

2. I

3. B

TEST DOCTOR

Question 3: Seat belts are not mentioned in the passage, so answer A is incorrect. Although the passage mentions that an airbag can prevent serious injuries, this point is not the main focus of the passage, so answer C is incorrect. And the focus of the passage is not chemical reactions, so D is incorrect. Answer B correctly reflects the whole of the passage.

READING

Read each of the passages below. Then, answer the questions that follow each passage.

Passage 1 The key to an air bag's success during a crash is the speed at which it inflates. Inside the bag is a gas generator that contains the compounds sodium azide, potassium nitrate, and silicon dioxide. At the moment of a crash, an electronic sensor in the car detects the sudden change in speed. The sensor sends a small electric current to the gas generator. This electric current provides the activation energy for the chemicals in the gas generator. The rate at which the reaction happens is very fast. In 1/25 of a second, the gas formed in the reaction inflates the bag. The air bag fills upward and outward. By filling the space between a person and the car's dashboard, the air bag protects him or her from getting hurt.

1. Which of the following events happens first?
   A The sensor sends an electric current to the gas generator.
   B The air bag inflates.
   C The air bag fills the space between the person and the dashboard.
   D The sensor detects a change in speed.

2. What provides the activation energy for the reaction to occur?
   F the speed of the car
   G the inflation of the air bag
   H the hot engine
   I the electric current from the sensor

3. What is the purpose of this passage?
   A to convince the reader to wear a seat belt
   B to describe the series of events that inflate an air bag
   C to explain why air bags are an important safety feature in cars
   D to show how chemical reactions protect pedestrians

Passage 2 An important tool in fighting forest fires is a slimy, red goop. This mixture of powder and water is a very powerful fire retardant. The burning of trees, grass, and brush is an exothermic reaction. The fire retardant slows or stops this self-feeding reaction by increasing the activation energy for the materials to which it sticks. A plane can carry between 4,500 and 11,000 L of the goop. The plane then drops it all in front of the raging flames of a forest fire when the pilot presses the button. Firefighters on the ground can gain valuable time when a fire is slowed with a fire retardant. This extra time allows the ground team to create a fire line that will finally stop the fire.

1. Which of the following sentences best summarizes the passage?
   A The burning of forests and other brush is an exothermic reaction.
   B Dropping fire retardants ahead of a flame can help firefighters on the ground stop a fire.
   C Firefighters on the ground create a fire line that will help stop the fire from spreading.
   D The slimy, red goop used as a fire retardant is made of a mixture of powder and water.

2. Based on the passage, which of the following statements is a fact?
   F Fire retardants are always successful in putting out fires.
   G No more than 4,500 L of red goop are loaded onto a plane.
   H A fire retardant works by increasing the activation energy for the materials that it sticks on.
   I The burning of trees is an endothermic reaction.

Passage 2

1. B

2. H

TEST DOCTOR

Question 1: The passage does mention that the burning of forests is an exothermic reaction but then goes on to discuss other details about forest fires, so A is not the correct answer. Creating a fire line is also mentioned, but the main idea of the passage does not relate directly to it, so C is not the answer. The composition of the fire retardant is not the main point of the passage, so D is not the answer. The role of fire retardants ahead of the flames in helping firefighters is the main point of the passage, so B is the answer.

Use the energy diagram below to answer the questions that follow.

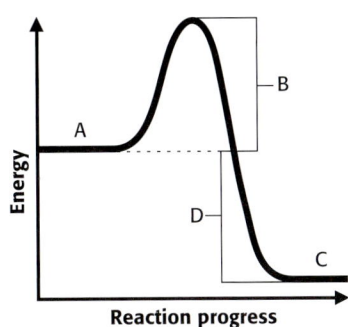

Energy

Reaction progress

1. Which letter represents the energy of the products?
   A   A
   B   B
   C   C
   D   D

2. Which letter represents the activation energy of the reaction?
   F   A
   G   B
   H   C
   I   D

3. Which of the following statements best describes the reaction represented by the graph?
   A   The reaction is endothermic because the energy of the products is greater than the energy of the reactants.
   B   The reaction is endothermic because the energy of the reactants is greater than the energy of the products.
   C   The reaction is exothermic because the energy of the products is greater than the energy of the reactants.
   D   The reaction is exothermic because the energy of the reactants is greater than the energy of the products.

Read each question below, and choose the best answer.

1. Nina has 15 pens in her backpack. She has 3 red pens, 10 black pens, and 2 blue pens. If Ben selects a pen to borrow at random, what is the probability that the pen selected is red?
   A   2/15
   B   1/5
   C   1/3
   D   2/3

2. How many atoms of nitrogen, N, are in the formula for calcium nitrate, $Ca(NO_3)_2$?
   F   3
   G   2
   H   6
   I   1

3. Which letter best represents the number 2 3/5 on the number line?

   A   P
   B   Q
   C   R
   D   S

4. According to the following chemical equation, how many reactants are needed to form water and carbon dioxide?

$$H_2CO_3 \longrightarrow H_2O + CO_2$$

   F   one
   G   two
   H   three
   I   four

Standardized Test Preparation

1. C
2. G
3. D

 TEST DOCTOR

Question 3: The correct answer is D. Students may be confused and think that if a reaction is exothermic, an energy diagram will show a higher energy after the reaction. But the diagram reflects the energy of the system: because energy is given off to the surroundings by an exothermic reaction, the system's energy will be lower after the reaction.

1. B
2. G
3. C
4. F

 **TEST DOCTOR**

Question 1: Correctly answering this question requires reducing the fraction 3/15 to 1/5.

Question 2: Calcium nitrate contains 2 formula units of nitrate, NO_3^-, each of which contains 1 atom of nitrogen. Therefore, there are $2 \times 1 = 2$ atoms of nitrogen per formula unit of calcium nitrate.

CHAPTER RESOURCES

Chapter Resource File
   • Standardized Test Preparation **GENERAL**

State Resources

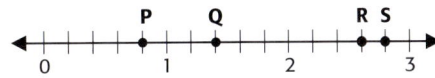 For specific resources for your state, visit **go.hrw.com** and type in the keyword **HSMSTR**.

Science in Action

Science, Technology, and Society

Background

Some students may be surprised to learn that in addition to being used in thermostats, mercury is also used in fluorescent bulbs. When an electric current is passed through mercury vapor, the vapor produces a low-heat, energy-efficient light.

Weird Science

Background

Students may be interested in the details of the chemical reaction that takes place in most light sticks. When the glass vial is broken, the hydrogen peroxide solution mixes with the phenyl oxalate ester and the fluorescent dye. The hydrogen peroxide oxidizes the phenyl oxalate ester. This reaction produces phenol and an unstable peroxyacid ester. The unstable peroxyacid ester then decomposes. This reaction produces additional phenol and a cyclic peroxy compound, which decomposes to carbon dioxide. This decomposition releases energy, which is absorbed by the fluorescent dye.

Science, Technology & Society

Bringing Down the House!

Have you ever watched a building being demolished? It takes only minutes to demolish it, but a lot of time was spent planning the demolition. And it takes time to remove hazardous chemicals from the building. For example, asbestos, which is found in insulation, can cause lung cancer. Mercury found in thermostats can cause brain damage, birth defects, and death. It is important to remove these substances because most of the rubble is sent to a landfill. If hazardous chemicals are not removed, they could leak into the groundwater and enter the water supply.

Math ACTIVITY

A city produces 4 million tons of waste in 1 year. Of this waste, 82% is solid waste. If 38% of the solid waste comes from the construction and demolition of buildings, how many tons of waste does this represent?

Weird Science

Light Sticks

Have you ever seen light sticks at a concert? Your family may even keep them in the car for emergencies. But how do light sticks work? To activate the light stick, you have to bend it. Most light sticks are made of a plastic tube that contains a mixture of two chemicals. Also inside the tube is a thin glass vial, which contains hydrogen peroxide. As long as the glass vial is unbroken, the two chemicals are kept separate. But bending the ends of the tube breaks the glass vial. This action releases the hydrogen peroxide into the other chemicals and a chemical reaction occurs, which makes the light stick glow.

Social Studies ACTIVITY

Who invented light sticks? What was their original purpose? Research the answers to these questions. Make a poster that shows what you have learned.

Answer to Math Activity

4,000,000 tons × 0.82 × 0.38 = 1,200,000 tons

Answer to Social Studies Activity

Chemist Michael M. Rauhut, manager of exploratory research at American Cyanamid in Stamford, Connecticut, and his colleague Laszlo J. Bollyky developed the reaction now used in light sticks. The chemists who were working on the reaction were trying to develop an artifical chemiluminescence that would mimic the natural chemiluminescence of fireflies. The trademark name of Cyalume® was given to the light sticks that Rauhut and Bollyky invented.

Careers

Larry McKee

Arson Investigator Once a fire dies down, you might see an arson investigator like Lt. Larry McKee on the scene. "After the fire is out, I can investigate the fire scene to determine where the fire started and how it started," says McKee, who questions witnesses and firefighters about what they have seen. He knows that the color of the smoke can indicate certain chemicals. He also has help detecting chemicals from an accelerant-sniffing dog, Nikki. Nikki has been trained to detect about 11 different chemicals. If Nikki finds one of these chemicals, she begins to dig. McKee takes a sample of the suspicious material to the laboratory. He treats the sample so that any chemicals present will dissolve in a liquid. A sample of this liquid is placed into an instrument called a *gas chromatograph* and tested. The results of this test are printed out in a graph, from which the suspicious chemical is identified. Next, McKee begins to search for suspects. By combining detective work with scientific evidence, fire investigators can help find clues that can lead to the conviction of the arsonist.

Language Arts ACTiViTY

WRITING SKILL Write a one-page story about an arson investigator. Begin the story at the scene of a fire. Take the story through the different steps that you think an investigator would have to go through to solve the crime.

To learn more about these Science in Action topics, visit **go.hrw.com** and type in the keyword **HP5REAF.**

Current Science
Check out Current Science® articles related to this chapter by visiting **go.hrw.com.** Just type in the keyword **HP5CS14.**

Answer to Language Arts Activity
Answers may vary but should reflect the information given in the passage about Larry McKee.

Careers

Background

In an arson investigator's laboratory, a mass spectrometer is sometimes used to test a vapor sample for accelerants. The spectrometer can provide detailed information about the chemical makeup of any accelerants present. Unlike the gas chromatograph, the mass spectrometer breaks down the chemical compounds in the vapor sample into their characteristic fragments according to their chemical composition. The fragments are then checked against data in a computer to see how closely they match known accelerants. Turner explains, "We can't always identify everything. The sample may contain a lot of chemicals that come from the carpeting or the synthetic products that are present in the household. A lot of the time, there are so many chemicals in a sample that we can't make a conclusive statement about what is there."

Teaching Strategy GENERAL

Invite a firefighter, arson investigator, or fire chief to visit your classroom and to bring visual aids, such as household items found after a fire or some of the equipment used to detect how and where a fire started. Prepare the class to ask questions about what the nature of fires is, how fires start, how to prevent them, and how fire investigators use science to determine the cause of fires.

Chemical Compounds
Chapter Planning Guide

Compression guide:
To shorten instruction because of time limitations, omit Section 4.

OBJECTIVES	LABS, DEMONSTRATIONS, AND ACTIVITIES	TECHNOLOGY RESOURCES
PACING • 90 min pp. 56–61 **Chapter Opener**	**SE** Start-up Activity, p. 57 `GENERAL`	**OSP** Parent Letter ■ `GENERAL` **CD** Student Edition on CD-ROM **CD** Guided Reading Audio CD ■ **TR** Chapter Starter Transparency* **VID** Brain Food Video Quiz
Section 1 Ionic and Covalent Compounds • Describe the properties of ionic and covalent compounds. • Classify compounds as ionic or covalent based on their properties.	**TE** Demonstration Observing Crystals, p. 59 ◆ `GENERAL` **SE** Connection to Language Arts Electrolyte Solutions, p. 60 `GENERAL` **SE** Science in Action Math, Social Studies, and Language Arts Activities, pp. 84–85 `GENERAL`	**TR** Lesson Plans* **TR** Bellringer Transparency*
PACING • 90 min pp. 62–67 **Section 2 Acids and Bases** • Describe four properties of acids. • Identify four uses of acids. • Describe four properties of bases. • Identify four uses of bases.	**TE** Activity Tasting a Weak Acid, p. 62 `GENERAL` **TE** Demonstration A Fruit Juice Indicator, p. 63 ◆ `GENERAL` **SE** Connection to Biology Acids Can Curl Your Hair!, p. 64 `GENERAL` **TE** Activity Comparing Acids and Bases, p. 64 `BASIC` **TE** Demonstration A Base From a Metal, p. 64 `GENERAL` **TE** Demonstration Making Soap, p. 65 `GENERAL` **SE** School-to-Home Activity Acids and Bases at Home, p. 66 `GENERAL` **SE** Quick Lab Blue to Red—Acid!, p. 66 `GENERAL` **SE** Skills Practice Lab Cabbage Patch Indicators, p. 78 `GENERAL` **LB** Calculator-Based Labs Cabbage Patch Indicators `ADVANCED` **LB** Labs You Can Eat Can You Say Seviche?* `GENERAL`	**CRF** Lesson Plans* **TR** Bellringer Transparency* **CRF** SciLinks Activity* `GENERAL` **VID** Lab Videos for Physical Science
PACING • 45 min pp. 68–71 **Section 3 Solutions of Acids and Bases** • Explain the difference between strong acids and bases and weak acids and bases. • Identify acids and bases by using the pH scale. • Describe the formation and uses of salts.	**TE** Connection Activity Math, p. 68 `GENERAL` **SE** Quick Lab pHast Relief!, p. 69 ◆ `GENERAL` **CRF** Datasheet for Quick Lab* **SE** Connection to Biology Blood and pH, p. 70 `GENERAL` **SE** Skills Practice Lab Making Salt, p. 114 `GENERAL` **CRF** Datasheet for LabBook* **LB** EcoLabs & Field Activities Greener Cleaners* `BASIC`	**CRF** Lesson Plans* **TR** Bellringer Transparency* **TR** pH Values of Common Materials*
PACING • 45 min pp. 72–77 **Section 4 Organic Compounds** • Explain why there are so many organic compounds. • Identify and describe saturated, unsaturated, and aromatic hydrocarbons. • Describe the characteristics of carbohydrates, lipids, proteins, and nucleic acids and their functions in the body.	**TE** Connection Activity Math, p. 73 `GENERAL` **SE** Quick Lab Food Facts, p. 75 `GENERAL` **TE** Demonstration Lipids and Water, p. 75 ◆ `GENERAL` **SE** Connection to Social Studies DNA "Fingerprinting" and Crime-Scene Investigation, p. 76 `GENERAL` **TE** Activity DNA, p. 77 ◆ `ADVANCED` **LB** Long-Term Projects & Research Ideas Tiny Plastic Factories* `ADVANCED`	**CRF** Lesson Plans* **TR** Bellringer Transparency* **TR** Structural Formulas* **TR** **LINK TO LIFE SCIENCE** Phospholipid Molecule and Cell Membrane*

PACING • 90 min

CHAPTER REVIEW, ASSESSMENT, AND STANDARDIZED TEST PREPARATION

CRF Vocabulary Activity* `GENERAL`
SE Chapter Review, pp. 80–81 `GENERAL`
CRF Chapter Review* ■ `GENERAL`
CRF Chapter Tests A* ■ `GENERAL`, B* `ADVANCED`, C* `SPECIAL NEEDS`
SE Standardized Test Preparation, pp. 82–83 `GENERAL`
CRF Standardized Test Preparation* `GENERAL`
CRF Performance-Based Assessment* `GENERAL`
OSP Test Generator `GENERAL`
CRF Test Item Listing* `GENERAL`

Online and Technology Resources

Visit **go.hrw.com** for a variety of free resources related to this textbook. Enter the keyword **HP5CMP**.

Holt Online Learning

Students can access interactive problem-solving help and active visual concept development with the *Holt Science and Technology* Online Edition available at **www.hrw.com**.

Guided Reading Audio CD

A direct reading of each chapter using instructional visuals as guideposts. For auditory learners, reluctant readers, and Spanish-speaking students. Available in English and Spanish.

SKILLS DEVELOPMENT RESOURCES	SECTION REVIEW AND ASSESSMENT	STANDARDS CORRELATIONS
SE Pre-Reading Activity, p. 56 `GENERAL` **OSP** Science Puzzlers, Twisters & Teasers* `GENERAL`		National Science Education Standards UCP 1; SAI 1; PS 3a
CRF Directed Reading A* ■ `BASIC`, B* `SPECIAL NEEDS` **CRF** Vocabulary and Section Summary* ■ `GENERAL` **SE** Reading Strategy Reading Organizer, p. 58 `GENERAL` **TE** Reading Strategy Prediction Guide, p. 59 `GENERAL`	**SE** Reading Checks, pp. 59, 60 `GENERAL` **TE** Reteaching, p. 60 `BASIC` **TE** Quiz, p. 60 `GENERAL` **TE** Alternative Assessment, p. 60 `GENERAL` **SE** Section Review,* p. 61 ■ `GENERAL` **CRF** Section Quiz* ■ `GENERAL`	UCP 1; SAI 2; PS 1a, 1b; *LabBook:* SAI 1; PS 1b
CRF Directed Reading A* ■ `BASIC`, B* `SPECIAL NEEDS` **CRF** Vocabulary and Section Summary* ■ `GENERAL` **SE** Reading Strategy Reading Organizer, p. 62 `GENERAL` **TE** Inclusion Strategies, p. 65 **MS** Math Skills for Science Creating Exponents* `GENERAL` **CRF** Reinforcement Worksheet A Simple Solution* `BASIC` **CRF** Critical Thinking Battle of the Breads* `ADVANCED`	**SE** Reading Checks, pp. 62, 64, 67 `GENERAL` **TE** Homework, p. 65 `GENERAL` **TE** Reteaching, p. 66 `BASIC` **TE** Quiz, p. 66 `GENERAL` **TE** Alternative Assessment, p. 66 `GENERAL` **SE** Section Review,* p. 67 ■ `GENERAL` **CRF** Section Quiz* ■ `GENERAL`	UCP 1; SAI 1; SPSP 1, 4; PS 1a, 1b; *Chapter Lab:* SAI 1, 2; PS 1a, 1b
CRF Directed Reading A* ■ `BASIC`, B* `SPECIAL NEEDS` **CRF** Vocabulary and Section Summary* ■ `GENERAL` **SE** Reading Strategy Discussion, p. 68 `GENERAL`	**SE** Reading Checks, pp. 68, 70 `GENERAL` **TE** Homework, p. 69 `GENERAL` **TE** Reteaching, p. 70 `BASIC` **TE** Quiz, p. 70 `GENERAL` **TE** Alternative Assessment, p. 70 `GENERAL` **SE** Section Review,* p. 71 ■ `GENERAL` **CRF** Section Quiz* ■ `GENERAL`	UCP 1, 2, 4; SAI 1; ST 2; SPSP 1, 3; PS 1b
CRF Directed Reading A* ■ `BASIC`, B* `SPECIAL NEEDS` **CRF** Vocabulary and Section Summary* ■ `GENERAL` **SE** Reading Strategy Paired Summarizing, p. 72 `GENERAL` **TE** Inclusion Strategies, p. 73	**SE** Reading Checks, pp. 72, 75, 76 `GENERAL` **TE** Reteaching, p. 76 `BASIC` **TE** Quiz, p. 76 `GENERAL` **TE** Alternative Assessment, p. 76 `GENERAL` **SE** Section Review,* p. 77 ■ `GENERAL` **CRF** Section Quiz* ■ `GENERAL`	UCP 1; SAI 1; ST 2; SPSP 1; PS 1a, 1b,1c

One-Stop Planner® CD-ROM

This convenient CD-ROM includes:
- **Lab Materials QuickList Software**
- **Holt Calendar Planner**
- **Customizable Lesson Plans**
- **Printable Worksheets**
- **ExamView® Test Generator**

cnnstudentnews.com

Find the latest news, lesson plans, and activities related to important scientific events.

www.scilinks.org

Maintained by the **National Science Teachers Association.** See Chapter Enrichment pages for a complete list of topics.

Check out *Current Science* articles and activities by visiting the HRW Web site at **go.hrw.com.** Just type in the keyword **HP5CS15T.**

Classroom Videos
- **Lab Videos** demonstrate the chapter lab.
- **Brain Food Video Quizzes** help students review the chapter material.

Visual Resources

CHAPTER STARTER TRANSPARENCY

Strange but True!

During World War II, the United States could not obtain natural rubber from Asian suppliers, who gathered it from rubber trees as shown below. Faced with a shortage of raw material, American scientists searched for other materials to use in truck tires and soldiers' boots.

James Wright, an engineer at General Electric, was working with silicone oil—a clear, gooey compound composed of silicon bonded to several other elements. By substituting silicon for carbon, the main element in rubber, Wright hoped to create a new compound with all the flexibility and bounce of rubber.

In 1943, Wright made a surprising discovery. He mixed boric acid with silicone oil in a test tube. Instead of forming the hard rubber material he was looking for, the compound remained slightly gooey to the touch. Disappointed with the results, Wright tossed a gob of the material from the test tube onto the floor. To his surprise, the gob bounced right back at him.

The new compound was very bouncy and could be stretched and pulled. However, it wasn't a good rubber substitute, so Wright and other General Electric scientists continued their search.

Seven years later, a toy seller named Peter Hodgson packaged some of Wright's creation in small plastic "eggs" and presented his new product at the 1950 International Toy Fair in New York. The material, called Silly Putty®, proved quite popular. Millions of eggs containing Silly Putty have been sold to kids of all ages since then.

Rubber and boric acid are substances with very different properties. In this chapter, you will learn about the properties that are used to classify many different compounds.

Natural rubber is collected from a rubber tree as it flows from cuts made in the bark.

BELLRINGER TRANSPARENCIES

Section: Ionic and Covalent Compounds
Take a rubber ball and stand in your group. Stand and face your partner. In Group 1, one student from each pair gives his or her ball to the other student. In Group 2, both students should hold both rubber balls, as in a tug of war. Which group represents a compound formed by ionic bonding and which represents a compound formed by covalent bonding?

Write a paragraph explaining the differences between the two types of bonds in your **science journal**.

Section: Acids and Bases
A lemon and a tomato are both fruits that contain citric acid, which gives them a tangy flavor. Make a list of other foods whose tanginess may be due to the presence of acids. What kinds of foods are non-acidic? Do you often eat these foods paired with an acidic food? Why or why not?

Record your thoughts in your **science journal**.

TEACHING TRANSPARENCIES

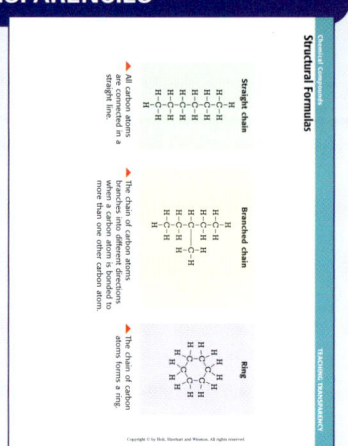

pH Values of Common Materials

Structural Formulas

TEACHING TRANSPARENCIES

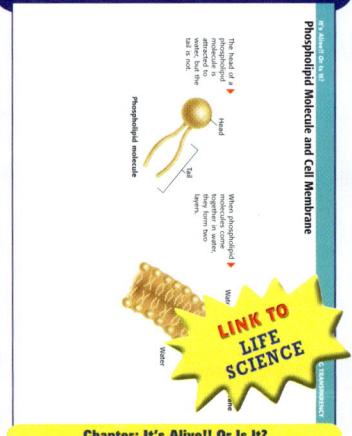

Phospholipid Molecule and Cell Membrane

LINK TO LIFE SCIENCE

Chapter: It's Alive!! Or Is It?

CONCEPT MAPPING TRANSPARENCY

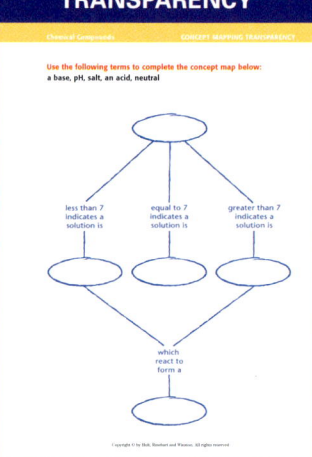

Use the following terms to complete the concept map below: a base, pH, salt, an acid, neutral

Planning Resources

LESSON PLANS

Lesson Plan   SAMPLE

Section: Waves

Pacing
Regular Schedule: with lab(s):2 days   without lab(s):2 days
Block Schedule: with lab(s):1 1/2 days   without lab(s):1 day

Objectives
1. Relate the seven properties of life to a living organism.
2. Describe seven themes that can help you to organize what you learn about biology.
3. Identify the tiny structures that make up all living organisms.
4. Differentiate between reproduction and heredity and between metabolism and homeostasis.

National Science Education Standards Covered
LSInter6:Cells have particular structures that underlie their functions.
LSMat1:Most cell functions involve chemical reactions.
LSBeh1:Cells store and use information to guide their functions.
UCP1:Cell functions are regulated.
ST1:Cells can differentiate and form complete multicellular organism.
PS1:Species evolve over time.
ESS1:The great diversity of organisms is the result of more than 3.5 billion years of evolution.
ESS2:Natural selection and its evolutionary consequences provide a scientific explanation for the fossil record of ancient life forms as well as for the striking molecular similarities observed among the diverse species of living organisms.
ST1:The millions of different species of plants, animals, and microorganisms that live on Earth today are related by descent from common ancestors.
ST2:The energy for life primarily comes from the sun.
SPSP1:The complexity and organization of organisms accommodates the need for obtaining, transforming, transporting, releasing, and eliminating the matter and energy used to sustain the organism.
SPSP6:As matter and energy flows through different levels of organization of living systems—cells, organs, communities—and between living systems and the physical environment, chemical elements are recombined in different ways.
HNS1:Organisms have behavioral responses to internal changes and to external stimuli.

PARENT LETTER

SAMPLE

Dear Parent,

Your son's or daughter's science class will soon begin exploring the chapter entitled "The World of Physical Science." In this chapter, students will learn about how the scientific method applies to the world of physical science and the role of physical science in the world. By the end of the chapter, students should demonstrate a clear understanding of the chapter's main ideas and be able to discuss the following topics:

1. physical science as the study of energy and matter (Section 1)
2. the role of physical science in the world around them (Section 1)
3. careers that rely on physical science (Section 1)
4. the steps used in the scientific method (Section 2)
5. examples of technology (Section 2)
6. how the scientific method is used to answer questions and solve problems (Section 2)
7. how our knowledge of science changes over time (Section 2)
8. how models represent real objects or systems (Section 3)
9. examples of different ways models are used in science (Section 3)
10. the importance of the International System of Units (Section 4)
11. the appropriate units to use for particular measurements (Section 4)
12. how area and density are derived quantities (Section 4)

Questions to Ask Along the Way

You can help your son or daughter learn about these topics by asking interesting questions such as the following:

• What are some surprising careers that use physical science?
• What is a characteristic of a good hypothesis?
• When is it a good idea to use a model?
• Why do Americans measure things in terms of inches and yards and meters?

ALSO IN SPANISH

TEST ITEM LISTING

TEST ITEM LISTING
The World of Science   SAMPLE

MULTIPLE CHOICE

1. A limitation of models is that
   a. they are large enough to see.
   b. they do not act exactly like the things that they model.
   c. they are smaller than the things that they model.
   d. they model unfamiliar things
   Answer: B   Difficulty: 1   Section: 3   Objective: 2

2. The length 10 m is equal to
   a. 100 cm.   c. 10,000 mm.
   b. 1,000 cm.   d. Both (b) and (c).
   Answer: D   Difficulty: 1   Section: 3   Objective: 2

3. To be valid, a hypothesis must be
   a. testable.   c. made into a law.
   b. supported by evidence.   d. Both (a) and (b).
   Answer: D   Difficulty: 1   Section: 1   Objective: 2

4. The statement "Sheila has a stain on her shirt" is an example of a(n)
   a. law.   c. observation.
   b. hypothesis.   d. prediction.
   Answer: C   Difficulty: 1   Section: 2   Objective: 2

5. A hypothesis is often developed out of
   a. observations   c. laws
   b. experiments   d. Both (a) and (b).
   Answer: B   Difficulty: 1   Section: 2   Objective: 2

6. How many milliliters are in 3.5 kL?
   a. 3,500 mL   c. 3,500,000 mL
   b. 0.0035 mL   d. 35,000 mL
   Answer: B   Difficulty: 1   Section: 3   Objective: 2

7. A map of Seattle is an example of a
   a. law.   c. model
   b. theory   d. unit
   Answer: B   Difficulty: 1   Section: 3   Objective: 2

8. A lab has the safety icons shown below. These icons mean that you should wear
   a. only safety goggles.   c. safety goggles and a lab apron.
   b. only a lab apron.   d. safety goggles, a lab apron, and gloves.
   Answer: B   Difficulty: 1   Section: 1   Objective: 2

9. The law of conservation of mass says the total of mass before a chemical change is
   a. more than the total mass after the change.
   b. less than the total mass after the change.
   c. the same as the total mass after the change.
   d. not the same as the total mass after the change.
   Answer: B   Difficulty: 1   Section: 3   Objective: 2

10. In which of the following areas might you find a geochemist at work?
    a. studying the chemistry of rocks   c. studying fishes
    b. studying humidity   d. studying the atmosphere
    Answer: A   Difficulty: 1   Section: 3   Objective: 2

One-Stop Planner® CD-ROM

This CD-ROM includes all of the resources shown here and the following time-saving tools:

- *Lab Materials QuickList Software*
- *Customizable lesson plans*
- *Holt Calendar Planner*
- *The powerful ExamView® Test Generator*

Meeting Individual Needs

DIRECTED READING A
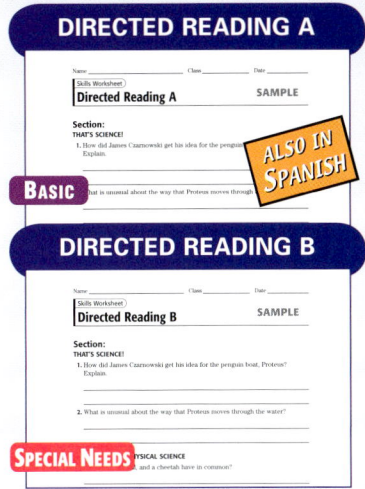
BASIC · ALSO IN SPANISH

DIRECTED READING B
SPECIAL NEEDS

VOCABULARY ACTIVITY
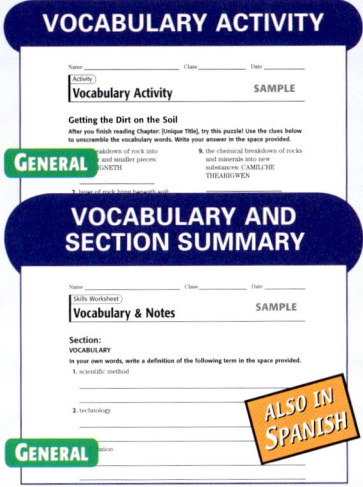
GENERAL

VOCABULARY AND SECTION SUMMARY
GENERAL · ALSO IN SPANISH

REINFORCEMENT
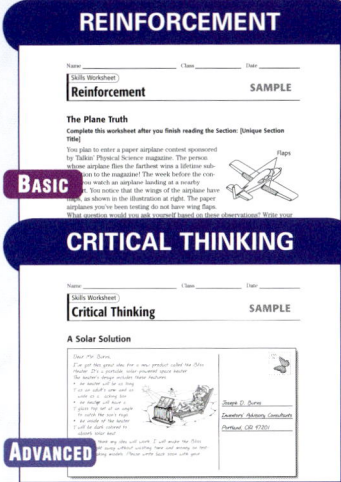
BASIC

CRITICAL THINKING
ADVANCED

SCILINKS ACTIVITY

GENERAL

SCIENCE PUZZLERS, TWISTERS & TEASERS
GENERAL

Labs and Activities

ECOLABS & FIELD ACTIVITIES
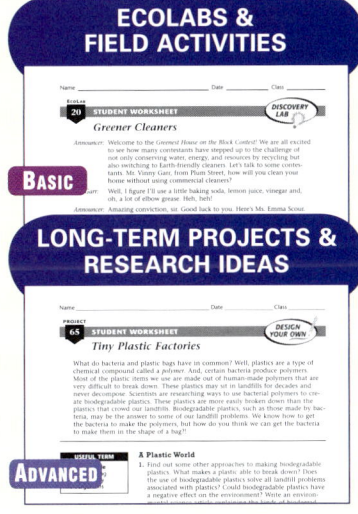
BASIC

LONG-TERM PROJECTS & RESEARCH IDEAS
ADVANCED

LABS YOU CAN EAT
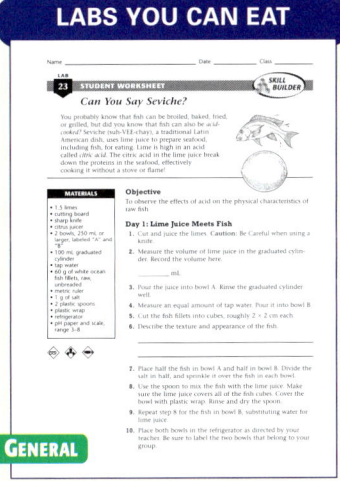
GENERAL

CALCULATOR-BASED LABS
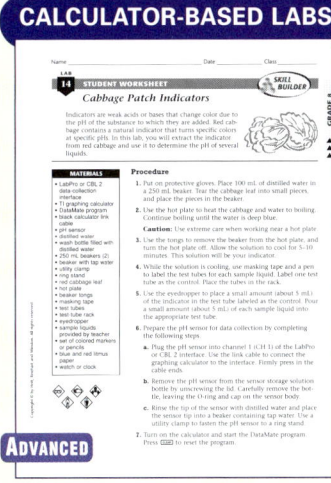
ADVANCED

DATASHEETS FOR QUICK LABS
DATASHEETS FOR CHAPTER LABS
DATASHEETS FOR LABBOOK
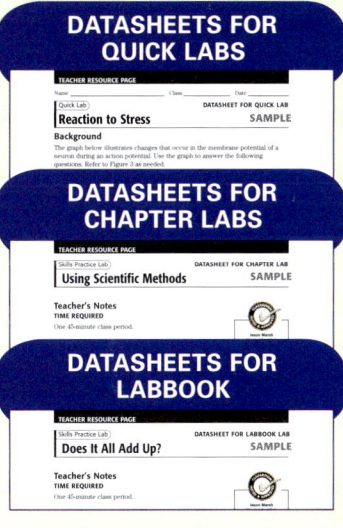

Review and Assessments

SECTION QUIZ
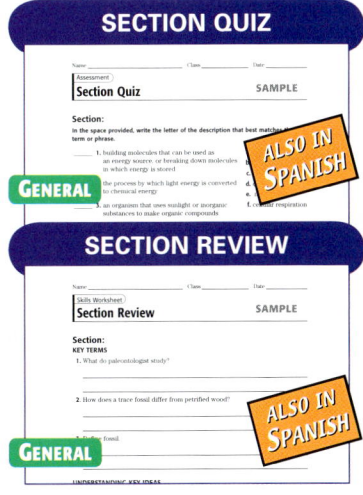
GENERAL · ALSO IN SPANISH

SECTION REVIEW
GENERAL · ALSO IN SPANISH

CHAPTER REVIEW
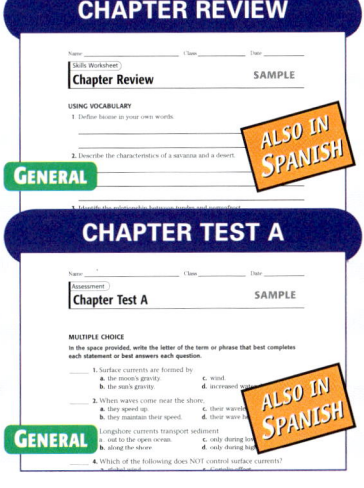
GENERAL · ALSO IN SPANISH

CHAPTER TEST A
GENERAL · ALSO IN SPANISH

CHAPTER TEST B
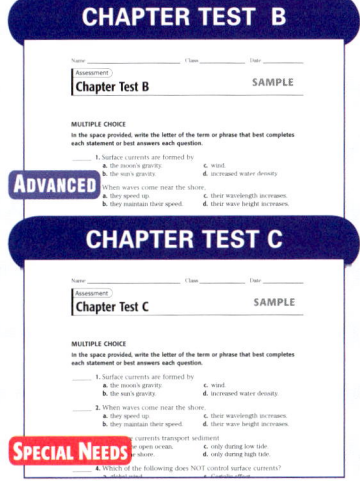
ADVANCED

CHAPTER TEST C
SPECIAL NEEDS

STANDARDIZED TEST PREPARATION
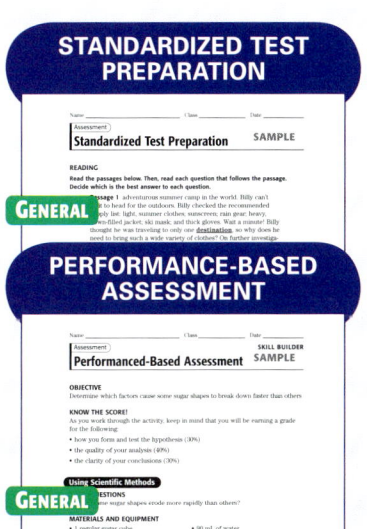
GENERAL

PERFORMANCE-BASED ASSESSMENT
GENERAL

This Chapter Enrichment provides relevant and interesting information to expand and enhance your presentation of the chapter material.

Ionic and Covalent Compounds

Network Crystals

- A crystal formed by covalent bonds between atoms becomes one large molecule called a *macromolecule* or *network crystal*. Network crystals are usually nonmetallic. Diamonds are examples of network crystals. In a diamond, each carbon atom is covalently bonded to four other carbon atoms, forming a network crystal. Silicon carbide also forms a network crystal with covalent bonds between its atoms. Network crystals are very hard, have high melting points, and do not conduct electric current.

Is That a Fact!

- ◆ Water is the most common and most important compound on Earth. Each of a water molecule's hydrogen atoms shares its single electron with the molecule's oxygen atom, forming a covalent bond.

Diamonds

- Diamonds can be used in several different ways, not just as jewelry. Diamonds are composed of carbon atoms that are covalently bonded in a complex network. This structure gives diamonds a unique combination of properties. Diamonds are very hard, have a high melting point, and do not conduct electric current. As a result of these properties, industry uses diamonds in tools that drill, cut, or grind. Furthermore, manufacturers can use them in semiconductor devices. To counteract the problems of expense and scarcity of diamonds, scientists are developing methods to produce high-quality synthetic diamonds.

Is That a Fact!

- ◆ Not all ionic compounds are soluble in water. Silver chloride, zinc sulfide, copper(II) oxide, and magnesium phosphate are some ionic compounds that are not very water soluble.

Acids and Bases

Neutralizing Stings

- Bee venom is acidic, while wasp venom is basic. You can reduce the pain of bee and wasp stings by neutralizing the venom. To neutralize a bee's acidic venom, apply a paste of sodium bicarbonate (baking soda and water) or a weak ammonia solution. To neutralize a wasp's venom, apply vinegar. Of course, stinging insects inject their venom beneath the skin's outermost layer, so a topical salve will neutralize venom only on or near the skin's surface.

Indicators

- The pH scale was introduced by S.P.L. Sørensen (1868–1939) to measure the concentration of hydrogen ions in solution. The more hydrogen ions there are in solution, the more acidic a solution is.

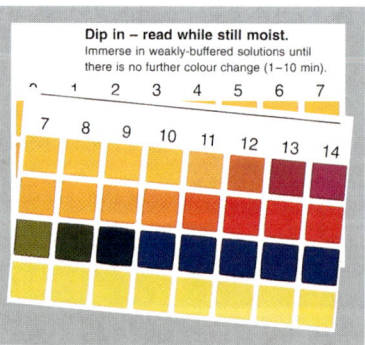

- The percentage or number of hydrogen ions in solution affects the color of certain natural dyes. These dyes can be used as indicators of acidity and basicity (alkalinity). Litmus, an indicator obtained from lichens, is red in acids and blue in bases. Phenol red is yellow in acids and red in bases; methyl red is red in acids and yellow in bases; and bromthymol blue is yellow in acids and blue in bases. However, most indicators do not change color at pH 7. For example, methyl red changes color around pH 5.

Section 3

Solutions of Acids and Bases

Salt Crystals

- The forces within a salt crystal hold the ions together in what is called a *crystal lattice.* Chemists define a crystal lattice as a repetitive, geometric packing arrangement. All of the ions in a crystal of table salt, NaCl, are part of one giant lattice. However, the smallest visible sodium chloride crystal still has more than a billion billion ions!

Is That a Fact!

- The mineral beryl is colorless. However, when beryl contains tiny amounts of the green salt chromium(III) oxide, the valuable green gemstone emerald is formed.

Section 4

Organic Compounds

Hodgkin and Vitamin B12

- In 1948, vitamin B12 was isolated as a red crystalline compound. Dorothy Hodgkin analyzed the vitamin's structure using X-ray crystallography. She won a Nobel Prize in 1964 for her work. Vitamin B12 is a complex, organic molecule containing 181 atoms.

Fredrich Wöhler

- In 1828, while attempting to prepare ammonium cyanate from cyanic acid and ammonia, Friedrich Wöhler (1800–1882) accidentally synthesized urea. This was the first artificially synthesized organic compound. Wöhler's work proved that a compound naturally produced by animals could be made in the laboratory from inorganic chemicals.

Marcelin Berthelot

- Marcelin Berthelot (1827–1907) further demonstrated that plants and animals are not the only sources of organic compounds. He synthesized many organic compounds from inorganic compounds and elements. In 1860, Berthelot synthesized acetylene from its elements, hydrogen and carbon. Berthelot also synthesized another commercially important organic compound, benzene.

Is That a Fact!

- Two compounds are considered structural isomers if they have the same molecular formula but different connections between atoms. Two compounds are stereoisomers if they have the same molecular formula and connections between atoms but different arrangements of atoms in three-dimensional space.

- An alkane containing 30 carbon atoms has 4,111,846,763 possible isomers.

SciLinks is maintained by the National Science Teachers Association to provide you and your students with interesting, up-to-date links that will enrich your classroom presentation of the chapter.

Visit www.scilinks.org and enter the SciLinks code for more information about the topic listed.

Developed and maintained by the National Science Teachers Association

Topic: **Ionic Compounds**	Topic: **Salts**
SciLinks code: **HSM0817**	SciLinks code: **HSM1347**
Topic: **Covalent Compounds**	Topic: **Aromatic Compounds**
SciLinks code: **HSM0365**	SciLinks code: **HSM0095**
Topic: **Acids and Bases**	Topic: **Organic Compounds**
SciLinks code: **HSM0013**	SciLinks code: **HSM1078**
Topic: **pH Scale**	
SciLinks code: **HSM1130**	

Overview

This chapter discusses chemical compounds. Students will learn about the properties of ionic and covalent compounds and acids and bases. The chapter discusses the pH scale and concludes with a discussion of organic compounds and their properties.

Assessing Prior Knowledge

Students should be familiar with the following topics:

• elements, compounds, and mixtures
• chemical bonding
• chemical reactions

Identifying Misconceptions

Students might assume that the properties of a chemical compound result from only the properties of the elements that compose the compound. Point out that the type of bond between atoms also determines the properties of a compound.

3

Chemical Compounds

About the PHOTO

The bean weevil feeds on bean seeds, which are rich in chemical compounds such as proteins, carbohydrates, and lipids. The bean weevil begins life as a tiny grub that lives in the seed where it eats starch and protein. The adult then cuts holes in the seed coat and crawls out, as you can see in this photo.

PRE-READING ACTIVITY

FOLDNOTES **Layered Book** Before you read the chapter, create the FoldNote entitled "Layered Book" described in the **Study Skills** section of the Appendix. Label the tabs of the layered book with "Ionic and covalent compounds," "Acids and bases," "Solutions of acids and bases," and "Organic compounds." As you read the chapter, write information you learn about each category under the appropriate tab.

Standards Correlations

National Science Education Standards

The following codes indicate the National Science Education Standards that correlate to this chapter. The full text of the standards is at the front of the book.

Chapter Opener
UCP 1; SAI 1; PS 3a

Section 1 Ionic and Covalent Compounds
UCP 1; SAI 2; PS 1a, 1b

Section 2 Acids and Bases
UCP 1; SAI 1; SPSP 1, 4; PS 1a, 1b

Section 3 Solutions of Acids and Bases
UCP 1, 2, 4; SAI 1; ST 2; SPSP 1, 3; PS 1b

Section 4 Organic Compounds
UCP 1; SAI 1; ST 2; SPSP 1; PS 1a, 1b, 1c

Chapter Lab
SAI 1, 2; PS 1a, 1b

Chapter Review
PS 1a, 1b, 1c

Science in Action
SAI 1; ST 2; SPSP 5; HNS 1, 2; PS 1b

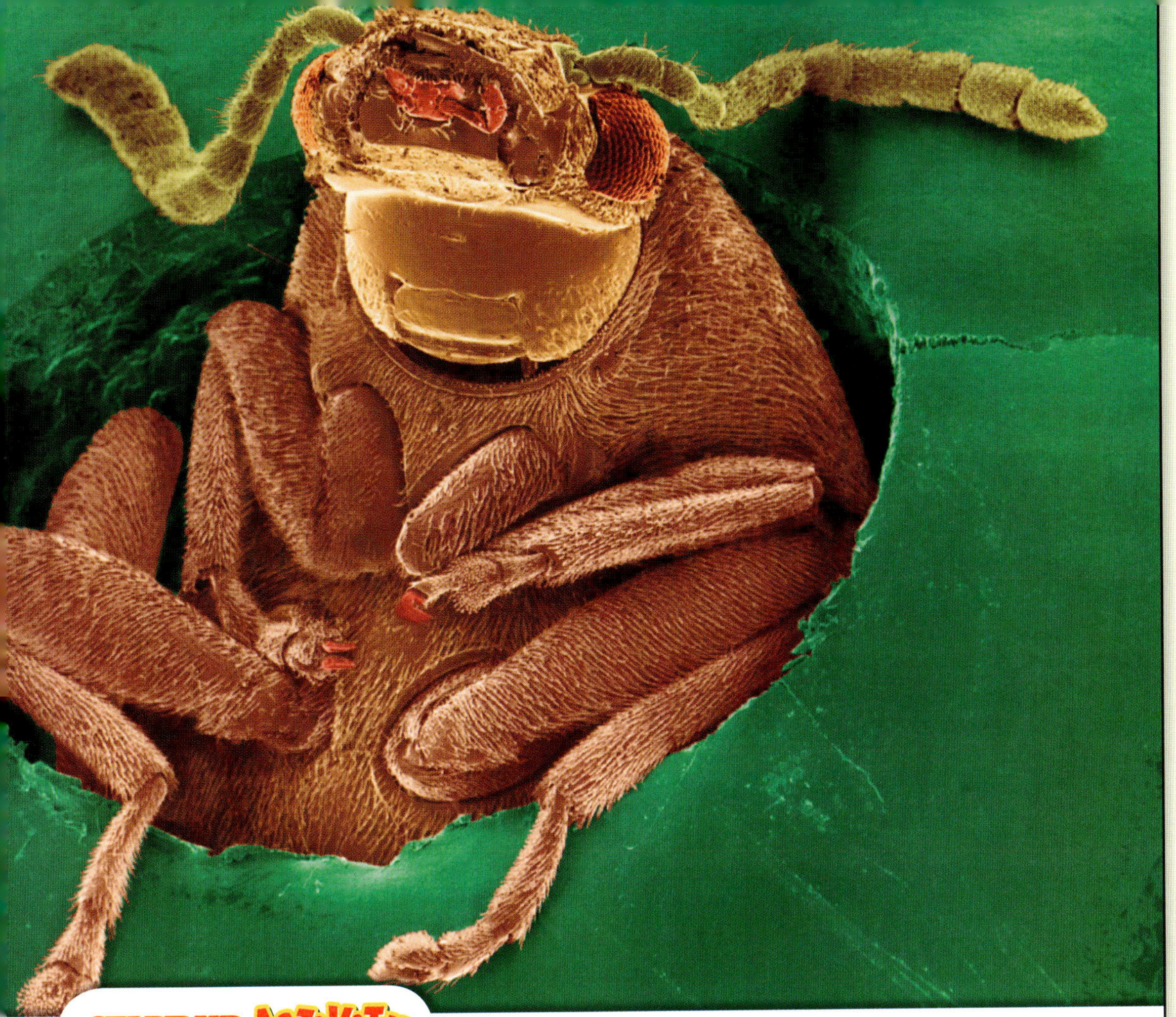

FOR EACH GROUP
- balloons, 2
- cloth, wool 1 piece

Teacher's Notes: This activity will work best on a dry day. Be sure students do not overinflate balloons. It may be necessary to have students rub the balloon with the wool cloth again before they place the balloon against the wall in step 2. Foam packing peanuts may be substituted for balloons in this activity.

Answers

1. Sample answer: The balloons have like charges because they repel each other.

2. The charge on the wall where the balloon is placed must be positive because the balloon was attracted to the wall.

3. Sample answer: The particles that make up compounds must have opposite charges to be attracted to one another.

START-UP ACTIVITY

Sticking Together

In this activity, you will demonstrate the force that keeps particles together in some compounds.

Procedure

1. Rub **two balloons** with a **wool cloth.** Move the balloons near each other. Describe what you see.

2. Put one balloon against a wall. Record your observations.

Analysis

1. The balloons are charged by rubbing them with the wool cloth. Like charges repel each other. Opposite charges attract each other. Do the balloons have like or opposite charges? Explain.

2. If the balloon that was placed against the wall has a negative charge, what is the charge on the wall? Explain your answer.

3. The particles that make up compounds are attracted to each other in the same way that the balloon is attracted to the wall. What can you infer about the particles that make up such compounds?

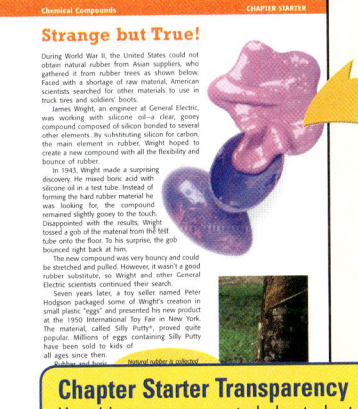

Strange but True!

During World War II, the United States could not obtain natural rubber from Asian suppliers, who gathered it from rubber trees as shown below. Faced with a shortage of raw material, American scientists searched for other materials to use in truck tires and soldiers' boots.

James Wright, an engineer at General Electric, was working with silicone oil—a clear, gooey compound composed of silicon bonded to several other elements. By substituting silicon for carbon, the main element in rubber, Wright hoped to create a new compound with all the flexibility and bounce of rubber.

In 1943, Wright made a surprising discovery. He mixed boric acid with silicone oil in a test tube. Instead of forming the hard rubber material he was looking for, the compound remained slightly gooey to the touch. Disappointed with the results, Wright tossed a gob of the material from the test tube onto the floor. To his surprise, the gob bounced right back at him.

The new compound was very bouncy and could be stretched and pulled. However, it wasn't a good rubber substitute, so Wright and other General Electric scientists continued their search.

Seven years later, a toy seller named Peter Hodgson packaged some of Wright's creation in small plastic "eggs" and presented his new product at the 1950 International Toy Fair in New York. The material, called Silly Putty®, proved quite popular. Millions of eggs containing Silly Putty have been sold to kids of all ages since then.

Chapter Starter Transparency
Use this transparency to help students begin thinking about the importance of chemical compounds.

CHAPTER RESOURCES

Technology

Transparencies
- Chapter Starter Transparency

READING SKILLS

Student Edition on CD-ROM

Guided Reading Audio CD
- English or Spanish

Classroom Videos
- Brain Food Video Quiz

Workbooks

Science Puzzlers, Twisters & Teasers
- Chemical Compounds GENERAL

Focus

Overview

In this section, students learn that chemical compounds can be classified by the bonds they contain: ionic bonds or covalent bonds. Students learn the distinguishing properties of each type of bond.

Bellringer

Give every student a foam ball. Organize the class into two groups, and organize each group into pairs. Tell partners to stand and face each other. In Group 1, have one student from each pair give his or her ball to the other student. In Group 2, tell both students to hold both foam balls, as in a tug of war. Explain that the students in Group 1 represent a compound formed by ionic bonding and that those in Group 2 represent a compound formed by covalent bonding. Ask students to write a paragraph in their **science journal** explaining the differences between the two types of bonds.

LS Kinesthetic   English Language Learners

Objectives

- Describe the properties of ionic and covalent compounds.
- Classify compounds as ionic or covalent based on their properties.

Terms to Learn

chemical bond
ionic compound
covalent compound

Reading Organizer As you read this section, create an outline of the section. Use the headings from the section in your outline.

 chemical bond the combining of atoms to form molecules or compounds

 ionic compound a compound made of oppositely charged ions

Ionic and Covalent Compounds

When ions or molecules combine, they form compounds. Because there are millions of compounds, it is helpful to organize them into groups. But how can scientists tell the difference between compounds?

One way to group compounds is by the kind of chemical bond they have. A **chemical bond** is the combining of atoms to form molecules or compounds. Bonding can occur between valence electrons of different atoms. *Valence electrons* are electrons in the outermost energy level of an atom. The behavior of valence electrons determines if an ionic compound or a covalent compound is formed.

Ionic Compounds and Their Properties

The properties of ionic compounds are a result of strong attractive forces called ionic bonds. An *ionic bond* is an attraction between oppositely charged ions. Compounds that contain ionic bonds are called **ionic compounds.** Ionic compounds can be formed by the reaction of a metal with a nonmetal. Metal atoms become positively charged ions when electrons are transferred from the metal atoms to the nonmetal atoms. This transfer of electrons also causes the nonmetal atom to become a negatively charged ion. Sodium chloride, commonly known as *table salt,* is an ionic compound.

Brittleness

Ionic compounds tend to be brittle solids at room temperature. So, they usually break apart when hit. This property is due to the arrangement of ions in a repeating three-dimensional pattern called a *crystal lattice,* shown in **Figure 1.** Each ion in a lattice is surrounded by ions of the opposite charge. And each ion is bonded to the ions around it. When an ionic compound is hit, the pattern of ions shifts. Ions that have the same charge line up and repel one another, which causes the crystal to break.

Figure 1 *The sodium ions, shown in purple, and the chloride ions, shown in green, are bonded in the crystal lattice structure of sodium chloride.*

CONNECTION to Life Science ——— GENERAL

The body's nerve cells contain a large number of sodium, potassium, and chloride ions. The movement of these ions into and out of the cells causes a type of electric current to exist in the nerve cells. This electric current allows "messages" to move very quickly through the body's nervous system.

Potassium dichromate
Melting point: 398°C

Magnesium oxide
Melting point: 2,800°C

Nickel(II) oxide
Melting point: 1,984°C

High Melting Points

Because of the strong ionic bonds that hold ions together, ionic compounds have high melting points. These high melting points are the reason that most ionic compounds are solids at room temperature. For example, solid sodium chloride must be heated to 801°C before it will melt. The melting points of three other ionic compounds are given in **Figure 2.**

Solubility and Electrical Conductivity

Many ionic compounds are highly soluble. So, they dissolve easily in water. Water molecules attract each of the ions of an ionic compound and pull the ions away from one another. The solution that forms when an ionic compound dissolves in water can conduct an electric current, as shown in **Figure 3.** The solution can conduct an electric current because the ions are charged and are able to move freely past one another. However, an undissolved crystal of an ionic compound does not conduct an electric current.

✓ Reading Check Why do ionic solutions conduct an electric current? (*See the Appendix for answers to Reading Checks.*)

Figure 3 *The pure water does not conduct an electric current. However, the solution of salt water conducts an electric current, so the bulb lights up.*

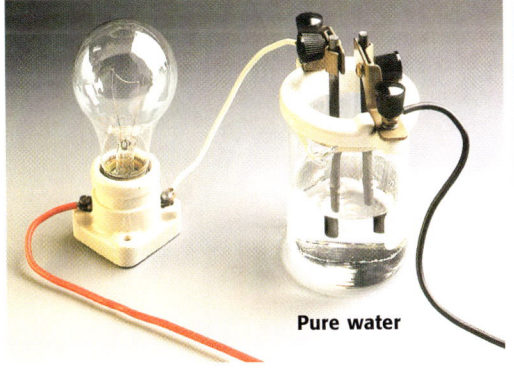

Pure water

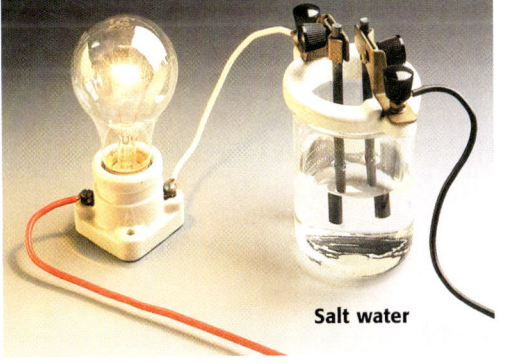

Salt water

Is That a Fact!

Some compounds contain both ionic and covalent bonds. For example, silver nitrate, $AgNO_3$, is made of Ag^+ (silver) and NO_3^- (nitrate) ions joined by ionic bonds. However, the atoms in each nitrate ion are held together by covalent bonds.

Answer to Reading Check
Ionic solutions conduct an electric current because the ions in the solution are charged and are able to move past each other easily.

Chemical Bond Review On the board, create a table with the headings "Ionic" and "Covalent." Ask students to help you fill in the chart with descriptions of each type of bond and to give examples of compounds formed by each type of bond. **LS Logical**

Quiz — GENERAL

1. How are ionic compounds formed? (by the transfer of electrons from metal atoms to nonmetal atoms)

2. Give two examples of covalent compounds. (Answers will vary but may include sugar, water, and carbon dioxide.)

3. Potassium chloride is a crystalline solid at room temperature. Is this compound more likely to be ionic or covalent? (ionic)

Alternative Assessment — GENERAL

Writing **Bonding Story** Have students write a one- or two-page story that uses the concept of ionic and covalent bonds. For example, the story could be a mystery that can be solved only by determining which of the two types of bonds a certain compound has. **LS Verbal**

PORTFOLIO

covalent compound a chemical compound formed by the sharing of electrons

CONNECTION TO Language Arts

WRITING SKILL **Electrolyte Solutions** Ionic compounds that conduct electricity when they are dissolved in water are called *electrolytes*. Some electrolytes play important roles in the functioning of living cells. Electrolytes can be lost by the body during intense physical activity or illness and must be replenished for cells to work properly. Research two electrolytes that your body cells need and the function that they serve. Present your findings in a one-page research paper.

Figure 4 *Olive oil, which is used in salad dressings, is made of very large covalent molecules that do not mix with water.*

Covalent Compounds and Their Properties

Most compounds are covalent compounds. **Covalent compounds** are compounds that form when a group of atoms shares electrons. This sharing of electrons forms a covalent bond. A *covalent bond* is a weaker attractive force than an ionic bond is. The group of atoms that make up a covalent compound is called a molecule. A *molecule* is the smallest particle into which a covalently bonded compound can be divided and still be the same compound. Properties of covalent compounds are very different from the properties of ionic compounds.

Low Solubility

Many covalent compounds are not soluble in water, which means that they do not dissolve well in water. You may have noticed this if you have ever left off the top of a soda bottle. The carbon dioxide gas that gives the soda its fizz eventually escapes, and your soda pop goes "flat." The attraction between water molecules is much stronger than their attraction for to the molecules of most other covalent compounds. So, water molecules stay together instead of mixing with the covalent compounds. If you have ever made salad dressing, you probably know that oil and water don't mix. Oils, such as the oil in the salad dressing in **Figure 4,** are made of covalent compounds.

✓ **Reading Check** Why won't most covalent compounds dissolve in water?

CONNECTION to History — ADVANCED

Bonding Theory Gilbert Newton Lewis (1875–1946) was an American chemist who developed the electron-pair bonding theory of atoms and molecules. He was the first chemist to describe covalent bonding. Lewis theorized that electrons in an atom usually form a tetrahedral arrangement of pairs around the nucleus. The octet rule is often associated with Lewis, although he never actually used the term "octet" for four pairs of electrons.

Answer to Reading Check

Most covalent compounds will not dissolve in water because the attraction of the water molecules to each other is much stronger than their attraction to the compound.

Low Melting Points

The forces of attraction between molecules of covalent compounds are much weaker than the bonds holding ionic solids together. Less heat is needed to separate the molecules of covalent compounds, so these compounds have much lower melting and boiling points than ionic compounds do.

Electrical Conductivity

Although most covalent compounds don't dissolve in water, some do. Most of the covalent compounds that dissolve in water form solutions that have uncharged molecules. Sugar is a covalent compound that dissolves in water and that does not form ions. So, a solution of sugar and water does not conduct an electric current, as shown in **Figure 5.** However, some covalent compounds do form ions when they dissolve in water. Many acids, for example, form ions in water. These solutions, like ionic solutions, conduct an electric current.

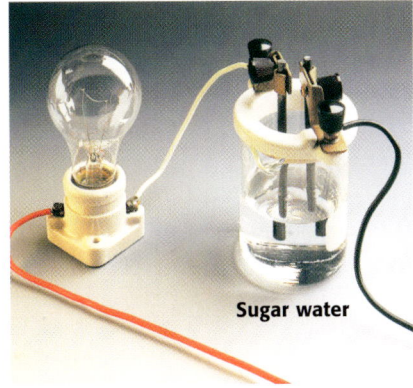

Sugar water

Figure 5 *This solution of sugar, a covalent compound, and water does not conduct an electric current because the molecules of sugar are not charged.*

SECTION Review

Summary

- Ionic compounds have ionic bonds between ions of opposite charges.
- Ionic compounds are usually brittle, have high melting points, dissolve in water, and often conduct an electric current.
- Covalent compounds have covalent bonds and consist of particles called molecules.
- Covalent compounds have low melting points, don't dissolve easily in water, and do not conduct an electric current.

Using Key Terms

1. Use each of the following terms in a separate sentence: *ionic compound, covalent compound,* and *chemical bond.*

Understanding Key Ideas

2. Which of the following describes an ionic compound?
   a. It has a low melting point.
   b. It consists of shared electrons.
   c. It conducts electric current in water solutions.
   d. It consists of two nonmetals.

3. List two properties of covalent compounds.

Math Skills

4. A compound contains 39.37% chromium, 38.10% oxygen, and potassium. What percentage of the compound is potassium?

Critical Thinking

5. **Making Inferences** Solid crystals of ionic compounds do not conduct an electric current. But when the crystals dissolve in water, the solution conducts an electric current. Explain.

6. **Applying Concepts** Some white solid crystals are dissolved in water. If the solution does not conduct an electric current, is the solid an ionic compound or a covalent compound? Explain.

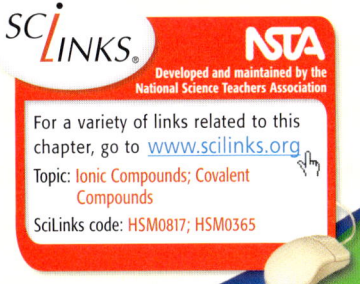

Developed and maintained by the National Science Teachers Association

For a variety of links related to this chapter, go to www.scilinks.org

Topic: Ionic Compounds; Covalent Compounds

SciLinks code: HSM0817; HSM0365

CHAPTER RESOURCES

Chapter Resource File

- Section Quiz **GENERAL**
- Section Review **GENERAL**
- Vocabulary and Section Summary **GENERAL**

Focus

Overview

This section describes the properties and uses of acids and bases. Students will learn the properties of acids and bases and will learn about the uses of acids and bases.

 Bellringer

Show students a lemon and a tomato. Tell them that these fruits contain citric acid, which gives them a tangy flavor. Ask students to suggest other foods whose tanginess may be due to the presence of acids. (Sample answer: dill pickles, grapefruits, strawberries, or vinegar.)

Motivate

ACTIVITY ——————— GENERAL

Tasting a Weak Acid Have groups of students taste samples of carbonated water and compare its flavor with that of regular water. Have students report their findings. Explain that carbonated water contains carbonic acid, a weak acid that forms when carbon dioxide dissolves in water. The acid is responsible for the tangy flavor. Ask students if they can think of other foods that may be flavored with acids. **LS** Kinesthetic

READING WARM-UP

Objectives

- Describe four properties of acids.
- Identify four uses of acids.
- Describe four properties of bases.
- Identify four uses of bases.

Terms to Learn

acid
indicator
base

READING STRATEGY

Reading Organizer As you read this section, make a table comparing acids and bases.

acid any compound that increases the number of hydronium ions when dissolved in water

Acids and Bases

Would you like a nice, refreshing glass of acid? This is just what you get when you have a glass of lemonade.

Lemons contain a substance called an *acid.* One property of acids is a sour taste. In this section, you will learn about the properties of acids and bases.

Acids and Their Properties

A sour taste is not the only property of an acid. Have you noticed that when you squeeze lemon juice into tea, the color of the tea becomes lighter? This change happens because acids cause some substances to change color. An **acid** is any compound that increases the number of hydronium ions, H_3O^+, when dissolved in water. Hydronium ions form when a hydrogen ion, H^+, separates from the acid and bonds with a water molecule, H_2O, to form a hydronium ion, H_3O^+.

✓ **Reading Check** How is a hydronium ion formed? (*See the Appendix for answers to Reading Checks.*)

Acids Have a Sour Flavor

Have you ever taken a bite of a lemon or lime? If so, like the boy in **Figure 1,** you know the sour taste of an acid. The taste of lemons, limes, and other citrus fruits is a result of citric acid. However, taste, touch, or smell should NEVER be used to identify an unknown chemical. Many acids are *corrosive,* which means that they destroy body tissue, clothing, and many other things. Most acids are also poisonous.

NEVER touch or taste a concentrated solution of a strong acid.

Figure 1 *Foods that have a sour taste usually contain acids.*

CHAPTER RESOURCES

Chapter Resource File

- Lesson Plan
- Directed Reading A **BASIC**
- Directed Reading B **SPECIAL NEEDS**

Technology

Transparencies
- Bellringer

Answer to Reading Check

A hydronium ion forms when a hydrogen ion bonds to a water molecule in a water solution.

Figure 2 Detecting Acids with Indicators

The indicator, bromthymol blue, is pale blue in water.

When acid is added, the color changes to yellow because of the presence of the indicator.

indicator a compound that can reversibly change color depending on conditions such as pH

Acids Change Colors in Indicators

A substance that changes color in the presence of an acid or base is an **indicator.** Look at **Figure 2.** The flask on the left contains water and an indicator called *bromthymol blue* (BROHM THIE MAWL BLOO). Acid has been added to the flask on the right. The color changes from pale blue to yellow because the indicator detects the presence of an acid.

Another indicator commonly used in the lab is litmus. Paper strips containing litmus are available in both blue and red. When an acid is added to blue litmus paper, the color of the litmus changes to red.

Acids React with Metals

Acids react with some metals to produce hydrogen gas. For example, hydrochloric acid reacts with zinc metal to produce hydrogen gas, as shown in **Figure 3.** The equation for the reaction is the following:

$$2HCl + Zn \longrightarrow H_2 + ZnCl_2$$

In this reaction, zinc displaces hydrogen in the compound, hydrochloric acid. This displacement happens because zinc is an active metal. But if the element silver were put into hydrochloric acid, nothing would happen. Silver is not an active metal, so no reaction would take place.

Figure 3 *Bubbles of hydrogen gas form when zinc metal reacts with hydrochloric acid.*

ACTIVITY — BASIC

Comparing Acids and Bases

Have students make a chart that compares the properties of acids and bases. Encourage them to refer to their chart as they read through this section. **LS** Logical

Demonstration — GENERAL

Safety Caution: Students should not look directly at the bright light produced by this reaction.

A Base from a Metal Explain to students that metal oxides react with water to form bases. Cut a piece of magnesium ribbon a few centimeters long. Hold the magnesium ribbon with a pair of tongs, and burn it over a plate or pie (tin). Explain to students that the residue left from the reaction is magnesium oxide. Place the residue in a test tube, and add several drops of water. Ask students what will form in the test tube. (magnesium hydroxide, a base)

Ask them what color of litmus paper they would use to test the solution. (red)

Then ask them to predict what will happen to the litmus paper. (The litmus paper will turn blue.)

Ask a volunteer to use litmus paper to check the predictions. **LS** Visual/Logical

CONNECTION TO Biology

Acids Can Curl Your Hair! Permanents contain acids. Acids make hair curly by denaturing a certain amino acid in hair proteins. Research how acids are used in products that either curl or straighten hair. Then, make a poster that demonstrates this process. Present your poster to your classmates.

ACTIVITY

Acids Conduct Electric Current

When acids are dissolved in water, they break apart and form ions in the solution. The ions make it possible for the solution to conduct an electric current. A car battery is one example of how an acid can be used to produce an electric current. The sulfuric acid in the battery conducts electricity to help start the car's engine.

Uses of Acids

Acids are used in many areas of industry and in homes. Sulfuric acid is the most widely made industrial chemical in the world. It is used to make many products, including paper, paint, detergents, and fertilizers. Nitric acid is used to make fertilizers, rubber, and plastics. Hydrochloric acid is used to make metals from their ores by separating the metals from the materials with which they are combined. It is also used in swimming pools to help keep them free of algae. Hydrochloric acid is even found in your stomach, where it aids in digestion. Hydrofluoric acid is used to etch glass, as shown in **Figure 4.** Citric acid and ascorbic acid (Vitamin C) are found in orange juice. And carbonic acid and phosphoric acid help give a sharp taste to soft drinks.

✓ **Reading Check** What are three uses of acids?

Figure 4 *The image of the swan was etched into the glass through the use of hydrofluoric acid.*

Is That a Fact!

Ammonia is a base even though it does not have hydroxide in its chemical formula. A solution of ammonia and water contains ammonium ions, NH_4^+, and hydroxide ions, OH^-.

Answer to Reading Check

Sulfuric acid is used in car batteries to conduct electric current. Hydrochloric acid is used as an algicide in swimming pools. Nitric acid is used to make fertilizers.

Figure 5 Examples of Bases

Soaps are made from sodium hydroxide, which is a base. Soaps remove dirt and oils from skin and feel slippery when you touch them.

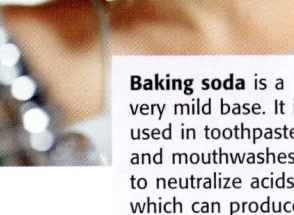

Baking soda is a very mild base. It is used in toothpastes and mouthwashes to neutralize acids, which can produce unpleasant odors.

Bleach and detergents contain bases and are used for removing stains from clothing. Detergents feel slippery like soap.

Bases and Their Properties

A **base** is any compound that increases the number of hydroxide ions, OH⁻, when dissolved in water. For example, sodium hydroxide breaks apart to form sodium ions and hydroxide ions as shown below.

$$NaOH \longrightarrow Na^+ + OH^-$$

Hydroxide ions give bases their properties. **Figure 5** shows examples of bases that you are probably familiar with.

Bases Have a Bitter Flavor and a Slippery Feel

The properties of a base solution include a bitter taste and a slippery feel. If you have ever accidentally tasted soap, you know the bitter taste of a base. Soap will also have the slippery feel of a base. However, taste, touch or smell should NEVER be used to identify an unknown chemical. Like acids, many bases are corrosive. If your fingers feel slippery when you are using a base in an experiment, you may have gotten the base on your hands. You should immediately rinse your hands with large amounts of water and tell your teacher.

base any compound that increases the number of hydroxide ions when dissolved in water

> **NEVER** touch or taste a concentrated solution of a strong base.

Homework —— GENERAL

Writing

Soapmaking Before soap was made commercially, people made their own soap. They combined an oil or fat with a strong base, such as lye (sodium hydroxide). Have students research some aspect of soapmaking and compile their findings into a short report.
LS Verbal/Logical

Demonstration —— GENERAL

Making Soap Place 40 g of lard or tallow in a beaker. Carefully add 10 mL of 20% sodium hydroxide solution (2 g of sodium hydroxide dissolved in 10 mL of water) to the beaker. Warm the mixture very gently on a hot plate for about 30 min. Stir the mixture periodically with a wooden spoon to prevent spattering. Let the solution cool for a few minutes, then stir in 10 g of salt. Skim the soap off the top with the spoon, and rinse the soap well in very cold water. Place the soap on a paper towel to dry. Place a small amount of the soap in a test tube with 10 mL of distilled water. Stopper the test tube, and shake it to form suds. *English Language Learners*
LS Visual

Close

Reteaching — BASIC

Acids and Bases Review Have students make an outline of this section using the headings and subheadings. Under each heading and subheading, have students add at least two facts. Review the outlines as a class, and compile students' outlines to create a master outline that can be used as a study guide.
LS Logical

Quiz — GENERAL

1. Classify each of the following compounds as acidic or basic: soap, vinegar, bleach, baking soda, ammonia, lemonade, and magnesium hydroxide.
(acidic: vinegar, lemonade; basic: soap, bleach, baking soda, ammonia, magnesium hydroxide)

2. When an acid is added to water, does the number of hydronium ions increase or decrease? (increase)

Alternative Assessment — GENERAL

 Writing **Acids and Bases Poster** Have students create a poster-board display using photographs from magazines. On one half of the poster board, students should compile photographs of acid compounds, and on the other half, students should compile photographs of basic compounds. Have students add information about each type of compound and display the posters for the class. **LS Visual** PORTFOLIO

SCHOOL to HOME

Acids and Bases at Home
Ask an adult to join in a contest with you. The object is to find products at home that contain an acid or a base. Each person will write the name of the product and the name of the acid or base that it contains. The person who finds the most products containing an acid or base is the winner.
ACTIVITY

Figure 6 Detecting Bases with Indicators

The indicator, bromthymol blue, is pale blue in water.

When a base is added to the indicator, the indicator turns dark blue.

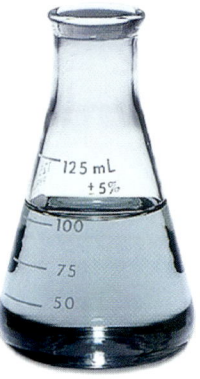 

Bases Change Color in Indicators

Like acids, bases change the color of an indicator. Most indicators turn a different color in the presence of bases than they do in the presence of acids. For example, bases change the color of red litmus paper to blue. And the indicator, bromthymol blue, turns blue when a base is added to it, as shown in **Figure 6.**

Bases Conduct Electric Current

Solutions of bases conduct an electric current because bases increase the number of hydroxide ions, OH^-, in a solution. A hydroxide ion is actually a hydrogen atom and an oxygen atom bonded together. The extra electron gives the hydroxide ion a negative charge.

Quick Lab

Blue to Red—Acid!

1. Take one strip of **red litmus paper** and one strip of **blue litmus paper,** and carefully dip the end of each strip into one of the **solutions** that you will be testing.

2. Note which strip changes color, and note the color.

3. Repeat this step with each of the other solutions. Use two new pieces of litmus paper each time.

4. Determine which solutions are acids and which are bases by observing the color changes that happen on the litmus paper.

Quick Lab

MATERIALS

FOR EACH GROUP
- litmus paper, red and blue
- spot plate
- test solutions

Safety Caution: Students should wear safety goggles, gloves, and aprons. Remind students that they should not taste any of the solutions.

Have students use the dropper and spot plate to test four acidic solutions and four basic solutions. Use common household solutions such as lemon juice, shampoo, baking soda, tap water, black coffee, milk, soft drinks, diluted household ammonia, milk of magnesia, and vinegar.

Answer
4. Blue litmus will turn red to indicate acids. Red litmus will turn blue to indicate bases.

Uses of Bases

Like acids, bases have many uses. Sodium hydroxide is a base used to make soap and paper. It is also used in oven cleaners and in products that unclog drains. Calcium hydroxide, $Ca(OH)_2$, is used to make cement and plaster. Ammonia is found in many household cleaners and is used to make fertilizers. And magnesium hydroxide and aluminum hydroxide are used in antacids to treat heartburn. **Figure 7** shows some of the many products that contain bases. Carefully follow the safety instructions when using these products. Remember that bases can harm your skin.

Reading Check What three ways can bases be used at home?

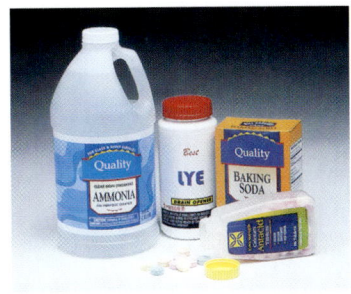

Figure 7 *Bases are common around the house. They are useful as cleaning agents, as cooking aids, and as medicines.*

SECTION Review

Summary

- An acid is a compound that increases the number of hydronium ions in solution.
- Acids taste sour, turn blue litmus paper red, react with metals to produce hydrogen gas, and may conduct an electric current when in solution.
- Acids are used for industrial purposes and in household products.
- A base is a compound that increases the number of hydroxide ions in solution.
- Bases taste bitter, feel slippery, and turn red litmus paper blue. Most solutions of bases conduct an electric current.
- Bases are used in cleaning products and acid neutralizers.

Using Key Terms

1. In your own words, write a definition for each of the following terms: *acid, base,* and *indicator.*

Understanding Key Ideas

2. A base is a substance that
   a. feels slippery.
   b. tastes sour.
   c. reacts with metals to produce hydrogen gas.
   d. turns blue litmus paper red.

3. Acids are important in
   a. making antacids.
   b. preparing detergents.
   c. keeping algae out of swimming pools.
   d. manufacturing cement.

4. What happens to red litmus paper when when it touches a base?

Math Skills

5. A cake recipe calls for 472 mL of milk. You don't have a metric measuring cup at home, so you need to convert milliliters to cups. You know that 1 L equals 1.06 quarts and that there are 4 cups in 1 quart. How many cups of milk will you need to use?

Critical Thinking

6. **Making Comparisons** Compare the properties of acids and bases.

7. **Applying Concepts** Why would it be useful for a gardener or a vegetable farmer to use litmus paper to test soil samples?

8. **Analyzing Processes** Suppose that your teacher gives you a solution of an unknown chemical. The chemical is either an acid or a base. You know that touching or tasting acids and bases is not safe. What two tests could you perform on the chemical to determine whether it is an acid or a base? What results would help you decide if the chemical was an acid or a base?

SCI LINKS

NSTA
Developed and maintained by the
National Science Teachers Association

For a variety of links related to this chapter, go to www.scilinks.org

Topic: Acids and Bases
SciLinks code: HSM0013

Answers to Section Review

1. Sample answer: An acid is a compound that increases the number of hydronium ions when dissolved in water. A base is a compound that increases the number of hydroxide ions when dissolved in water. An indicator is something that changes color in the presence of an acid or a base.

2. a

3. c

4. Red litmus paper turns blue when it touches a base.

5. 472 mL = 0.472 L × 1.06 qt × 4 cups = 2 cups

6. Sample answer: Acids have a sour flavor, change the color of indicators, react with metals, and conduct electric current. Bases have a bitter flavor, feel slippery, change the color of indicators, and conduct electric current.

7. Sample answer: A gardener or a farmer might use litmus paper to test soil samples to see if the soil is acidic or basic.

8. Sample answer: I would test the solution with red and blue litmus paper, and I would put zinc in the solution. If the solution is an acid, it should turn the blue litmus paper red and should react with the zinc. If the solution is a base, it should turn the red litmus paper blue and should not react with the zinc.

Answer to Reading Check

Bases can be used at home in the form of soap, oven cleaner, or antacid.

CHAPTER RESOURCES

Chapter Resource File

- Section Quiz **GENERAL**
- Section Review **GENERAL**
- Vocabulary and Section Summary **GENERAL**
- Reinforcement Worksheet **BASIC**
- Critical Thinking **ADVANCED**
- SciLinks Activity **GENERAL**
- Datasheet for Quick Lab

Overview

This section discusses the strength of acids and bases, the way that acids and bases neutralize each other, and the ways that acids and bases affect the environment. This section also discusses the properties and uses of salts.

 Bellringer

Bring in product labels from vinegar products, citrus products, soaps, cleaning agents, and other household products containing acids or bases. Ask students to work in pairs and examine the labels to see if they can identify which ingredients are acids and which are bases.

Motivate

CONNECTION ACTIVITY
Math ———————— GENERAL

pH Scale Each one-point step on the pH scale represents a tenfold difference in acidity. Thus, a solution of pH 3 is 10 times more acidic than one of pH 4, and a solution of pH 9 is 10 times more basic than one of pH 8. Ask students: "How much more acidic is a solution of pH 2 than one of pH 6?" $(6 - 2 = 4,$ and $10^4 = 10,000$ times more acidic)

 Logical

READING WARM-UP

Objectives

● Explain the difference between strong acids and bases and weak acids and bases.

● Identify acids and bases by using the pH scale.

● Describe the formation and uses of salts.

Terms to Learn

neutralization reaction
pH
salt

READING STRATEGY

Discussion Read this section silently. Write down questions that you have about this section. Discuss your questions in a small group.

Solutions of Acids and Bases

Suppose that at your friend's party, you ate several large pieces of pepperoni pizza followed by cake and ice cream. Now, you have a terrible case of indigestion.

If you have ever had an upset stomach, you may have felt very much like the boy in **Figure 1.** And you may have taken an antacid. But do you know how antacids work? An antacid is a weak base that neutralizes a strong acid in your stomach. In this section, you will learn about the strengths of acids and bases. You will also learn about reactions between acids and bases.

Strengths of Acids and Bases

Acids and bases can be strong or weak. The strength of an acid or a base is not the same as the concentration of an acid or a base. The concentration of an acid or a base is the amount of acid or base dissolved in water. But the strength of an acid or a base depends on the number of molecules that break apart when the acid or base is dissolved in water.

Strong Versus Weak Acids

As an acid dissolves in water, the acid's molecules break apart and produce hydrogen ions, H^+. If all of the molecules of an acid break apart, the acid is called a *strong acid*. Strong acids include sulfuric acid, nitric acid, and hydrochloric acid. If only a few molecules of an acid break apart, the acid is a weak acid. Weak acids include acetic (uh SEET ik) acid, citric acid, and carbonic acid.

✓ **Reading Check** What is the difference between a strong acid and a weak acid? (*See the Appendix for answers to Reading Checks.*)

Figure 1 Antacids may help relieve your stomachache by reacting with the acid in your stomach.

CHAPTER RESOURCES

Chapter Resource File

• Lesson Plan
• Directed Reading A **BASIC**
• Directed Reading B **SPECIAL NEEDS**

Technology

Transparencies
• Bellringer
• pH Values of Common Materials

Answer to Reading Check

In a strong acid, all of the molecules of the acid break apart when the acid is dissolved in water. In a weak acid, only a few of the acid molecules break apart when the acid is dissolved in water.

Strong Versus Weak Bases

When all molecules of a base break apart in water to produce hydroxide ions, OH⁻, the base is a strong base. Strong bases include sodium hydroxide, calcium hydroxide, and potassium hydroxide. When only a few molecules of a base break apart, the base is a weak base, such as ammonium hydroxide and aluminum hydroxide.

Acids, Bases, and Neutralization

When the base in an antacid meets stomach acid, a reaction occurs. The reaction between acids and bases is a **neutralization reaction** (NOO truhl i ZA shuhn ree AK shuhn). Acids and bases neutralize one another because the hydrogen ions (H^+), which are present in an acid, and the hydroxide ions (OH^-), which are present in a base, react to form water, H_2O, which is neutral. Other ions from the acid and base dissolve in the water. If the water evaporates, these ions join to form a compound called a *salt*.

The pH Scale

An *indicator*, such as litmus, can identify whether a solution contains an acid or base. To describe how acidic or basic a solution is, the pH scale is used. The **pH** of a solution is a measure of the hydronium ion concentration in the solution. A solution that has a pH of 7 is neutral, which means that the solution is neither acidic nor basic. Pure water has a pH of 7. Basic solutions have a pH greater than 7. Acidic solutions have a pH less than 7. **Figure 2** shows the pH values for many common materials.

Figure 2 pH Values of Common Materials

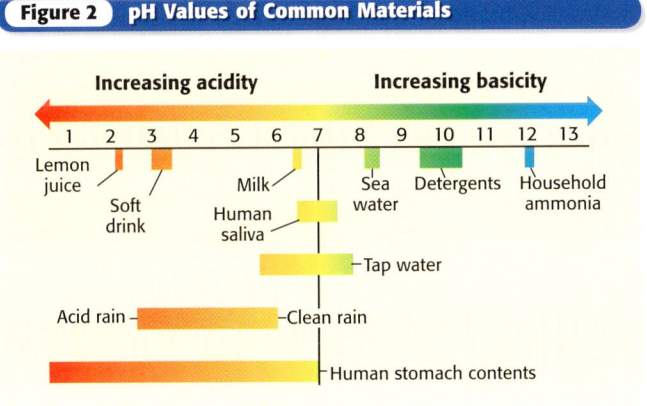

neutralization reaction the reaction of an acid and a base to form a neutral solution of water and a salt

pH a value that is used to express the acidity or basicity (alkalinity) of a system

pHast Relief!

1. Pour **vinegar** into a **small plastic cup** until the cup is half full. Test the vinegar with **red and blue litmus paper.** Record your results.

2. Crush one **antacid tablet,** and mix it with the vinegar. Test the mixture with litmus paper. Record your results.

3. Compare the acidity of the solution before the antacid was added with the acidity of the solution after it was added.

Is That a Fact!

The pH-sensitive compound cyanidin is present in many plants, including cornflowers, poppies, and rhubarb. Cornflowers are blue because their sap is basic, and cyanidin is blue in the presence of bases. Poppies have red flowers because their sap is acidic, and cyanidin is red in the presence of acids. The oxalic acid in rhubarb affects the cyanidin and is responsible for the vegetable's red color.

Homework ——— GENERAL

Dissolving Marble Have students research the effect of acid precipitation on statues, gravestones, and buildings. Tell students to make a small poster describing their findings and to display it to the class. **LS Visual**

Teach

MATERIALS

FOR EACH STUDENT
• antacid tablet
• cup, plastic, small
• litmus paper, red and blue
• vinegar

Safety Caution: Remind students to review all safety cautions and icons before beginning this lab activity.

Answer

3. The vinegar was more acidic before the reaction than the mixture was after the reaction.

Using the Figure – GENERAL

pH Values Direct students' attention to **Figure 2.** Ask what a pH of 7 indicates. (A solution is neither acidic nor basic; it is neutral.) Ask what all acidic substances have in common. (a pH of less than 7)

Discussion ——— GENERAL

Strong and Weak Acids To reinforce the difference between strong and weak acids, place two Petri dishes on an overhead projector. In each dish, place a small piece of magnesium ribbon. Place several drops of 0.8 M HCl on one piece of magnesium and several drops of vinegar on the other piece. Emphasize to students that the concentrations of the two acids are the same. Have students compare the action of the strong acid with that of the weak acid. **LS Visual**   English Language Learners

Quiz — GENERAL

1. Is the compound H_3PO_4 an acid or a base? How do you know? (It is an acid because it will form hydrogen ions if dissolved in water.)

2. Would you expect the pH of a sample of acid rain to be 4 or 9? Why? (4; because it is acidic)

3. What products would form when hydrochloric acid, HCl, and sodium hydroxide, NaOH, react? (The products would be water and the salt sodium chloride.)

Figure 3   Using Indicators to Find pH

pH Indicator Scale

pH 4

pH 10

CONNECTION TO Biology

WRITING SKILL **Blood and pH** Human blood has a pH between 7.38 and 7.42. If the blood pH is lower or higher, the body cannot function properly. Research what can cause the pH of blood to rise above or fall below normal ranges. Write a one-page paper that details your findings.

Figure 4 *To grow blue flowers, plant hydrangeas in soil that has a low pH. To grow pink flowers, use soil that has a high pH.*

Using Indicators to Determine pH

A combination of indicators can be used to find out how basic or how acidic a solution is. This can be done if the colors of the indicators are known at different pH values. **Figure 3** shows strips of pH paper, which contains several different indicators. These strips were dipped into two different solutions. The pH of each solution is found by comparing the colors on each strip with the colors on the indicator scale provided. This kind of indicator is often used to test the pH of water in pools and aquariums. Another way to find the pH of a solution is to use a pH meter. These meters can detect and measure hydronium ion concentration electronically.

✔ **Reading Check** How can indicators determine pH?

pH and the Environment

Living things depend on having a steady pH in their environment. Some plants, such as pine trees, prefer acidic soil that has a pH between 4 and 6. Other plants, such as lettuce, need basic soil that has a pH between 8 and 9. Plants may also have different traits under different growing conditions. For example, the color of hydrangea flowers varies when the flowers are grown in soils that have different pH values. These differences are shown in **Figure 4**. Many organisms living in lakes and streams need a neutral pH to survive.

Most rain has a pH between 5.5 and 6. When rainwater reacts with compounds found in air pollution, acids are formed and the rainwater's pH decreases. In the United States, most acid rain has a pH between 4 and 4.5, but some precipitation has a pH as low as 3.

Answer to Reading Check

Indicators turn different colors at different pH levels. The color on the pH strip can be compared with the colors on the indicator scale to determine the pH of the solution being tested.

Salts

When an acid neutralizes a base, a salt and water are produced. A **salt** is an ionic compound formed from the positive ion of a base and the negative ion of an acid. When you hear the word *salt*, you probably think of the table salt you use to season your food. But the sodium chloride found in your salt shaker is only one example of a large group of compounds called *salts*.

Uses of Salts

Salts have many uses in industry and in homes. You already know that sodium chloride is used to season foods. It is also used to make other compounds, including lye (sodium hydroxide) and baking soda. Sodium nitrate is a salt that is used to preserve food. And calcium sulfate is used to make wallboard, which is used in construction. Another use of salt is shown in **Figure 5.**

Figure 5 *Salts help keep roads free of ice by decreasing the freezing point of water.*

salt an ionic compound that forms when a metal atom replaces the hydrogen of an acid

SECTION Review

Summary

- Every molecule of a strong acid or base breaks apart to form ions. Few molecules of weak acids and bases break apart to form ions.

- An acid and a base can neutralize one another to make salt and water.

- pH is a measure of hydronium ion concentration in a solution.

- A salt is an ionic compound formed in a neutralization reaction. Salts have many industrial and household uses.

Using Key Terms

1. Use the following terms in the same sentence: *neutralization reaction* and *salt.*

Understanding Key Ideas

2. A neutralization reaction
   a. includes an acid and a base.
   b. produces a salt.
   c. forms water.
   d. All of the above

3. Explain the difference between a strong acid and a weak acid.

Math Skills

4. For each point lower on the pH scale, the hydrogen ions in solution increase tenfold. For example, a solution of pH 3 is not twice as acidic as a solution of pH 6 but is 1,000 times as acidic. How many times more acidic is a solution of pH 2 than a solution of pH 4?

Critical Thinking

5. **Analyzing Processes** Predict what will happen to the hydrogen ion concentration and the pH of water if hydrochloric acid is added to the water.

6. **Analyzing Relationships** Would fish be healthy in a lake that has a low pH? Explain.

7. **Applying Concepts** Soap is made from a strong base and oil. Would you expect the pH of soap to be 4 or 9? Explain.

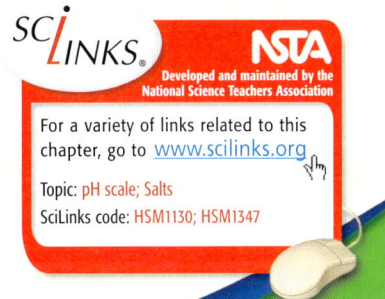

For a variety of links related to this chapter, go to www.scilinks.org

Topic: pH scale; Salts
SciLinks code: HSM1130; HSM1347

Answers to Section Review

1. Sample answer: A salt and water are produced during a neutralization reaction.

2. d

3. A strong acid is one in which all of the molecules of the acid break apart when dissolved in water. A weak acid is one in which only a few molecules break apart when dissolved in water.

4. 100 times more acidic (Students may solve this by noting that there is a two point difference between pH 2 and pH 4 and thinking that each point is considered to be 10. So, 10 × 10 = 100.)

5. The hydrogen ion concentration will increase when HCl is added to water. The pH will decrease below pH 7. due to the increased hydrogen ion concentration.

6. no; Fish would not be healthy in a lake with a low pH because fish need water that is near pH 7. Fish may die in a lake with a low pH.

7. I would expect the pH of soap to be 9 rather than 4 because soap is made from a base and bases have pHs higher than 7.

Is That a Fact!

Explain to students that the indicator litmus is actually a pigment derived from lichens. Lichens are organisms formed from the union of fungal and algal cells that appear as colored scales on trees and rocks. Litmus paper is paper that has been impregnated with litmus.

CHAPTER RESOURCES

Chapter Resource File

- Section Quiz **GENERAL**
- Section Review **GENERAL**
- Vocabulary and Section Summary **GENERAL**
- Datasheet for Quick Lab

Focus

Overview

This section discusses organic compounds—covalent compounds that contain carbon. This section discusses the way carbon forms bonds and introduces hydrocarbons and biochemicals.

 Bellringer

Ask students to list as many carbon-containing items as they can think of. Give them 3 minutes to complete their lists. Once time is up, ask students to share some of their ideas. Point out that researchers have discovered millions of different carbon compounds.

Motivate

Discussion ——— GENERAL

Amino Acids and Vegetarians Tell students that meat contains protein. Ask students how they think people with vegetarian diets can get all the building blocks they need to make proteins. (Sample answer: Vegetarians can get the building blocks of proteins from foods such as grains and legumes.) **LS Verbal/Auditory**

READING WARM-UP

Objectives

- Explain why there are so many organic compounds.
- Identify and describe saturated, unsaturated, and aromatic hydrocarbons.
- Describe the characteristics of carbohydrates, lipids, proteins, and nucleic acids and their functions in the body.

Terms to Learn

organic compound   lipid
hydrocarbon   protein
carbohydrate   nucleic acid

READING STRATEGY

Paired Summarizing Read this section silently. In pairs, take turns summarizing the material. Stop to discuss ideas that seem confusing.

Organic Compounds

Can you believe that more than 90% of all compounds are members of a single group of compounds? It's true!

Most compounds are members of a group called organic compounds. **Organic compounds** are covalent compounds composed of carbon-based molecules. Fuel, rubbing alcohol, and sugar are organic compounds. Even cotton, paper, and plastic belong to this group. Why are there so many kinds of organic compounds? Learning about the carbon atom can help you understand why.

The Four Bonds of a Carbon Atom

All organic compounds contain carbon. Each carbon atom has four valence electrons. So, each carbon atom can make four bonds with four other atoms.

Carbon Backbones

The models in **Figure 1** are called *structural formulas*. They are used to show how atoms in a molecule are connected. Each line represents a pair of electrons that form a covalent bond. Many organic compounds are based on the types of carbon backbones shown in **Figure 1.** Some compounds have hundreds or thousands of carbon atoms as part of their backbone! Organic compounds may also contain hydrogen, oxygen, sulfur, nitrogen, and phosphorus.

✓ Reading Check What is the purpose of structural formulas? (*See the Appendix for answers to Reading Checks.*)

Figure 1   Three Models of Carbon Backbones

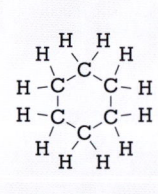

Straight chain	Branched chain	Ring

▲ All carbon atoms are connected in a straight line.

▲ The chain of carbon atoms branches into different directions when a carbon atom is bonded to more than one other carbon atom.

▲ The chain of carbon atoms forms a ring.

CHAPTER RESOURCES

Chapter Resource File

 • Lesson Plan
- Directed Reading A **BASIC**
- Directed Reading B **SPECIAL NEEDS**

Technology

 Transparencies
- Bellringer
- Structural Formulas

Answer to Reading Check

Structural formulas show how atoms in a molecule are connected.

Figure 2 Three Types of Hydrocarbons

Alkane

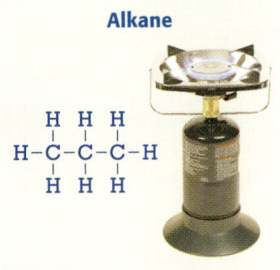

H—C—C—C—H (with H atoms)

The **propane** in this camping stove is a saturated hydrocarbon.

Alkene

H $C=C$ H (ethene structure)

Fruits make **ethene,** which is a compound that helps ripen the fruit.

Alkyne

$H-C\equiv C-H$

Ethyne is better known as acetylene. It is burned in this miner's lamp and in welding torches.

Hydrocarbons and Other Organic Compounds

Although many organic compounds contain several kinds of atoms, some contain only two. Organic compounds that contain only carbon and hydrogen are called **hydrocarbons.**

Saturated Hydrocarbons

The propane shown in **Figure 2** is a saturated hydrocarbon. A *saturated hydrocarbon,* or *alkane,* is a hydrocarbon in which each carbon atom in the molecule shares a single bond with each of four other atoms. A single bond is a covalent bond made up of one pair of shared electrons.

Unsaturated Hydrocarbons

An *unsaturated hydrocarbon,* such as ethene or ethyne shown in **Figure 2,** is a hydrocarbon in which at least one pair of carbon atoms shares a double bond or a triple bond. A double bond is a covalent bond made up of two pairs of shared electrons. A triple bond is a covalent bond made up of three pairs of shared electrons. Hydrocarbons that contain double or triple bonds are unsaturated because these bonds can be broken and more atoms can be added to the molecules.

Compounds that contain two carbon atoms connected by a double bond are called *alkenes.* Hydrocarbons that contain two carbon atoms connected by a triple bond are called *alkynes.*

Aromatic Hydrocarbons

Most aromatic (AR uh MAT ik) compounds are based on benzene. As shown in **Figure 3,** benzene has a ring of six carbons that have alternating double and single bonds. Aromatic hydrocarbons often have strong odors.

organic compound a covalently bonded compound that contains carbon

hydrocarbon an organic compound composed only of carbon and hydrogen

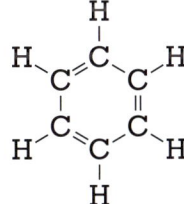

Figure 3 *Benzene is the starting material for manufacturing many products, including medicines.*

Is That a Fact!

Alkanes will burn and can be used as fuels. Natural gas is a mixture of alkanes. About 75% of natural gas is methane; the rest is a mixture of ethane, propane, and butane.

Nutrition Labels Give each student a photocopy of nutrition labels from several different foods, such as milk, cereal, and peanut butter. Ask each student to find and list the carbohydrate, lipid (fat), and protein content of each food. **LS** Logical

BRAIN FOOD

Carbohydrates Tell students that carbohydrate molecules are composed of carbon, hydrogen, and oxygen atoms. Carbohydrates can be classified as simple or complex carbohydrates. Sugars are simple carbohydrates, while starches and cellulose are complex. Most of the fiber we get from plants is in the form of cellulose. The sugar units in cellulose are bonded together to form a structure that gives plants strength and shape. Because of the way cellulose is structured, cellulose is impossible for humans to digest. However, high-fiber foods, such as plums and oatmeal, are healthy because they slow the absorption of calories and help move food through the digestive system.

Table 1 Types and Uses of Organic Compounds		
Type of compound	**Uses**	**Examples**
Alkyl halides	starting material for Teflon™ refrigerant (Freon™)	chloromethane, CH_3Cl bromoethane, C_2H_5Br
Alcohols	rubbing alcohol gasoline additive antifreeze	methanol, CH_3OH ethanol, C_2H_5OH
Organic acids	food preservatives flavorings	ethanoic acid, CH_3COOH propanoic acid, C_2H_5COOH
Esters	flavorings fragrances clothing (polyester)	methyl ethanoate, CH_3COOCH_3 ethyl propanoate, $C_2H_5COOC_2H_5$

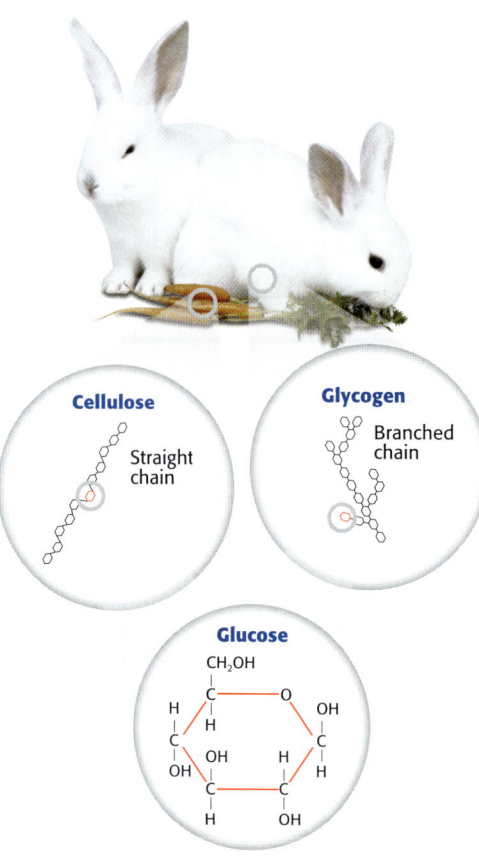

Figure 4 *Glucose molecules, represented by hexagons, can bond to form complex carbohydrates, such as cellulose and glycogen.*

Other Organic Compounds

There are many other kinds of organic compounds. Some have atoms of halogens, oxygen, sulfur, and phosphorus in their molecules. A few of these compounds and their uses are listed in **Table 1.**

Biochemicals: The Compounds of Life

Organic compounds that are made by living things are called *biochemicals*. Biochemicals are divided into four categories: carbohydrates, lipids, proteins, and nucleic acids (noo KLEE ik AS idz).

Carbohydrates

Carbohydrates are biochemicals that are composed of one or more simple sugar molecules bonded together. Carbohydrates are used as a source of energy. There are two kinds of carbohydrates: simple carbohydrates and complex carbohydrates.

Simple carbohydrates include simple sugars, such as glucose. **Figure 4** shows how glucose molecules can bond to form different complex carbohydrates. Complex carbohydrates may be made of hundreds or thousands of sugar molecules bonded together. *Cellulose* gives plant cell walls their rigid structure, and *glycogen* supplies energy to muscle cells.

MISCONCEPTION ALERT

Sugar By Any Other Name When it comes to your health, sugar is sugar. Many packaged foods claim that they are healthful because they contain glucose or fructose (fruit sugar) instead of sucrose (refined table sugar). All of these are concentrated sugars; all they provide is calories. It is much healthier to eat sugars in their truly natural form—as part of whole foods, such as fruits and vegetables.

Is That a Fact!

The main difference between fats and oils is that fats are solid at room temperature, while oils are liquid at room temperature. Most fats, such as chicken fat and lard, come from animal sources. Most oils, such as olive oil and corn oil, come from plant sources.

Lipids

Lipids are biochemicals that do not dissolve in water. Fats, oils, and waxes are kinds of lipids. Lipids have many functions, including storing energy and making up cell membranes. Although too much fat in your diet can be unhealthy, some fat is important to good health. The foods in **Figure 5** are sources of lipids.

Lipids store excess energy in the body. Animals tend to store lipids as fats, while plants store lipids as oils. When an organism has used up most of its carbohydrates, it can obtain energy by breaking down lipids. Lipids are also used to store some vitamins.

Proteins

Most of the biochemicals found in living things are proteins. In fact, after water, proteins are the most common molecules in your cells. **Proteins** are biochemicals that are composed of "building blocks" called *amino acids*.

Amino acids are small molecules made up of carbon, hydrogen, oxygen, and nitrogen atoms. Some amino acids also include sulfur atoms. Amino acids bond to form proteins of many shapes and sizes. The shape of a protein determines the function of the protein. If even a single amino acid is missing or out of place, the protein may not function correctly or at all. Proteins have many functions. They regulate chemical activities, transport and store materials, and provide structural support.

 **Reading Check** What are proteins made of?

Figure 5 *Vegetable oil, meat, cheese, nuts, eggs, and milk are sources of lipids in your diet.*

carbohydrate a class of energy-giving nutrients that includes sugars, starches, and fiber; composed of one or more simple sugars bonded together

lipid a type of biochemical that does not dissolve in water; fats and steroids are lipids

protein a molecule that is made up of amino acids and that is needed to build and repair body structures and to regulate processes in the body

Food Facts

1. Select **four empty food packages.**

2. Without reading the Nutrition Facts labels, rank the items from most carbohydrate content to least carbohydrate content.

3. Rank the items from most fat content to least fat content.

4. Read the Nutrition Facts labels, and compare your rankings with the real rankings.

5. Why do you think your rankings were right, or why were they wrong? Explain your answers.

Figure 6 *Spider webs are made up of proteins that are shaped like long fibers.*

nucleic acid a molecule made up of subunits called *nucleotides*

Examples of Proteins

Proteins have many roles in your body and in living things. Enzymes (EN ZIEMZ) are proteins that are catalysts. *Catalysts* regulate chemical reactions in the body by increasing the rate at which the reactions occur. Some hormones are proteins. For example, insulin is a protein hormone that helps regulate your blood-sugar level. Another kind of protein, called *hemoglobin,* is found in red blood cells and delivers oxygen throughout the body. There are also large proteins that extend through cell membranes. These proteins help control the transport of materials into and out of cells. Some proteins, such as those in your hair, provide structural support. The structural proteins of silk fibers make the spider web shown in **Figure 6** strong and lightweight.

Nucleic Acids

The largest molecules made by living organisms are nucleic acids. **Nucleic acids** are biochemicals made up of *nucleotides* (NOO klee oh TIEDZ). Nucleotides are molecules made of carbon, hydrogen, oxygen, nitrogen, and phosphorus atoms. There are only five kinds of nucleotides. But nucleic acids may have millions of nucleotides bonded together. The only reason living things differ from each other is that each living thing has a different order of nucleotides.

Nucleic acids have several functions. One function of nucleic acids is to store genetic information. They also help build proteins and other nucleic acids. Nucleic acids are sometimes called *the blueprints of life,* because they contain all the information needed for a cell to make all of its proteins.

Reading Check What are two functions of nucleic acids?

CONNECTION TO Social Studies

DNA "Fingerprinting" and Crime-Scene Investigation The chemical structure of all human DNA is the same. The only difference between one person's DNA and another's is the order, or sequence, of the building blocks in the DNA. The number of ways these building blocks can be sequenced are countless.

DNA fingerprinting is new process. However, it has changed the way that criminal investigations are carried out. Research DNA fingerprinting. Find out when DNA fingerprinting was first used, who developed the process, and how DNA fingerprinting is used in crime-scene investigations. Present your findings in an oral presentation to your class. Include a model or a poster to help explain the process to your classmates.

DNA and RNA

There are two kinds of nucleic acids: DNA and RNA. A model of DNA (**d**eoxyribo**n**ucleic **a**cid) is shown in **Figure 7.** DNA is the genetic material of the cell. DNA molecules can store a huge amount of information because of their length. The DNA molecules in a single human cell have a length of about 2 m—which is more than 6 ft long! When a cell needs to make a certain protein, it copies a certain part of the DNA. The information copied from the DNA directs the order in which amino acids are bonded to make that protein. DNA also contains information used to build the second type of nucleic acid, RNA (**r**ibo**n**ucleic **a**cid). RNA is involved in the actual building of proteins.

Figure 7 *Two strands of DNA are twisted in a spiral shape. Four different nucleotides make up the rungs of the DNA ladder.*

SECTION Review

Summary

- Organic compounds contain carbon, which can form four bonds.
- Hydrocarbons are composed of only carbon and hydrogen.
- Hydrocarbons may be saturated, unsaturated, or aromatic hydrocarbons.
- Carbohydrates are made of simple sugars.
- Lipids store energy and make up cell membranes.
- Proteins are composed of amino acids.
- Nucleic acids store genetic information and help cells make proteins.

Using Key Terms

1. Use the following terms in the same sentence: *organic compound, hydrocarbon,* and *biochemical.*

2. In your own words, write a definition for each of the following terms: *carbohydrate, lipid, protein,* and *nucleic acid.*

Understanding Key Ideas

3. A saturated hydrocarbon has
   a. only single bonds.
   b. double bonds.
   c. triple bonds.
   d. double and triple bonds.

4. List two functions of proteins.

5. What is an aromatic hydrocarbon?

Critical Thinking

6. **Identifying Relationships** Hemoglobin is a protein that is in blood and that transports oxygen to the tissues of the body. Information stored in nucleic acids tells a cell how to make proteins. What might happen if there is a mistake in the information needed to make hemoglobin?

7. **Making Comparisons** Compare saturated hydrocarbons with unsaturated hydrocarbons.

Interpreting Graphics

Use the structural formula of this organic compound to answer the questions that follow.

$$\begin{array}{ccc} H & H & H \\ | & | & | \\ H-C-C-C-H \\ | & | & | \\ H & H & H \end{array}$$

8. What type of bonds are present in this molecule?

9. Can you determine the shape of the molecule from this structural formula? Explain your answer.

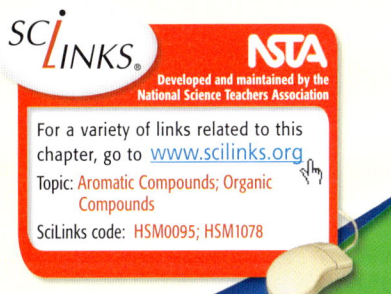

Developed and maintained by the National Science Teachers Association

For a variety of links related to this chapter, go to www.scilinks.org
Topic: Aromatic Compounds; Organic Compounds
SciLinks code: HSM0095; HSM1078

ACTIVITY — ADVANCED

Writing **DNA** Suggest that interested students learn about how DNA stores the information that cells need to make proteins. Tell students to research the topic in the library and on the Internet and to compile their findings into a short report. **LS Verbal**

CHAPTER RESOURCES

Chapter Resource File

- Section Quiz GENERAL
- Section Review GENERAL
- Vocabulary and Section Summary GENERAL

Skills Practice Lab

Cabbage Patch Indicators

Teacher's Notes

Time Required

One 45-minute class period

Lab Ratings

EASY ————————————→ HARD

Teacher Prep 🧪🧪
Student Set-Up 🧪🧪
Concept Level 🧪🧪
Clean Up 🧪🧪

MATERIALS

Materials listed are for groups of 2–3 students. Choose a wide variety of sample liquids, including bleach, ammonia, clear soda pop, lemon juice, milk, and baking soda (dissolved in water). Either red or blue litmus paper (or both) will work.

Safety Caution

Remind students to review all safety cautions and icons before beginning this lab activity. Caution students to use care when using the hot plate. Have students use tongs when handling the beaker with the hot water. Tell students that even diluted acids and bases can irritate the skin. Students should wash the affected area immediately if any sample liquid touches their skin.

OBJECTIVES

Make a natural acid-base indicator solution.

Determine the pH of various common substances.

MATERIALS

- beaker, 250 mL
- beaker tongs
- eyedropper
- hot plate
- litmus paper
- pot holder
- red cabbage leaf
- sample liquids provided by teacher
- tape, masking
- test tubes
- test-tube rack
- water, distilled

SAFETY

Cabbage Patch Indicators

Indicators are weak acids or bases that change color due to the pH of the substance to which they are added. Red cabbage contains a natural indicator. It turns specific colors at specific pHs. In this lab you will extract the indicator from red cabbage. Then, you will use it to determine the pH of several liquids.

Procedure

1. Copy the table below. Be sure to include one line for each sample liquid.

Data Collection Table			
Liquid	Color with indicator	pH	Effect on litmus paper
Control			
	DO NOT WRITE IN BOOK		

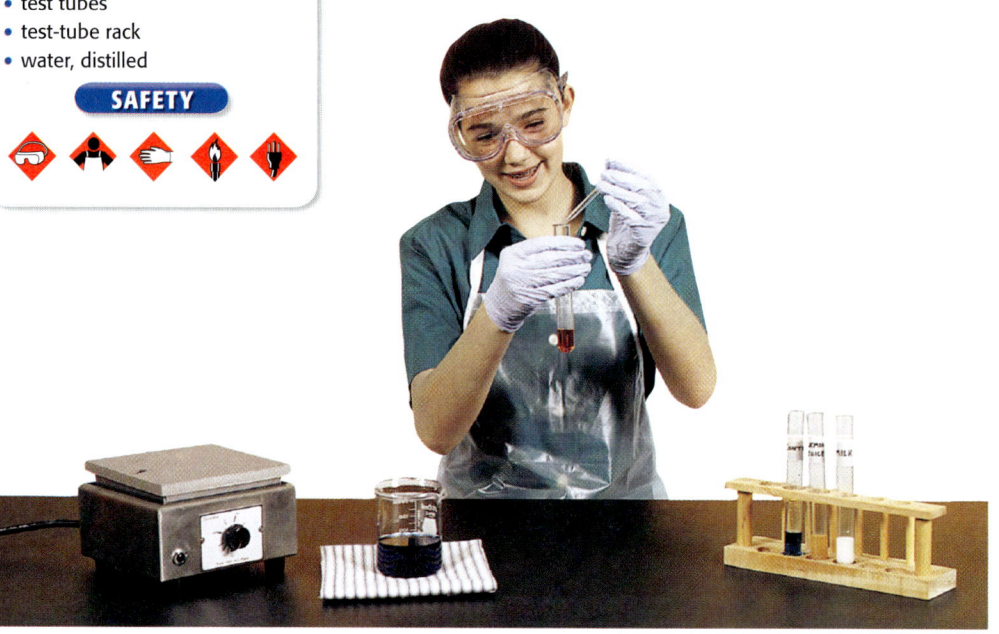

Dennis Hanson
Big Bear Middle School
Big Bear Lake, California

CHAPTER RESOURCES

Chapter Resource File

- Datasheet for Chapter Lab
- Lab Notes and Answers

Technology

💿 **Classroom Videos**
- Lab Video

LabBook

- Making Salt

2 Put on protective gloves. Place 100 mL of distilled water in the beaker. Tear the cabbage leaf into small pieces. Place the pieces in the beaker.

3 Use the hot plate to heat the cabbage and water to boiling. Continue boiling until the water is deep blue. **Caution:** Use extreme care when working near a hot plate.

4 Use tongs to remove the beaker from the hot plate. Turn the hot plate off. Allow the solution to cool on a pot hoder for 5 to 10 minutes.

5 While the solution is cooling, use masking tape and a pen to label the test tubes for each sample liquid. Label one test tube as the control. Place the tubes in the rack.

6 Use the eyedropper to place a small amount (about 5 mL) of the indicator (cabbage juice) in the test tube labeled as the control.

7 Pour a small amount (about 5 mL) of each sample liquid into the appropriate test tube.

8 Using the eyedropper, place several drops of the indicator into each test tube. Swirl gently. Record the color of each liquid in the table.

9 Use the chart below to the find the pH of each sample. Record the pH values in the table.

10 Litmus paper has an indicator that turns red in an acid and blue in a base. Test each liquid with a strip of litmus paper. Record the results.

Analyze the Results

1 **Analyzing Data** What purpose does the control serve? What is the pH of the control?

2 **Examining Data** What colors in your samples indicate the presence of an acid? What colors indicate the presence of a base?

3 **Analyzing Results** Why is red cabbage juice considered a good indicator?

Draw Conclusions

4 **Interpreting Information** Which do you think would be more useful to help identify an unknown liquid—litmus paper or red cabbage juice? Why?

Applying Your Data

Unlike distilled water, rainwater has some carbon dioxide dissolved in it. Is rainwater acidic, basic, or neutral? To find out, place a small amount of the cabbage juice indicator (which is water-based) in a clean test tube. Use a straw to gently blow bubbles in the indicator. Continue blowing bubbles until you see a color change. What can you conclude about the pH of your "rainwater?" What is the purpose of blowing bubbles in the cabbage juice?

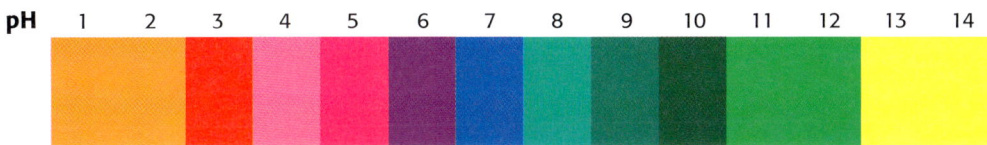

pH 1 2 3 4 5 6 7 8 9 10 11 12 13 14

Analyze the Results

1. The control serves as a color comparison for test tubes containing the sample liquids. The control is reddish blue, which indicates that it is neutral, so the pH is about 7.

2. Reddish colors indicate acids; Bluish colors indicate bases.

3. Red cabbage juice is a good indicator because its color can indicate many pH values. It can be used to identify the relative strengths of acids and bases if their concentrations are the same.

Draw Conclusions

4. red cabbage juice; It would give you an approximate idea of the pH of the unknown liquid. The pH could then be used to help identify the substance. Litmus paper can indicate only whether the unknown liquid is acidic or basic.

Applying Your Data

Students should find that rainwater is slightly acidic. Blowing bubbles dissolves carbon dioxide in the cabbage juice.

CHAPTER RESOURCES

Workbooks

Labs You Can Eat
• Can You Say Seviche? **GENERAL**

EcoLabs & Field Activities
• Greener Cleaners **BASIC**

Long-Term Projects & Research Ideas
• Tiny Plastic Factories **ADVANCED**

Calculator-Based Labs
• Cabbage Patch Indicators **ADVANCED**

Disposal Information

Use the appropriate disposal technique for each sample liquid.

Chapter Review

Assignment Guide

SECTION	QUESTIONS
1	1, 8, 16, 21
2	2, 3, 14–15, 18, 22
3	9, 11–12, 19–20
4	4–7, 10, 13, 17, 21, 23–26

ANSWERS

Using Key Terms

1. An ionic compound contains ionic bonds formed by atoms gaining or losing one or more electrons. A covalent compound contains covalent bonds formed by atoms sharing electrons.

2. An acid increases the number of hydronium ions when dissolved in water, and a base increases the number of hydroxide ions when dissolved in water.

3. pH is the measure of the hydronium ion concentration in a solution, and an indicator is a substance that changes color in the presence of an acid or a base.

4. A hydrocarbon is a compound composed only of hydrogen and carbon, and an organic compound is a carbon-based compound that may contain hydrogen, oxygen, sulfur, nitrogen, or phosphorus.

USING KEY TERMS

For each pair of terms, explain how the meanings of the terms differ.

1. *ionic compound* and *covalent compound*

2. *acid* and *base*

3. *pH* and *indicator*

4. *hydrocarbon* and *organic compound*

5. *carbohydrate* and *lipid*

6. *protein* and *nucleic acid*

UNDERSTANDING KEY IDEAS

Multiple Choice

7. Which of the following statements describes lipids?
   a. Lipids are used to store energy.
   b. Lipids do not dissolve in water.
   c. Lipids make up part of the cell membrane.
   d. All of the above

8. Ionic compounds
   a. have a low melting point.
   b. are often brittle.
   c. do not conduct electric current in water.
   d. do not dissolve easily in water.

9. An increase in the concentration of hydronium ions in solution
   a. raises the pH.
   b. lowers the pH.
   c. does not affect the pH.
   d. doubles the pH.

10. The compounds that store information for building proteins are
    a. lipids.
    b. hydrocarbons.
    c. nucleic acids.
    d. carbohydrates.

Short Answer

11. What type of compound would you use to neutralize a solution of potassium hydroxide?

12. Explain why the reaction of an acid with a base is called *neutralization*.

13. What characteristic of carbon atoms helps to explain the wide variety of organic compounds?

14. What kind of ions are produced when an acid is dissolved in water and when a base is dissolved in water?

Math Skills

15. Most of the vinegar used to make pickles is 5% acetic acid. So, in 100 mL of vinegar, 5 mL is acid diluted with 95 mL of water. If you bought a 473 mL bottle of 5% vinegar, how many milliliters of acetic acid would be in the bottle? How many milliliters of water were used to dilute the acetic acid?

16. If you dilute a 75 mL can of orange juice with enough water to make a total volume of 300 mL, what is the percentage of juice in the mixture?

5. A carbohydrate is composed of sugar molecules, and a lipid is a biochemical that does not dissolve in water.

6. A protein is composed of amino acids, and a nucleic acid is made of nucleotides.

Understanding Key Ideas

7. d
8. b
9. b
10. c
11. Potassium hydroxide is a base, so I would use an acid to neutralize it.

12. When an acid reacts with a base, the pH of the solution gets closer to pH 7, which indicates a neutral solution.

13. Each carbon atom has four valence electrons and can form four bonds. These bonds can be made to atoms of carbon or to atoms of other elements.

14. Acids produce hydronium ions in water, and bases produce hydroxide ions in water.

15. 473 mL × 5% = 473 mL × 0.05 = 23.65 mL acetic acid
    473 mL − 23.65 mL = 449.35 mL water

16. 75 mL ÷ 300 mL × 100% = 25% juice

CRITICAL THINKING

17. Concept Mapping Use the following terms to create a concept map: *acid, base, salt, neutral,* and *pH.*

18. Applying Concepts Fish give off the base, ammonia, NH_3, as waste. How does the release of ammonia affect the pH of the water in the aquarium? What can be done to correct the pH of the water?

19. Analyzing Methods Many insects, such as fire ants, inject formic acid, a weak acid, when they bite or sting. Describe the type of compound that should be used to treat the bite.

20. Making Comparisons Organic compounds are also covalent compounds. What properties would you expect organic compounds to have as a result?

21. Applying Concepts Farmers have been known to taste their soil to determine whether the soil has the correct acidity for their plants. How would taste help the farmer determine the acidity of the soil?

22. Analyzing Ideas A diet that includes a high level of lipids is unhealthy. Why is a diet containing no lipids also unhealthy?

INTERPRETING GRAPHICS

Use the structural formulas below to answer the questions that follow.

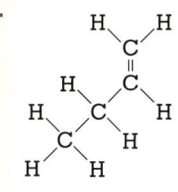

23 A saturated hydrocarbon is represented by which structural formula(s)?

24 An unsaturated hydrocarbon is represented by which structural formula(s)?

25 An aromatic hydrocarbon is represented by which structural formula(s)?

 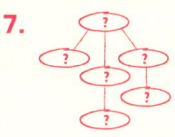
CHAPTER RESOURCES

Chapter Resource File

- Chapter Review **GENERAL**
- Chapter Test A **GENERAL**
- Chapter Test B **SPECIAL NEEDS**
- Chapter Test C **ADVANCED**
- Vocabulary Activity **GENERAL**

Workbooks

Study Guide
- Assessment resources are also available in Spanish.

Standardized Test Preparation

Teacher's Note

To provide practice under more realistic testing conditions, give students 20 minutes to answer all of the questions in this Standardized Test Preparation.

MISCONCEPTION ALERT

Answers to the standardized test preparation can help you identify student misconceptions and misunderstandings.

READING

Passage 1

1. C
2. H
3. D

✚ TEST DOCTOR

Question 2: Students may select choice G because the passage states that the silk fiber is embedded in a glycine-rich substance. However, the passage also states that the silk fiber is made of two strands of alanine-rich protein. So, choice G is incorrect. The correct answer is choice H.

READING

Read each of the passages below. Then, answer the questions that follow each passage.

Passage 1 Spider webs often resemble a bicycle wheel. The "spokes" of the web are made of a silk thread called *dragline silk*. The sticky, stretchy part of the web is called *capture silk* because the spiders use this silk to <u>capture</u> their prey. Spider silk is made of proteins, and proteins are made of amino acids. There are 20 naturally occurring amino acids, but spider silk has only 7 of them. Scientists used a technique called *nuclear magnetic resonance* (NMR) to see the structure of dragline silk. The silk fiber is made of two tough strands of alanine-rich protein embedded in a glycine-rich substance. This protein resembles tangled spaghetti. Scientists believe that this tangled part makes the silk springy and that a repeating sequence of 5 amino acids makes the protein stretchy.

1. According to the passage, how many types of amino acids does spider silk contain?
   - **A** 20 amino acids
   - **B** 5 amino acids
   - **C** 7 amino acids
   - **D** all naturally occurring amino acids

2. Based on the passage, which of the following statements is a fact?
   - **F** Capture silk makes up the "spokes" of the web.
   - **G** The silk fiber is made of two strands of glycine-rich protein.
   - **H** Proteins are made of amino acids.
   - **I** Spider webs are strong because of a repeating sequence of 5 amino acids.

3. In this passage, what does *capture* mean?
   - **A** to kill
   - **B** to eat
   - **C** to free
   - **D** to trap

Passage 2 The earliest evidence of soapmaking dates back to 2,800 BCE. A soaplike material was found in clay cylinders in ancient Babylon. According to Roman legend, soap was named after Mount Sapo, where animals were sacrificed. A soaplike substance was made when rain washed the melted animal fat and wood ashes into the clay soil along the Tiber River. In 1791, a major step toward large scale <u>commercial</u> soapmaking began when Nicholas Leblanc, a French chemist, patented a process for making soda ash from salt. About 20 years later, the science of modern soapmaking was born. At that time, Michel Chevreul, another French chemist, discovered how fats, glycerin, and fatty acids interact. This interaction is the basis of saponification, or soap chemistry, today.

1. In this passage, what does *commercial* mean?
   - **A** for advertising purposes
   - **B** for public sale
   - **C** from French manufacturers
   - **D** for a limited time period

2. Based on the passage, which of the following statements is a fact?
   - **F** Saponification is a process used to make soap.
   - **G** The word *soap* probably originated from the French.
   - **H** Modern soapmaking began around 1791.
   - **I** Soapmaking began 2,000 years ago.

3. Which of the following statements is the best summary for the passage?
   - **A** The process of soapmaking has a history of at least 4,000 years.
   - **B** Most of the scientists responsible for soapmaking were from France.
   - **C** Commercial soapmaking began in 1791.
   - **D** Soap chemistry is called *saponification*.

Passage 2

1. B
2. H
3. A

✚ TEST DOCTOR

Question 3: Students may be confused by this question because all the choices appear to summarize the passage. However, students should note that the passage is about the entire history of soapmaking. Therefore, the best summary is choice A.

The diagram below shows a model of a water molecule (H_2O). Use the diagram to answer the questions that follow.

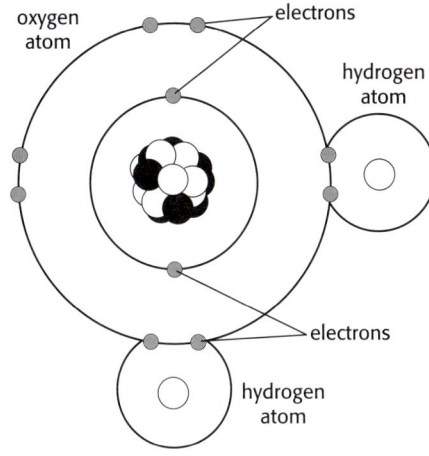

1. Which statement best describes the oxygen atom?

   A The oxygen atom has four valence electrons.

   B The oxygen atom is sharing two electrons with two other atoms.

   C The oxygen atom lost two electrons.

   D The oxygen atom has two valence electrons.

2. Which of the following features cannot be determined by looking at the model?

   F the number of atoms in the molecule

   G the number of electrons in each atom

   H the type of bonds joining the atoms

   I the physical state of the substance

3. Which statement best describes each of the smaller atoms in the molecule?

   A Each atom has eight total electrons.

   B Each atom has two protons.

   C Each atom has two valence electrons.

   D Each atom has lost two electrons.

Read each question below, and choose the best answer.

1. Marty is making 8 gal of lemonade. How much sugar does he need if 1 1/2 cups of sugar are needed for every 2 gal of lemonade?

   A 4 cups

   B 6 cups

   C 3 cups

   D 8 cups

2. The Jimenez family went to the science museum. They bought four tickets at $12.95 each, four snacks at $3 each, and two souvenirs at $7.95 each. What is the best estimate of the total cost of tickets, snacks, and souvenirs?

   F $24

   G $50

   H $63

   I $80

3. Each whole number on the pH scale represents a tenfold change in the concentration of hydronium ions. An acid that has a pH of 2 has how many times more hydronium ions than an acid that has a pH of 5?

   A 30 times more hydronium ions

   B 100 times more hydronium ions

   C 1,000 times more hydronium ions

   D 10,000 times more hydronium ions

Standardized Test Preparation

1. B

2. I

3. C

TEST DOCTOR

Question 1: To answer this question, students must remember that pairs of electrons drawn between two atoms in this kind of model represent a covalent bond. A covalent bond is a bond that forms when atoms share electrons. So, the oxygen atom is sharing one electron with each of the two hydrogen atoms, for a total of two electrons shared. The correct choice is B.

1. B

2. I

3. C

TEST DOCTOR

Question 2: To answer this question, students can estimate the cost of the tickets as $13 and the cost of the souvenirs as $8. Therefore, ($13 × 4) + ($3 × 4) + ($8 × 2) = $80.

CHAPTER RESOURCES

Chapter Resource File

 • Standardized Test Preparation GENERAL

State Resources

 For specific resources for your state, visit **go.hrw.com** and type in the keyword **HSMSTR**.

Science in Action

Science, Technology, and Society

Background

Getting tissue samples from archeological specimens isn't easy. Museum curators often won't give permission to study the DNA of a museum specimen because part of the specimen must be destroyed to study its genetic material. Even if DNA can be obtained, experiments can be difficult. Because there is so little DNA to work with, repeating experiments in order to verify results is not always possible.

Weird Science

Background

Silly Putty is a polymeric material with special properties that make it different from other polymers. The polymers in Silly Putty have covalent bonds within the molecules but hydrogen bonds between the molecules. The hydrogen bonds can be easily broken. When small amounts of stress are slowly applied to the putty, some of the hydrogen bonds break and the putty starts to "flow." When more stress is applied quickly, many hydrogen bonds break, causing the putty to tear or break off.

Science, Technology, and Society

Molecular Photocopying

To learn about our human ancestors, scientists can use DNA from mummies. Well-preserved DNA can be copied using a technique called polymerase chain reaction (PCR). PCR uses enzymes called *polymerases,* which make new strands of DNA using old strands as templates. Thus, PCR is called molecular photocopying. However, scientists have to be very careful when using this process. If just one of their own skin cells falls into the PCR mixture, it will contaminate the ancient DNA with their own DNA.

Social Studies ACTIVITY

WRITING SKILL DNA analysis of mummies is helping archeologists study human history. Write a research paper about what scientists have learned about human history through DNA analysis.

Weird Science

Silly Putty™

During World War II, the supply of natural rubber was very low. So, James Wright, at General Electric, tried to make a synthetic rubber. The putty he made could be molded, stretched, and bounced. But it did not work as a rubber substitute and was ignored. Then, Peter Hodgson, a consultant for a toy company, had a brilliant idea. He marketed the putty as a toy in 1949. It was an immediate success. Hodgson created the name Silly Putty™. Although Silly Putty™ was invented more than 50 years ago, it has not changed much. More than 300 million eggs of Silly Putty have been sold since 1950.

Math ACTIVITY

In 1949, Mr. Hodgson bought 9.5 kg of putty for $147. The putty was divided into balls, each having a mass of 14 g. What was his cost for one 14 g ball of putty?

Answer to Social Studies Activity

Answers may vary. Students' papers should outline what has been learned about human history through DNA analysis.

Answer to Math Activity

$147 ÷ 9,500 g = $0.01547/g

$0.01547/g × 14 g = $.022 /14 g ball

Jeannie Eberhardt

Forensic Scientist Jeannie Eberhardt says that her job as a forensic scientist is not really as glamorous as it may seem on popular TV shows. "If they bring me a garbage bag from the crime scene, then my job is to dig through the trash and look for evidence," she laughs. Jeannie Eberhardt explains that her job is to "search for, collect, and analyze evidence from crime scenes." Eberhardt says that one of the most important qualities a forensic scientist can have is the ability to be unbiased. She says that she focuses on the evidence and not on any information she may have about the alleged crime or the suspect. Eberhardt advises students who think they might be interested in a career as a forensic scientist to talk to someone who works in the field. She also recommends that students develop a broad science background. And she advises students that most of these jobs require extensive background checks. "Your actions now could affect your ability to get a job later on," she points out.

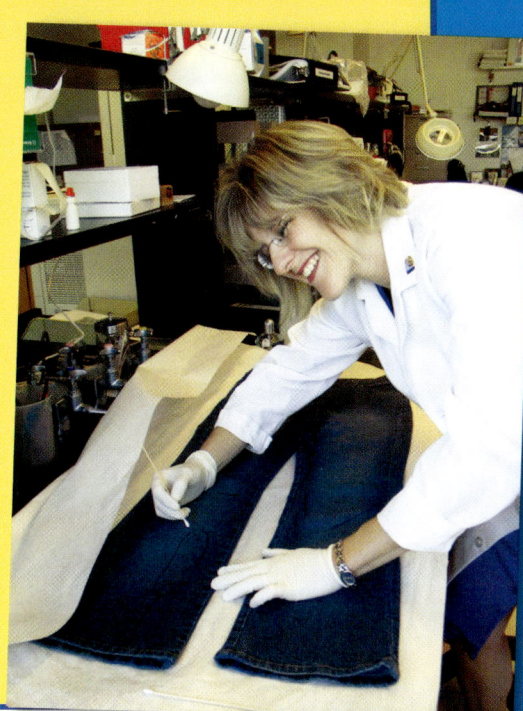

Language Arts ACTIVITY

WRITING SKILL Jeannie Eberhardt says that it is very important to be unbiased when analyzing a crime scene. Write a one-page essay explaining why it is necessary to focus on the evidence in a crime and not on personal feelings or news reports.

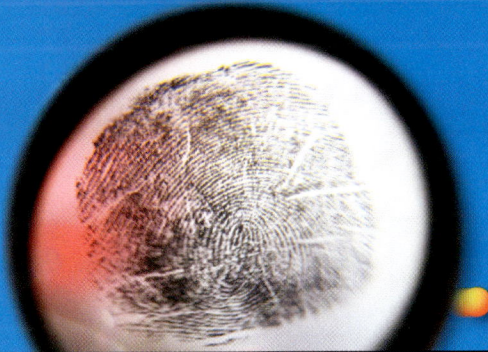

go.hrw.com

To learn more about these Science in Action topics, visit **go.hrw.com** and type in the keyword **HP5CMPF**

Current Science

Check out Current Science® articles related to this chapter by visiting **go.hrw.com**. Just type in the keyword **HP5CS15**

Careers

Background

After Jeannie Eberhardt earned her masters degree in forensic science, she began working in a crime lab. "You have to be working and training for at least six months in a crime lab before you can do any work on real cases," she explained. For the first six months of her first job, Eberhardt went through an intensive on-the-job training program. She tested different materials and analyzed mock samples from an invented crime scene. Then she had to testify at a mock trial to defend her findings. In the trial, her co-workers and supervisors from the crime lab played the parts of the judge and the attorneys. They invited everyone who worked in the building to watch the trial and videotaped Eberhardt's testimony. After she successfully analyzed the samples she was given and showed that she could defend her work in court, she was finally allowed to work on real cases.

Answer to Language Arts Activity

Answers may vary. Students' essays should explain that focusing on evidence is important to determine who committed a crime. The essays may explain that personal feelings and news reports might interfere with a criminal investigation.

Compression guide:
To shorten instruction because of time limitations, omit the Chapter Lab.

OBJECTIVES	LABS, DEMONSTRATIONS, AND ACTIVITIES	TECHNOLOGY RESOURCES
PACING • 90 min pp. 86–95 **Chapter Opener**	SE **Start-up Activity**, p. 87 **GENERAL**	OSP **Parent Letter** ■ **GENERAL** CD **Student Edition CD-ROM** CD **Guided Reading Audio CD** ■ TR **Chapter Starter Transparency*** VID **Brain Food Video Quiz**
Section 1 Radioactivity • Describe how radioactivity was discovered. • Compare alpha, beta, and gamma decay. • Describe the penetrating power of the three kinds of nuclear radiation. • Calculate ages of objects using half-life. • Identify uses of radioactive materials.	TE **Demonstration** Blocking Radiation, p. 88 **GENERAL** TE **Activity** X rays and Gamma Rays, p. 90 **BASIC** TE **Connection Activity** Language Arts, p. 91 **ADVANCED** TE **Connection Activity** Math, p. 93 **GENERAL** SE **Science in Action** Math, Social Studies, and Language Arts Activities, pp. 108–109 **GENERAL**	CRF **Lesson Plans*** TR **Bellringer Transparency*** TR Alpha Decay of Radium-226* TR Beta Decay of Carbon-14* TR The Penetrating Abilities of Nuclear Radiation* TR Radioactive Decay and Half-Life* SE **Internet Activity**, p. 94 **GENERAL** CRF **SciLinks Activity*** **GENERAL**
PACING • 90 min pp. 96–101 **Section 2 Energy from the Nucleus** • Describe nuclear fission. • Identify advantages and disadvantages of nuclear fission. • Describe nuclear fusion. • Identify advantages and disadvantages of nuclear fusion.	TE **Demonstration** Mousetrap Fission, p. 96 **GENERAL** TE **Connection Activity** History, p. 97 **GENERAL** TE **Connection Activity** Environmental Science, p. 98 **ADVANCED** SE **Quick Lab** Gone Fission, p. 99 **GENERAL** CRF **Datasheet for Quick Lab*** SE **Connection to Astronomy** Elements of the Stars, p. 100 **GENERAL** SE **Model-Making Lab** Domino Chain Reactions, p. 102 **GENERAL** CRF **Datasheet for Chapter Lab*** LB **Long-Term Projects & Research Ideas** Meltdown!* **ADVANCED**	CRF **Lesson Plans*** TR **Bellringer Transparency*** TR Fission of a Uranium-235 Nucleus* TR How a Nuclear Power Plant Works* TR Fusion of Hydrogen-1 Nuclei* TR **LINK TO EARTH SCIENCE** Fusion of Hydrogen in the Sun* VID **Lab Videos for Physical Science**

PACING • 90 min

CHAPTER REVIEW, ASSESSMENT, AND STANDARDIZED TEST PREPARATION

CRF **Vocabulary Activity*** **GENERAL**
SE **Chapter Review**, pp. 104–105 **GENERAL**
CRF **Chapter Review*** ■ **GENERAL**
CRF **Chapter Tests A*** **GENERAL**, B* **ADVANCED**, C* **SPECIAL NEEDS**
SE **Standardized Test Preparation**, pp. 106–107 **GENERAL**
CRF **Standardized Test Preparation*** **GENERAL**
CRF **Performance-Based Assessment*** **GENERAL**
OSP **Test Generator** **GENERAL**
CRF **Test Item Listing*** **GENERAL**

Online and Technology Resources

Visit **go.hrw.com** for a variety of free resources related to this textbook. Enter the keyword **HP5RAD**.

Students can access interactive problem-solving help and active visual concept development with the *Holt Science and Technology* Online Edition available at **www.hrw.com**.

Guided Reading Audio CD

A direct reading of each chapter using instructional visuals as guideposts. For auditory learners, reluctant readers, and Spanish-speaking students. Available in English and Spanish.

SKILLS DEVELOPMENT RESOURCES	SECTION REVIEW AND ASSESSMENT	STANDARDS CORRELATIONS
SE Pre-Reading Activity, p. 86 `GENERAL` **OSP** Science Puzzlers, Twisters & Teasers `GENERAL`		National Science Education Standards UCP 1, 2, 3; SAI 1
CRF Directed Reading A* ■ `BASIC`, B* `SPECIAL NEEDS` **CRF** Vocabulary and Section Summary* ■ `GENERAL` **SE** Reading Strategy Reading Organizer, p. 88 `GENERAL` **TE** Reading Strategy Prediction Guide, p. 89 `GENERAL` **SE** Connection to Environmental Science Radon in the Home, p. 92 `GENERAL` **TE** Reading Strategy Prediction Guide, p. 92 `GENERAL` **SE** Math Practice How Old Is It?, p. 93 `GENERAL` **TE** Inclusion Strategies, p. 93 **MS** Math Skills for Science Radioactive Decay and the Half-Life* `GENERAL` **CRF** Reinforcement Worksheet The Decay of a Nucleus* `BASIC`	**SE** Reading Checks, pp. 89, 91, 93, 94 `GENERAL` **TE** Homework, p. 89 `GENERAL` **TE** Reteaching, p. 94 `BASIC` **TE** Quiz, p. 94 `GENERAL` **TE** Alternative Assessment, p. 94 `GENERAL` **SE** Section Review,* p. 95 ■ `GENERAL` **CRF** Section Quiz* ■ `GENERAL`	UCP 1, 2, 3; SAI 2; ST 2; SPSP 1, 4, 5; PS 3a, 3e
CRF Directed Reading A* ■ `BASIC`, B* `SPECIAL NEEDS` **CRF** Vocabulary and Section Summary* ■ `GENERAL` **SE** Reading Strategy Reading Organizer, p. 96 `GENERAL` **SE** Connection to Language Arts Storage Site, p. 99 `GENERAL` **TE** Inclusion Strategies, p. 99 **CRF** Critical Thinking The Blue Flame* `ADVANCED` **CRF** Reinforcement Worksheet Fission or Fusion?* `BASIC`	**SE** Reading Checks, pp. 96, 99, 100 `GENERAL` **TE** Reteaching, p. 100 `BASIC` **TE** Quiz, p. 100 `GENERAL` **TE** Alternative Assessment, p. 100 `GENERAL` **SE** Section Review,* p. 101 ■ `GENERAL` **CRF** Section Quiz* ■ `GENERAL`	UCP 2, 3; ST 1, 2; SPSP 3, 4, 5; PS 3a, 3e; *Chapter Lab:* UCP 2, 3; SAI 1; PS 3a, 3e

One-Stop
Planner® CD-ROM

This convenient CD-ROM includes:
- **Lab Materials QuickList Software**
- **Holt Calendar Planner**
- **Customizable Lesson Plans**
- **Printable Worksheets**
- **ExamView® Test Generator**

cnnstudentnews.com

Find the latest news, lesson plans, and activities related to important scientific events.

www.scilinks.org

Maintained by the **National Science Teachers Association.** See Chapter Enrichment pages for a complete list of topics.

Check out *Current Science* articles and activities by visiting the HRW Web site at **go.hrw.com.** Just type in the keyword **HP5CS16T.**

Classroom Videos
- **Lab Videos** demonstrate the chapter lab.
- **Brain Food Video Quizzes** help students review the chapter material.

Visual Resources

CHAPTER STARTER TRANSPARENCY

Atomic Energy — CHAPTER STARTER

Would You Believe . . . ?

Replacing the battery in your camera can be a problem when you're traveling—especially 1.5 billion kilometers (932 million miles) from home! That's how far from Earth the Cassini spacecraft will be when it reaches Saturn in 2004. Cassini's camera and other equipment need an energy source that will work after a trip of nearly 7 years. Otherwise, the trip to study the atmosphere and moons of Saturn would be wasted. The answer? The radioactive element plutonium!

The nuclei (plural of nucleus) of plutonium atoms are radioactive, meaning they are unstable. They become stable by giving off radiation in the form of particles and energy. This process heats the materials surrounding the plutonium, and the thermal energy of the materials is converted into electrical energy by a Radioisotope Thermoelectric Generator (RTG). Earlier spacecraft, including Voyager, Galileo, and Ulysses, have also depended on RTGs for electrical energy. RTGs are also used on Earth. The deep ocean and the Arctic icecap are two places where RTGs are used in scientific

equipment. Because electrical energy can be produced using a sample of plutonium for 10 or more years, RTGs provide energy longer than any battery known, as described below.

In this chapter you will learn more about the energy given off by radioactive materials. In addition, you will learn about the energy associated with the splitting and joining of atomic nuclei. Read on to continue your own journey!

These RTGs on Cassini might provide energy even longer than RTGs on Voyager, which were still providing energy after 20 years!

BELLRINGER TRANSPARENCIES

Atomic Energy — BELLRINGER TRANSPARENCY

Section: Radioactivity
In your **science journal** write a few sentences about the term *nuclear radiation*. Include what you know about nuclear radiation, any benefits, and any dangers you can think of. For example, when is radiation used to help people? When is radiation harmful?

Section: Energy from the Nucleus
Define each of the following terms in your own words in your **science journal**:
fission
fusion
Are the terms opposites, or are they similar? How is energy involved in each? Discuss your ideas with the group.

TEACHING TRANSPARENCIES

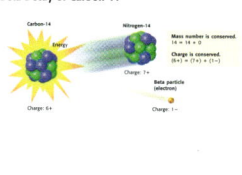

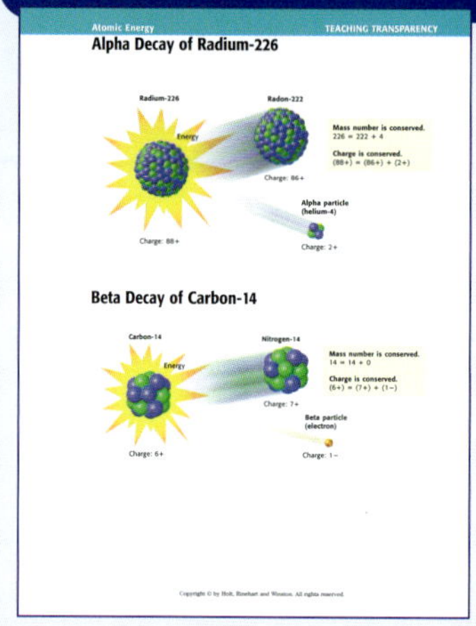

Atomic Energy — TEACHING TRANSPARENCY

Alpha Decay of Radium-226
Radium-226 → Radon-222
Mass number is conserved. 226 = 222 + 4
Charge is conserved. (88+) = (86+) + (2+)
Alpha particle (helium-4)

Beta Decay of Carbon-14
Carbon-14 → Nitrogen-14
Mass number is conserved. 14 = 14 + 0
Charge is conserved. (6+) = (7+) + (1−)
Beta particle (electron)

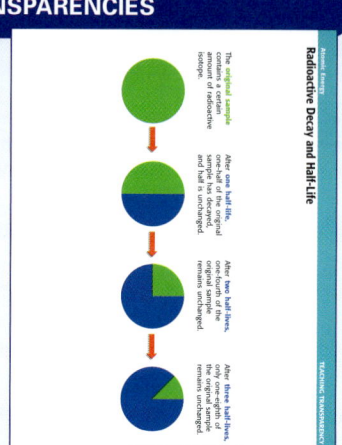

Radioactive Decay and Half-Life

The original sample contains a certain amount of radioactive isotope.

After one half-life, one-half of the original sample has decayed, and half is unchanged.

After two half-lives, one-fourth of the original sample remains unchanged.

After three half-lives, only one-eighth of the original sample remains unchanged.

TEACHING TRANSPARENCIES

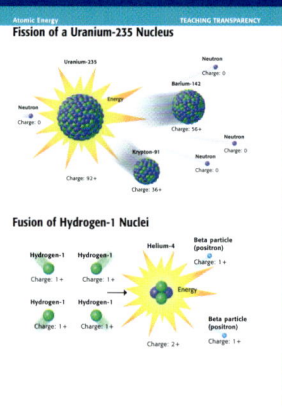

Atomic Energy — TEACHING TRANSPARENCY

Fission of a Uranium-235 Nucleus
Fusion of Hydrogen-1 Nuclei

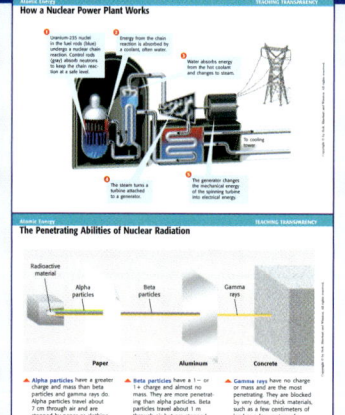

How a Nuclear Power Plant Works

The Penetrating Abilities of Nuclear Radiation

Alpha particles have a greater charge and mass than beta particles and gamma rays do. Alpha particles travel about 7 cm through air and are stopped by paper or clothing.

Beta particles have a 1− or 1+ charge and almost no mass. They are more penetrating than alpha particles. Beta particles travel about 1 m through air but are stopped by 3 mm of aluminum.

Gamma rays have no charge or mass and are the most penetrating. They are blocked by very dense, thick materials, such as a few centimeters of lead or a few meters of concrete.

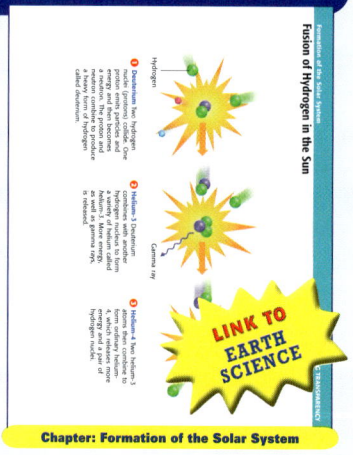

Fusion of Hydrogen in the Sun

Formation of the Solar System

LINK TO EARTH SCIENCE

Chapter: Formation of the Solar System

CONCEPT MAPPING TRANSPARENCY

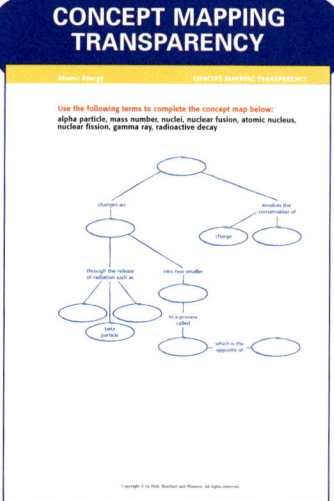

Atomic Energy — CONCEPT MAPPING TRANSPARENCY

Use the following terms to complete the concept map below:
alpha particle, mass number, nuclei, nuclear fusion, nuclear fission, gamma ray, radioactive decay

Planning Resources

LESSON PLANS

Lesson Plan SAMPLE

Section: Waves

Pacing
Regular Schedule: with lab(s):2 days without lab(s):2 days
Block Schedule: with lab(s): 1 1/2 days without lab(s):1 day

Objectives
1. Relate the seven properties of life to a living organism.
2. Describe seven themes that can help you to organize what you learn about biology.
3. Identify the tiny structures that make up all living organisms.
4. Differentiate between reproduction and heredity and between metabolism and homeostasis.

National Science Education Standards Covered
LSInter6:Cells have particular structures that underlie their functions.
LSMat1:Most cell functions involve chemical reactions.
LSBeh1:Cells store and use information to guide their functions.
UCP1:Cell functions are regulated.
SI1: Cells can differentiate and form complete multicellular organisms.
PS1: Species evolve over time.
ESS1: The great diversity of organisms is the result of more than 3.5 billion years of evolution.
ESS2: Natural selection and its evolutionary consequences provide a scientific explanation for the fossil record of ancient life forms as well as for the striking molecular similarities observed among the diverse species of living organisms.
ST1: The millions of different species of plants, animals, and microorganisms that live on Earth today are related by descent from common ancestors.
ST2: The energy for life primarily comes from the sun.
SPSP1: The complexity and organization of organisms accommodates the need for obtaining, transforming, transporting, releasing, and eliminating the matter and energy used to sustain the organism.
SPSP6: As matter and energy flows through different levels of organization of living systems—cells, organs, communities—and between living and the physical environment, chemical elements are recombined in different ways.
HNS1: Organisms have behavioral responses to internal changes and to external stimuli.

PARENT LETTER

SAMPLE

Dear Parent,

Your son's or daughter's science class will soon begin exploring the chapter entitled "The World of Physical Science." In this chapter, students will learn about how the scientific method applies to the world of physical science and the role of physical science in the world. By the end of the chapter, students should demonstrate a clear understanding of the chapter's main ideas and be able to discuss the following topics:

1. physical science is the study of energy and matter (Section 1)
2. the role of physical science in the world around them (Section 1)
3. careers that rely on physical science (Section 1)
4. the steps used in the scientific method (Section 2)
5. examples of technology (Section 2)
6. how the scientific method is used to answer questions and solve problems (Section 2)
7. how our knowledge of science changes over time (Section 2)
8. how models represent real objects or systems (Section 3)
9. examples of different ways models are used in science (Section 3)
10. the importance of the International System of Units (Section 4)
11. the appropriate units to use for particular measurements (Section 4)
12. how area and density are derived quantities (Section 4)

Questions to Ask Along the Way

You can help your son or daughter learn about these topics by asking interesting questions such as the following:

• What are some surprising careers that use physical science?
• What is a characteristic of a good hypothesis?
• When is it a good idea to use a model?
• Why do Americans measure things in terms of inches and yards and meters?

ALSO IN SPANISH

TEST ITEM LISTING

TEST ITEM LISTING
The World of Science SAMPLE

MULTIPLE CHOICE

1. A limitation of models is that
   a. they are large enough to see
   b. they do not act exactly like the things that they model.
   c. they are smaller than the things that they model.
   d. they model unfamiliar things.
   Answer: B   Difficulty: 1   Section: 1   Objective: 2
2. The length 10 m is equal to
   a. 100 cm.   c. 10,000 mm.
   b. 1,000 cm.   d. Both (b) and (c)
   Answer: B   Difficulty: 1   Section: 1   Objective: 2
3. To be valid, a hypothesis must be
   a. testable   c. made into a law.
   b. supported by evidence. d. Both (a) and (b)
   Answer: B   Difficulty: 1   Section: 2   Objective: 2 1
4. The statement "Sheila has a stain on her shirt" is an example of a(n)
   a. law.   c. observation.
   b. hypothesis.   d. prediction.
   Answer: B   Difficulty: 1   Section: 3   Objective: 2
5. A hypothesis is often developed out of
   a. observations.   c. laws.
   b. experiments.   d. Both (a) and (b)
   Answer: B   Difficulty: 1   Section: 2   Objective: 2
6. How many milliliters are in 3.5 kL?
   a. 3,500 mL   c. 3,500, 000 mL
   b. 0.0035 mL.   d. 35,000 mL
   Answer: B   Difficulty: 1   Section: 2   Objective: 2
7. A map of Seattle is an example of a
   a. law.   c. model.
   b. theory.   d. unit.
   Answer: B   Difficulty: 1   Section: 3   Objective: 2
8. A lab has the safety icons shown below. These icons mean that you should wear
   a. only safety goggles.   c. safety goggles and a lab apron.
   b. only a lab apron.   d. safety goggles, a lab apron, and gloves.
   Answer: B   Difficulty: 1   Section: 3   Objective: 3
9. The law of conservation of mass says the tot al mass before a chemical change is
   a. more than the total mass after the change
   b. less than the total mass after the change
   c. the same as the total mass after the change
   d. not the same as the total mass after the change
   Answer: B   Difficulty: 1   Section: 3   Objective: 2
10. In which of the following areas might you find a geochemist at work?
   a. studying the chemistry of rocks   c. studying fishes
   b. studying forestry   d. studying the atmosphere
   Answer: B   Difficulty: 1   Section: 1   Objective: 2

One-Stop Planner® CD-ROM

This CD-ROM includes all of the resources shown here and the following time-saving tools:

• **Lab Materials QuickList Software**
• **Customizable lesson plans**
• **Holt Calendar Planner**
• **The powerful ExamView® Test Generator**

Meeting Individual Needs

DIRECTED READING A
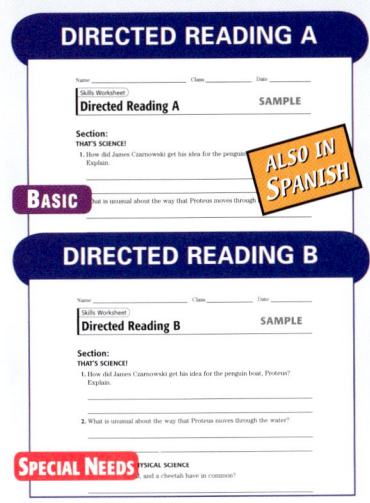
BASIC · ALSO IN SPANISH

DIRECTED READING B
SPECIAL NEEDS

VOCABULARY ACTIVITY
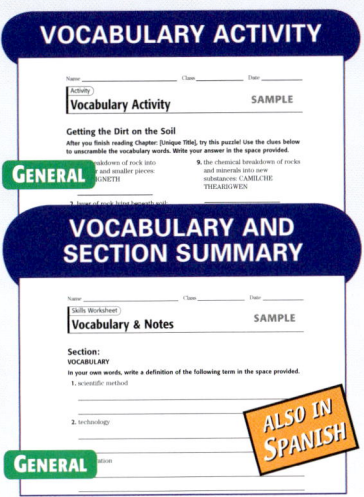
GENERAL

VOCABULARY AND SECTION SUMMARY
GENERAL · ALSO IN SPANISH

REINFORCEMENT
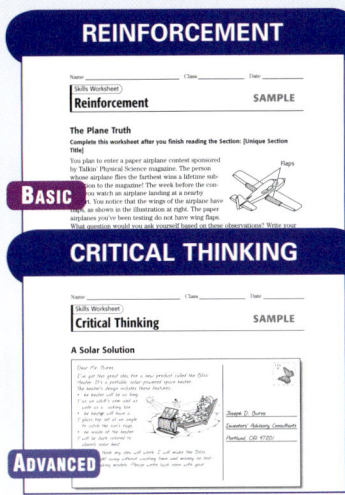
BASIC

CRITICAL THINKING
ADVANCED

SCILINKS ACTIVITY
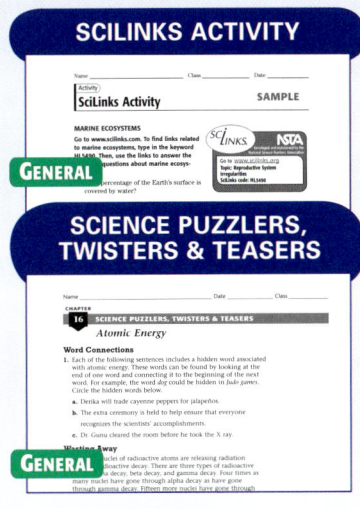
GENERAL

SCIENCE PUZZLERS, TWISTERS & TEASERS
GENERAL

Labs and Activities

LONG-TERM PROJECTS & RESEARCH IDEAS
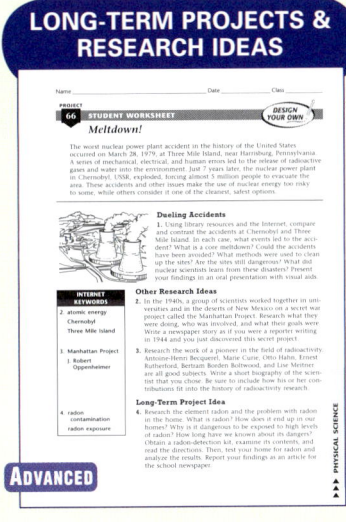
ADVANCED

DATASHEETS FOR QUICK LABS
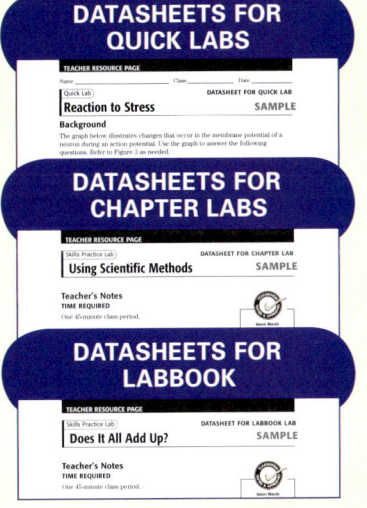

DATASHEETS FOR CHAPTER LABS

DATASHEETS FOR LABBOOK

Review and Assessments

SECTION QUIZ
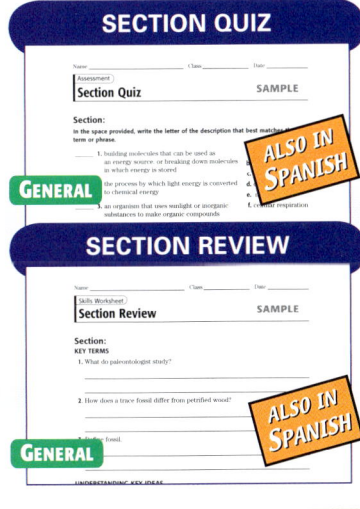
GENERAL · ALSO IN SPANISH

SECTION REVIEW
GENERAL · ALSO IN SPANISH

CHAPTER REVIEW
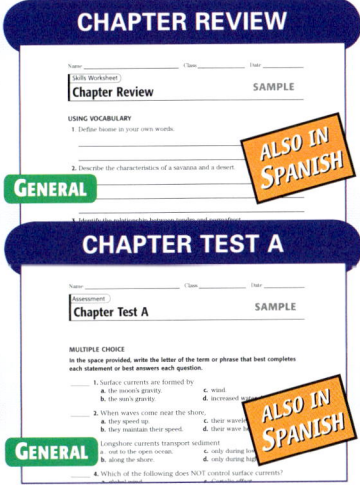
GENERAL · ALSO IN SPANISH

CHAPTER TEST A
GENERAL · ALSO IN SPANISH

CHAPTER TEST B
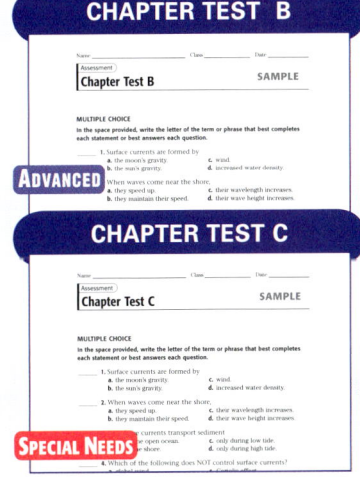
ADVANCED

CHAPTER TEST C
SPECIAL NEEDS

STANDARDIZED TEST PREPARATION
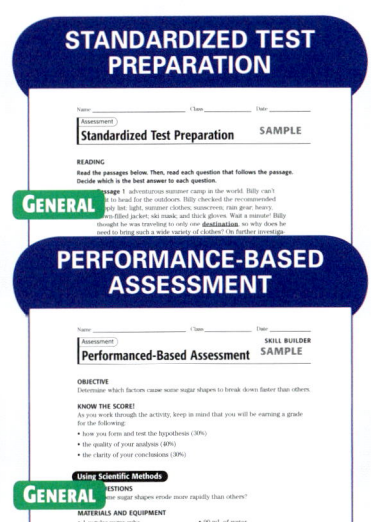
GENERAL

PERFORMANCE-BASED ASSESSMENT
GENERAL

This Chapter Enrichment provides relevant and interesting information to expand and enhance your presentation of the chapter material.

Section 1

Radioactivity

Before and After Radioactivity

- Radioactivity was discovered serendipitously by the French physicist Henri Becquerel (1852–1908) in 1896 while he was searching for evidence of X rays. X rays also had been discovered serendipitously by Wilhelm Conrad Roentgen (1845–1923) the previous year while he was studying cathode-ray tubes.

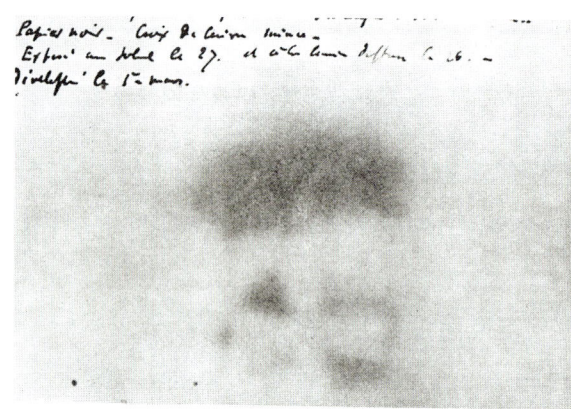

- Soon after Becquerel's discovery, Marie Curie (1867–1934) determined that the element thorium is radioactive. In 1898, Marie and her husband, Pierre (1859–1906), discovered polonium and radium, two new radioactive elements. Eventually, the Curies were able to quantify the amount of energy given off each hour by 1 g of radium—about 418 J.

- In 1899, André-Louis Debierne (1874–1949) discovered the radioactive element actinium. In 1908, Ernest Rutherford (1871–1934) helped discover that alpha particles were helium nuclei.

Is That a Fact!

◆ Antoine-César Becquerel (1788–1878), Henri Becquerel's grandfather, was one of the founders of electrochemistry. Alexandre-Edmond Becquerel (1820–1891), Henri's father, studied light and phosphorescence and invented the phosphoroscope. In fact, Henri Becquerel discovered radioactivity by using the minerals his father had collected and studied.

Alpha, Beta, Gamma

- Although there are many types of radiation, the three types discussed in this chapter are alpha particles, beta particles, and gamma rays.

- Alpha particles consist of two protons and two neutrons, which makes their mass much greater than that of beta particles. Alpha particles are identical to helium nuclei. Because they have no electrons, they have a 2+ charge. The large size and charge of alpha particles make alpha particles easy to block. Alpha particles can be stopped by a rubber glove or thick paper.

- Beta particles can be positively charged (positrons) or negatively charged (electrons). They can be stopped by wood only a few centimeters thick and by thin sheets of aluminum, iron, and lead.

- Gamma rays are high-energy photons that are similar to X rays in their effect. They have no mass or charge. All but the highest-energy gamma rays are stopped by thick concrete or thin sheets of lead.

Is That a Fact!

◆ It is fortunate that alpha particles are so easily stopped because they are potentially the most damaging if they enter the body.

Radioactive Decay

- The decay of naturally occurring isotopes is used to date fossils and rocks. The decay of artificially made isotopes has led to discoveries in nuclear energy, chemistry, and medicine.

- The decay of uranium-238 to lead-206 is a naturally occurring decay series. Uranium-238 decays through both alpha and beta decays accompanied by gamma decay to its end product, a stable isotope of lead.

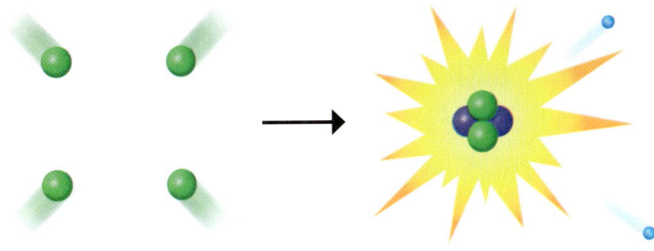

Is That a Fact!

◆ The alpha decay of radium-226 forms radon's most abundant isotope, radon-222, which undergoes alpha decay with a half-life of 3.8 days. The alpha particle and the radioactive polonium-198 made by the decay of radon-222 create a serious lung cancer risk.

Section 2

Energy from the Nucleus

Fission

● The German chemists Otto Hahn (1879–1968) and Fritz Strassmann (1902–1980) were the first to demonstrate the splitting of a nucleus through fission when they bombarded a sample of uranium with neutrons and detected nuclei much smaller than uranium.

● In 1939, this process was identified as nuclear fission by Austrian physicist Lise Meitner (1878–1968) and her nephew Otto Frisch.

● A large amount of energy can be generated from a continuous series of nuclear fissions called a *chain reaction*.

Is That a Fact!

◆ Controlled chain reactions occur in the reactors of nuclear power plants. Uncontrolled chain reactions occur during the detonation of atomic bombs.

Fusion

● Usually, two nuclei cannot collide because they are positively charged and repel each other. However, at very high temperature and pressure, nuclei can collide and fuse together to form a more massive nucleus.

● The fusion of hydrogen nuclei to produce helium is the principal source of the energy that is released as light by the sun and other stars.

● Nuclear fusion releases less energy per nuclear reaction than nuclear fission does because less matter is converted into energy. However, the energy released per gram of fuel is far greater in the fusion process because there are many more hydrogen atoms than uranium atoms in equal masses of the two fuels.

● Matter exists only in the plasma state at the extremely high temperatures required to make fusion occur. A chamber called a *tokamak* was invented in the 1960s to magnetically confine the plasma used for fusion reactions. The confinement prevents the plasma from destroying the tokamak.

Is That a Fact!

◆ In 1991, a controlled nuclear fusion reaction at the Joint European Torus Laboratory in England generated 1.7 million watts. In 1993, the Tokamak Fusion Test Reactor at Princeton University generated a controlled fusion reaction of 5.6 million watts. But both events required more energy than they generated, so fusion as an energy source is not yet practical.

SciLINKS

NSTA
Developed and maintained by the
National Science Teachers Association

SciLinks is maintained by the National Science Teachers Association to provide you and your students with interesting, up-to-date links that will enrich your classroom presentation of the chapter.

Visit www.scilinks.org and enter the SciLinks code for more information about the topic listed.

Topic: Discovering Radioactivity
SciLinks code: HSM0412

Topic: Radioactive Isotopes
SciLinks code: HSM1256

Topic: Nuclear Fission
SciLinks code: HSM1048

Topic: Nuclear Fusion
SciLinks code: HSM1050

Topic: Nuclear Reactors
SciLinks code: HSM1054

Overview

Tell students that this chapter will help them learn about radioactivity and three different kinds of radioactive decay. The chapter discusses nuclear fission and nuclear fusion and the advantages and disadvantages of each.

Assessing Prior Knowledge

Students should be familiar with the following topics:

• atomic structure
• isotopes

Identifying Misconceptions

As students learn the material in this chapter, some of them may be confused by differences between nuclear changes and chemical changes with which students are more familiar. You may want to stress the differences between these concepts by explaining that nuclear changes involve changing the identity of the atoms themselves, while chemical changes involve changing the connections between atoms. Students might also have trouble understanding that because matter is converted into energy during nuclear changes, mass is not conserved in these changes as it is in chemical changes.

Atomic Energy

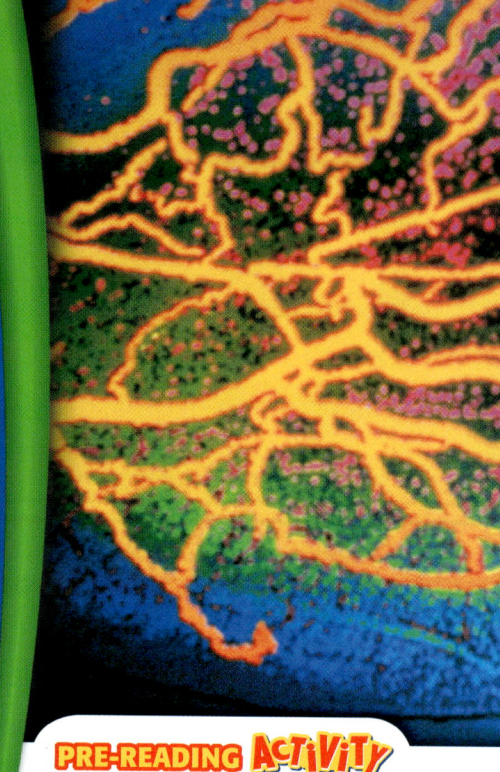

About the PHOTO

Look closely at the blood vessels that show up clearly in this image of a human hand. Doctors sometimes inject radioactive substances into a patient's body to help locate tumors and measure the activity of certain organs. Radioactive emissions from the substances are measured using a scanning device. Then, computers turn the data into an image.

PRE-READING ACTIVITY

Graphic Organizer

Spider Map Before you read the chapter, create the graphic organizer entitled "Spider Map" described in the **Study Skills** section of the Appendix. Label the circle "Radioactive Decay." Create a leg for each type of radioactive decay. As you read the chapter, fill in the map with details about each type of decay.

Standards Correlations

National Science Education Standards

The following codes indicate the National Science Education Standards that correlate to this chapter. The full text of the standards is at the front of the book.

Chapter Opener
UCP 1, 2, 3; SAI 1

Section 1 Radioactivity
UCP 1, 2, 3; SAI 2; ST 2; SPSP 1, 4, 5; HNS 1, 2, 3; PS 3a, 3e

Section 2 Energy from the Nucleus
UCP 2, 3; ST 1, 2; SPSP 3, 4, 5; PS 3a, 3e

Chapter Lab
UCP 2, 3; SAI 1; PS 3a, 3e

Chapter Review
UCP 3; SPSP 1, 3, 4; PS 3a, 3e

Science in Action
ST 2; SPSP 5; HNS 1, 3; PS 3a, 3e

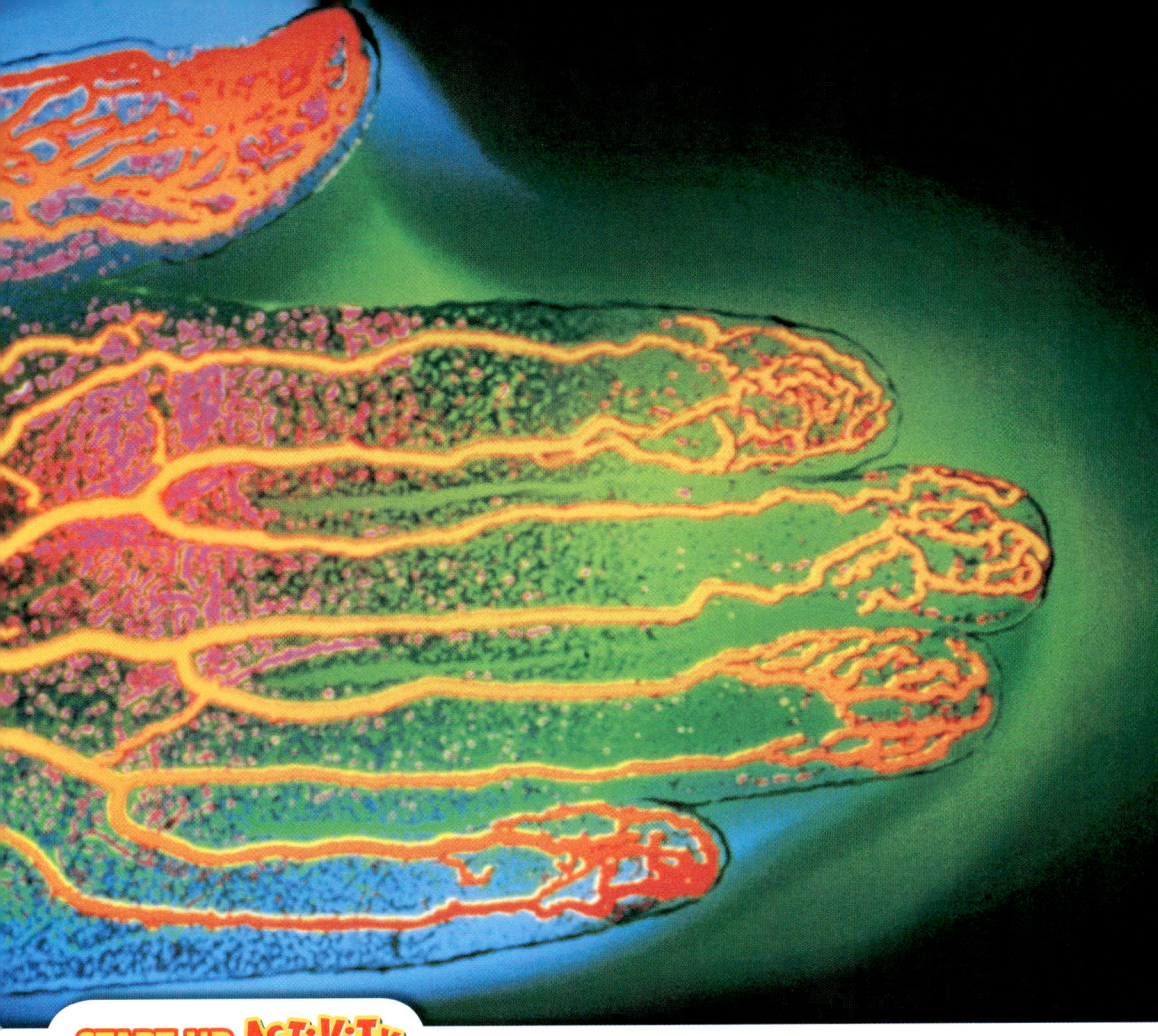

MATERIALS

FOR EACH GROUP
• box, with lid
• paper, graph
• pennies (100)

Safety Caution: Remind students to review all safety cautions and icons before beginning this lab activity.

Teacher's Notes: Because the process of radioactive decay modeled by this activity relies on the presence of a large number of nuclei, you might want students to enter their data into a class data table and graph the totals for the class.

Answers

1. Sample graph:

The number of headsium nuclei decreases over time. After each trial, the number of headsium nuclei is about half of the previous number.

2. The graphs have the same shape; however, the specific numbers of headsium nuclei at each trial are slightly different.

Watch Your Headsium!

In this activity, you will model the decay of unstable nuclei into stable nuclei.

Procedure

1. Place **100 pennies** with the heads' side up in a **box with a lid.** The pennies represent radioactive nuclei. Record 100 "headsium" nuclei as "Trial 0."

2. Close the box. Shake it up and down for 5 s.

3. Open the box. Remove the stable tails-up nuclei, or "tailsium" nuclei. Count the number of headsium nuclei remaining, and record it as "Trial 1."

4. Perform trials until you don't have any more pennies in the box or until you have finished five trials. Record your results.

Analysis

1. On a piece of **graph paper,** graph your data by plotting "Number of headsium nuclei" on the y-axis and "Trial number" on the x-axis. What trend do you see in the number of headsium nuclei?

2. Compare your graph with the graphs made by the other students in your class.

Would You Believe . . . ?

Replacing the battery in your camera can be a problem when you're traveling—especially 1.5 billion kilometers (932 million miles) from home! That's how far from Earth the *Cassini* spacecraft is as it studies Saturn. *Cassini's* camera and other equipment need an energy source that will work after a trip of nearly 7 years. Otherwise, the trip to study the atmosphere and moons of Saturn would be wasted. The answer? the radioactive element plutonium!

The nuclei (plural of nucleus) of plutonium atoms are radioactive, meaning they are unstable. They become stable by giving off radiation in the form of particles and energy. This process heats the materials surrounding the plutonium, and the thermal energy of the materials is converted into electrical energy by a

equipment. Because electrical energy can be produced using a sample of plutonium for 10 or more years, RTGs provide energy longer than any battery known, as described below.

In this chapter you will learn more about the energy given off by radioactive materials. In addition, you will learn about the energy associated with the splitting and joining of atomic nuclei. Read on to continue your own journey!

Chapter Starter Transparency
Use this transparency to help students begin thinking about the use of radioactive material to generate energy.

CHAPTER RESOURCES

Technology

Transparencies
• Chapter Starter Transparency

READING SKILLS

Student Edition on CD-ROM

Guided Reading Audio CD
• English or Spanish

Classroom Videos
• Brain Food Video Quiz

Workbooks

Science Puzzlers, Twisters & Teasers
• Atomic Energy **GENERAL**

Focus

Overview

In this section, students learn about alpha, beta, and gamma radiation and the penetrating ability of each. Students learn how to use an isotope's half-life to calculate the age of an object and learn other uses for radioactive materials.

 **Bellringer**

On the board write the following statement, "In a few sentences, write what you know about the term *nuclear radiation*. Include any benefits and any dangers you can think of."

Motivate

Demonstration — GENERAL

Blocking Radiation Display three items for students:

- piece of paper
- sheet of aluminum
- concrete cinder block

Explain to students that they are going to learn about three types of radiation and that one type can be stopped by paper, one by aluminum, and the third only by concrete or lead. Ask students what different characteristics of the radiation or of the barrier might be important.

LS Visual/Logical

READING WARM-UP

Objectives

- Describe how radioactivity was discovered.
- Compare alpha, beta, and gamma decay.
- Describe the penetrating power of the three kinds of nuclear radiation.
- Calculate ages of objects using half-life.
- Identify uses of radioactive materials.

Terms to Learn

radioactivity   isotope
mass number   half-life

READING STRATEGY

Reading Organizer As you read this section, create an outline of the section. Use the headings from the section in your outline.

Radioactivity

When scientists do experiments, they don't always find what they expect to find.

In 1896, a French scientist named Henri Becquerel found much more than he expected. He found a new area of science.

Discovering Radioactivity

Becquerel's hypothesis was that fluorescent minerals give off X rays. (*Fluorescent* materials glow when light shines on them.) To test his idea, he put a fluorescent mineral on top of a photographic plate wrapped in paper. After putting his setup in bright sunlight, he developed the plate and saw the strong image of the mineral he expected, as shown in **Figure 1**.

An Unexpected Result

Becquerel tried to do the experiment again, but the weather was cloudy. So, he put his materials in a drawer. He developed the plate anyway a few days later. He was shocked to see a strong image. Even without light, the mineral gave off energy. The energy passed through the paper and made an image on the plate. After more tests, Becquerel concluded that this energy comes from uranium, an element in the mineral.

Naming the Unexpected

This energy is called *nuclear radiation,* high-energy particles and rays that are emitted by the nuclei of some atoms. Marie Curie, a scientist working with Becquerel, named the process by which some nuclei give off nuclear radiation. She named the process **radioactivity,** which is also called *radioactive decay.*

Figure 1 *Sunlight could not pass through the paper. So, the image on the plate must have been made by energy given off by the mineral.*

CHAPTER RESOURCES

Chapter Resource File

- Lesson Plan
- Directed Reading A **BASIC**
- Directed Reading B **SPECIAL NEEDS**

Technology

Transparencies
- Bellringer
- Alpha Decay of Radium-226

WEIRD SCIENCE

A thorough scientist makes sure experimental results can be repeated. Becquerel tried to confirm his results, but cloudy weather stopped him. So, why did he develop the photographic plate anyway? One idea is that he was scheduled to speak about his research the next night and even a weak image would support his hypothesis. But when he saw a strong image, he realized that something special had happened.

Figure 2 Alpha Decay of Radium-226

Radium-226   Radon-222

Energy

Charge: 88+   Charge: 86+

Alpha particle
(helium-4)

Charge: 2+

Mass number is conserved.
226 = 222 + 4

Charge is conserved.
(88+) = (86+) + (2+)

Kinds of Radioactive Decay

During *radioactive decay,* an unstable nucleus gives off particles and energy. Three kinds of radioactive decay are alpha decay, beta decay, and gamma decay.

Alpha Decay

The release of an alpha particle from a nucleus is called *alpha decay.* An *alpha particle* is made up of two protons and two neutrons. It has a mass number of 4 and a charge of 2+. The **mass number** is the sum of the numbers of protons and neutrons in the nucleus of an atom. An alpha particle is the same as the nucleus of a helium atom. Many large radioactive nuclei give off alpha particles and become nuclei of atoms of different elements. One example of nuclei that give off alpha particles is radium-226. (The number that follows the name of an element is the mass number of the atom.)

Conservation in Decay

Look at the model of alpha decay in **Figure 2.** This model shows two important things about radioactive decay. First, the mass number is conserved. The sum of the mass numbers of the starting materials is always equal to the sum of the mass numbers of the products. Second, charge is conserved. The sum of the charges of the starting materials is always equal to the sum of the charges of the products.

✓ Reading Check What two things are conserved in radioactive decay? (*See the Appendix for answers to Reading Checks.*)

radioactivity the process by which an unstable nucleus gives off nuclear radiation

mass number the sum of the numbers of protons and neutrons in the nucleus of an atom

SCIENCE HUMOR

Q: Why did the unstable nucleus use toothpaste?

A: It wanted to prevent decay.

Concept Mapping Review **Figures 2, 3,** and **4** with students, and discuss the differences between alpha, beta, and gamma radiation. Have students create a concept map showing the differences between these three types of radiation. **LS Visual**

English Language Learners

ACTIVITY ——— BASIC

X rays and Gamma Rays Ask a local radiology clinic for some sample X-ray images to show students. Remove identifying marks from the films before using the films in class. Remind students that X rays are similar to gamma rays but that gamma rays have higher penetrating ability and energy than X rays do. Ask students why some structures appear brighter than others on an X-ray image.
(Denser structures and tissues absorb more X rays and therefore appear brighter in the image.)
LS Visual

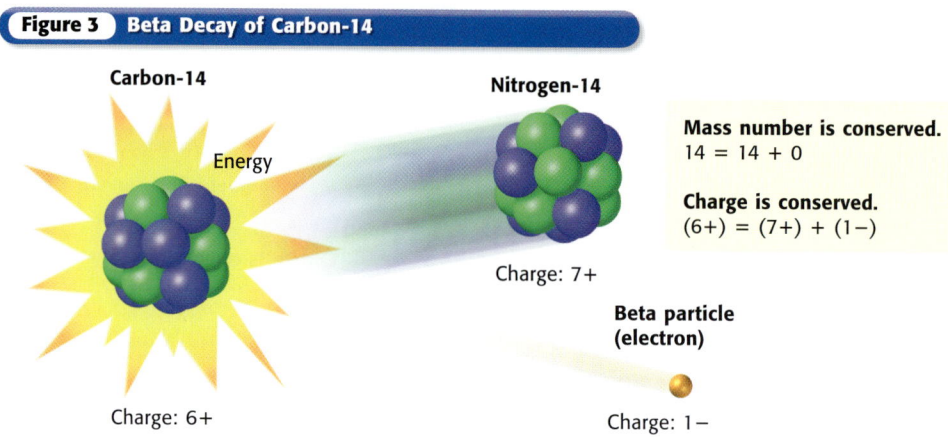

Figure 3 Beta Decay of Carbon-14

Carbon-14

Energy

Nitrogen-14

Mass number is conserved.
14 = 14 + 0

Charge is conserved.
(6+) = (7+) + (1−)

Charge: 7+

Beta particle (electron)

Charge: 6+

Charge: 1−

isotope an atom that has the same number of protons (or the same atomic number) as other atoms of the same element do but that has a different number of neutrons (and thus a different atomic mass)

Beta Decay

The release of a beta particle from a nucleus is called *beta decay*. A *beta particle* can be an electron or a positron. An electron has a charge of 1−. A positron has a charge of 1+. But electrons and positrons have a mass of almost 0. The mass number of a beta particle is 0 because it has no protons or neutrons.

Two Types of Beta Decay

A carbon-14 nucleus undergoes beta decay, as shown in the model in **Figure 3.** During this kind of decay, a neutron breaks into a proton and an electron. Notice that the nucleus becomes a nucleus of a different element. And both mass number and charge are conserved.

Not all isotopes of an element decay in the same way. **Isotopes** are atoms that have the same number of protons as other atoms of the same element do but that have different numbers of neutrons. A carbon-11 nucleus undergoes beta decay when a proton breaks into a positron and a neutron. But during any beta decay, the nucleus changes into a nucleus of a different element. And both mass number and charge are conserved.

Gamma Decay

Energy is also given off during alpha decay and beta decay. Some of this energy is in the form of light that has very high energy called *gamma rays*. The release of gamma rays from a nucleus is called *gamma decay*. This decay happens as the particles in the nucleus shift places. Gamma rays have no mass or charge. So, gamma decay alone does not cause one element to change into another element.

CHAPTER RESOURCES

Technology

 **Transparencies**
- Beta Decay of Carbon-14
- The Penetrating Abilities of Nuclear Radiation

Cultural Awareness — GENERAL

Scientists Everywhere Early discoveries of radiation and nuclear energy were made by scientists all over the world. For example, Wilhelm Conrad Roentgen did his research in Germany; the Curies and Henri Becquerel worked in France; and Ernest Rutherford and Frederick Soddy experimented in England. Other discoveries were made in Canada, Sweden, New Zealand, Japan, and the United States.

The Penetrating Power of Radiation

The three forms of nuclear radiation have different abilities to penetrate, or go through, matter. This difference is due to their mass and charge, as you can see in **Figure 4.**

Effects of Radiation on Matter

Atoms that are hit by nuclear radiation can give up electrons. Chemical bonds between atoms can break when hit by nuclear radiation. Both of these things can cause damage to living and nonliving matter.

Damage to Living Matter

When an organism absorbs radiation, its cells can be damaged. Radiation can cause burns like those caused by touching something that is hot. A single large exposure to radiation can lead to *radiation sickness*. Symptoms of this sickness include fatigue, loss of appetite, and hair loss. Destruction of blood cells and even death can result. Exposure to radiation can also increase the risk of cancer because of the damage done to cells. People who work near radioactive materials often wear a film badge. Radiation will make an image on the film to warn the person if the levels of radiation are too high.

✓ Reading Check Name three symptoms of radiation sickness.

Figure 4 The Penetrating Abilities of Nuclear Radiation

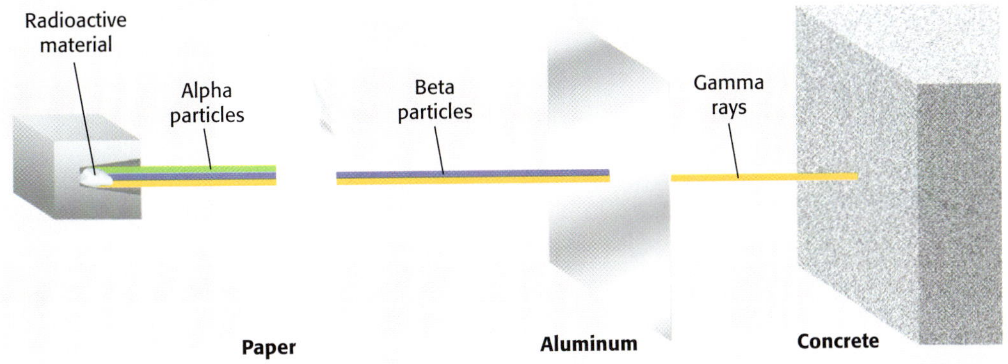

Radioactive material

Alpha particles

Beta particles

Gamma rays

Paper

Aluminum

Concrete

▲ **Alpha particles** have a greater charge and mass than beta particles and gamma rays do. Alpha particles travel about 7 cm through air and are stopped by paper or clothing.

▲ **Beta particles** have a 1− or 1+ charge and almost no mass. They are more penetrating than alpha particles. Beta particles travel about 1 m through air but are stopped by 3 mm of aluminum.

▲ **Gamma rays** have no charge or mass and are the most penetrating. They are blocked by very dense, thick materials, such as a few centimeters of lead or a few meters of concrete.

CONNECTION ACTIVITY
Language Arts — ADVANCED

Why Greek and Latin? Alpha particles, beta particles, and gamma rays are named for the first three letters of the Greek alphabet (α, β, and γ). Ask students to research why so many scientific phenomena are given Greek or Latin names. **LS Logical**

Answer to Reading Check
fatigue, loss of appetite, and hair loss

Is That a Fact!

Wilhelm Conrad Roentgen, Henri Becquerel, and the Curies all have units associated with radiation named after them.

- A *curie* (Ci) equals 3.7×10^{10} decays per second—approximately the activity of 1 g of radium.

- The SI designation for the curie is the *becquerel* (Bq), which equals one disintegration per second.

- The *roentgen* or *radiation unit* (rad) is used to express the dose of energy absorbed from radiation per kilogram of material.

BRAIN FOOD

Radiation in Food Natural radioactivity exists all around you—even in the food you eat. Potassium-40 and radium-226 are two radioactive isotopes found in food. Although these isotopes exist in extremely small amounts in foods, the variation between different foods is astounding; for example, 1 kg of fruit has an activity of 620 to 3,700 pCi of potassium-40. Brazil nuts have an activity of about 5,600 pCi of potassium-40 per kilogram!

READING STRATEGY — GENERAL

Prediction Guide Before students read the text under the Finding a Date by Decay head, ask them whether the following statements are true or false.

- Scientists measure the amount of uranium in human remains to determine how old they are. (false)

- We receive a steady supply of carbon-14 from the food we eat. (true)

- The rate of decay of carbon-14 does not change with changes in temperature and pressure. (true)

LS Logical

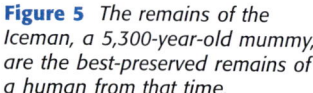

CONNECTION TO Environmental Science

WRITING SKILL **Radon in the Home** Radioactive radon-222 forms from the radioactive decay of uranium found in soil and rocks. Because radon is a gas, it can enter buildings through gaps in the walls and floors. Research the hazards of radon. Identify methods used to detect it and to prevent exposure to it. Present your findings by writing a pamphlet in the form of a public service announcement.

Damage to Nonliving Matter

Radiation can also damage nonliving matter. When metal atoms lose electrons, the metal is weakened. For example, radiation can cause the metal structures of buildings, such as nuclear power plants, to become unsafe. High levels of radiation from the sun can damage spacecraft.

Damage at Different Depths

Gamma rays go through matter easily. They can cause damage deep within matter. Beta particles cause damage closer to the surface. Alpha particles cause damage very near the surface. But alpha particles are larger and more massive than the particles of other kinds of radiation. So, if a source of alpha particles enters an organism, the particles can cause the most damage.

Finding a Date by Decay

Finding a date for someone can be tough—especially if the person is several thousand years old! Hikers in the Italian Alps found the remains of the Iceman, shown in **Figure 5,** in 1991. Scientists were able to estimate the time of death—about 5,300 years ago! How did the scientists do this? The decay of radioactive carbon was the key.

Carbon-14—It's in You!

Carbon atoms are found in all living things. A small percentage of these atoms is radioactive carbon-14 atoms. During an organism's life, the percentage of carbon-14 in the organism stays about the same. Any atoms that decay are replaced. Plants take in carbon from the atmosphere. Animals take in carbon from food. But when an organism dies, the carbon-14 is no longer replaced. Over time, the level of carbon-14 in the remains of the organism drops because of radioactive decay.

Figure 5 The remains of the Iceman, a 5,300-year-old mummy, are the best-preserved remains of a human from that time.

SCIENTISTS AT ODDS

Lung Cancer Most scientists think that the majority of lung cancer cases in the United States can be attributed to cigarette smoking. However, some scientists think that radon is responsible for some cases of lung cancer. Because some people are exposed to both cigarette smoke and radon, it may be impossible to determine exactly how many instances of lung cancer are caused by exposure to radon. Scientific estimates vary widely.

WEIRD SCIENCE

The Iceman was frozen in a glacier in the European Alps at an altitude of 3,200 m and was preserved for about 5,300 years in the ice. To protect the Iceman, the Institute for Anatomy in Innsbruck, Austria, keeps the Iceman in a cooler that simulates the temperature and environmental conditions of the glacier. A second cooler is ready as a backup if needed. If the Iceman warms up, he will begin to decompose.

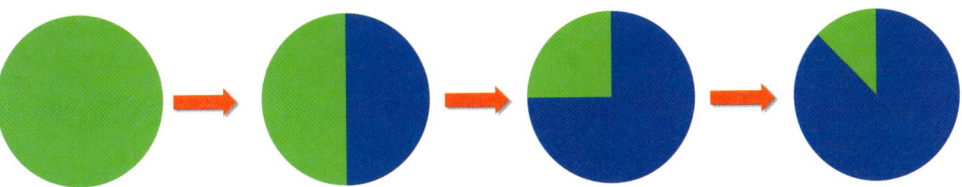

Figure 6 Radioactive Decay and Half-Life

The **original sample** contains a certain amount of radioactive isotope.

After **one half-life,** one-half of the original sample has decayed, and half is unchanged.

After **two half-lives,** one-fourth of the original sample remains unchanged.

After **three half-lives,** only one-eighth of the original sample remains unchanged.

A Steady Rate of Decay

Scientists have found that every 5,730 years, half of the carbon-14 in a sample decays. The rate of decay is constant. The rate is not changed by other conditions, such as temperature or pressure. Each radioactive isotope has its own rate of decay, called half-life. A **half-life** is the amount of time it takes one-half of the nuclei of a radioactive isotope to decay. **Figure 6** is a model of this process. **Table 1** lists some isotopes that have a wide range of half-lives.

half-life the time needed for half of a sample of a radioactive substance to undergo radioactive decay

✓ **Reading Check** What is the half-life of carbon-14?

Determining Age

Scientists measured the number of decays in the Iceman's body each minute. They found that a little less than half of the carbon-14 in the body had changed. In other words, not quite one half-life of carbon-14 (5,730 years) had passed since the Iceman died.

Carbon-14 can be used to find the age of objects up to 50,000 years old. To find the age of older things, other elements must be used. For example, potassium-40 has a half-life of 1.3 billion years. It is used to find the age of dinosaur fossils.

Table 1 Examples of Half-Lives

Isotope	Half-life	Isotope	Half-life
Uranium-238	4.5 billion years	Polonium-210	138 days
Oxygen-21	3.14 s	Nitrogen-13	10 min
Hydrogen-3	12.3 years	Calcium-36	0.1 s

MATH PRACTICE

How Old Is It?

One-fourth of the original carbon-14 of an antler is unchanged. As shown in **Figure 6,** two half-lives have passed. To determine the age of the antler, multiply the number of half-lives that have passed by the half-life of carbon-14. The antler's age is 2 times the half-life of carbon-14:

age = 2 × 5,730 years
age = 11,460 years

Determine the age of a wooden spear that contains one-eighth of its original amount of carbon-14.

BRAIN FOOD

Really Short Lived The half-life of some isotopes is only a few seconds, or even a tiny fraction of a second. Some of these isotopes occur naturally as a result of the radioactive decay of other nuclei.

CHAPTER RESOURCES

Technology

 Transparencies
• Radioactive Decay and Half-Life

Workbooks

 Math Skills for Science
• Radioactive Decay and the Half-Life **GENERAL**

Answer to Reading Check
5,730 years

CONNECTION ACTIVITY
Math ——————— GENERAL

Half-Life Practice Tell students that the half-life of nitrogen-13 is 10 min. Then, present the following problems to students:

1. A 20 g nitrogen-13 sample is prepared for an experiment. If a scientist begins the experiment 20 min later, how many grams of nitrogen-13 remain? (5 g)

2. If only 2.5 g of nitrogen-13 remain at the end, how much time has passed from the time the sample was prepared? (One-eighth of the original sample remains, so three half-lives have passed. The total time since the sample was prepared is 30 min (3 × 10 min).

LS Logical

Answer to Math Practice
3 × 5,730 years = 17,190 years

INCLUSION
Strategies

• *Developmentally Delayed*
• *Hearing impaired*
• *Learning Disabled*

Modeling half-life will help students better understand it.

1. Place students into groups of three, and assign the roles of timer, puller, and recorder. Give 16 chips to each group.

2. Every 15 s, a timer says "Pull," and the puller takes half of the chips and places them in a pile. (There should be one pile for each pull.) The recorder records that a half-life has happened with each pull.

4. After 45 s, the timer calls, "Stop!"

5. Help students notice that the number of chips pulled got smaller but that the time between pulls didn't change.

LS Interpersonal Co-op Learning

Uses of Radioactivity

You have learned how radioactive isotopes are used to determine the age of objects. But radioactivity is used in many areas for many things. The smoke detectors in your home might even use a small amount of radioactive material! Some isotopes can be used as tracers. *Tracers* are radioactive elements whose paths can be followed through a process or reaction.

✓ **Reading Check** What is a tracer?

Radioactivity in Healthcare

Doctors use tracers to help diagnose medical problems. Radioactive tracers that have short half-lives are fed to or injected into a patient. Then, a detector is used to follow the tracer as it moves through the patient's body. The image in **Figure 7** shows an example of the results of a tracer study. Radioactive materials are also used to treat illnesses, including cancer. Radioactive materials can even help prevent illness. For example, many food and healthcare products are sterilized using radiation.

Radioactivity in Industry

Radioactive isotopes can also help detect defects in structures. For example, radiation is used to test the thickness of metal sheets as they are made. Another way radioactive isotopes are used to test structures is shown in **Figure 7.**

Some space probes have been powered by radioactive materials. The energy given off as nuclei decay is converted into electrical energy for the probe.

INTERNET ACTIVITY

For another activity related to this chapter, go to **go.hrw.com** and type in the keyword **HP5RADW.**

Figure 7 **Uses of Radioactivity in Healthcare and in Industry**

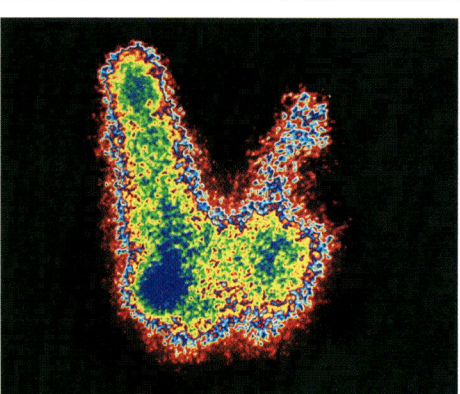 

Radioactive iodine-131 was used to make this scan of a thyroid gland. The dark area shows the location of a tumor.

Tracers are used to find weak spots in materials and leaks in pipes. A Geiger counter is often used to detect the tracer.

CONNECTION to Life Science — GENERAL

Tracers in Plants Scientists inject water composed of molecules that contain radioactive oxygen-15 or oxygen-19 atoms into a plant's root system. The plant uses these water molecules just as it uses water molecules that contain stable oxygen-16 atoms. A radiation counter detects the radioactive oxygen. Scientists can tell how much of the oxygen gas made by the plant through photosynthesis is made with the oxygen atoms from water molecules.

SECTION Review

Summary

- Henri Becquerel discovered radioactivity while trying to study X rays. Radioactivity is the process by which a nucleus gives off nuclear radiation.
- An alpha particle is composed of two protons and two neutrons. A beta particle can be an electron or a positron. Gamma rays are a form of light with very high energy.
- Gamma rays penetrate matter better than alpha or beta particles do. Beta particles penetrate matter better than alpha particles do.

- Nuclear radiation can damage living and nonliving matter.
- Half-life is the amount of time it takes for one-half of the nuclei of a radioactive isotope to decay. The age of some objects can be determined using half-lives.
- Uses of radioactive materials include detecting defects in materials, sterilizing products, diagnosing illness, and generating electrical energy.

Using Key Terms

1. Use the following terms in the same sentence: *mass number* and *isotope*.

Understanding Key Ideas

2. Which of the following statements correctly describes the changes that happen in radioactive decay?
   a. Alpha decay changes the atomic number and the mass number of a nucleus.
   b. Gamma decay changes the atomic number but not the mass number of a nucleus.
   c. Gamma decay changes the mass number but not the atomic number of a nucleus.
   d. Beta decay changes the mass number but not the atomic number of a nucleus.

3. Describe the experiment that led to the discovery of radioactivity.

4. Give two examples of how radioactivity is useful and two examples of how it is harmful.

Math Skills

5. A rock contains one-fourth of its original amount of potassium-40. The half-life of potassium-40 is 1.3 billion years. Calculate the rock's age.

6. How many half-lives have passed if a sample contains one-sixteenth of its original amount of radioactive material?

Critical Thinking

7. **Making Comparisons** Compare the penetrating power of the following nuclear radiation: alpha particles, beta particles, and gamma rays.

8. **Making Inferences** Why would uranium-238 not be useful in determining the age of a spear that is thought to be 5,000 years old? Explain your reasoning.

Interpreting Graphics

9. Look at the figure below. Which nucleus could not undergo alpha decay? Explain your answer.

Beryllium-10   **Hydrogen-3**

6 neutrons   2 neutrons
4 protons   1 proton

CHAPTER RESOURCES

Chapter Resource File

- Section Quiz `GENERAL`
- Section Review `GENERAL`
- Vocabulary and Section Summary `GENERAL`
- Reinforcement Worksheet `BASIC`
- SciLinks Activity `GENERAL`

Answers to Section Review

1. Sample answer: The mass number of each isotope of an element is different.

2. a

3. Sample answer: A fluorescent mineral was placed on a photographic plate wrapped in paper. Because of cloudy weather, Becquerel left the materials in a drawer. After a few days, he developed the plate and saw a strong image of the mineral.

4. Sample answer: Radioactivity is useful in determining the age of objects and in diagnosing medical problems. Radioactivity is harmful because it can weaken structures and can damage cells in living things.

5. If one-fourth of the original amount of potassium-40 remains, two half-lives have passed.
   $2 \times 1{,}300{,}000{,}000$ years $= 2{,}600{,}000{,}000$ years, or 2.6 billion years

6. four half-lives

7. Alpha particles are the least penetrating and are stopped by paper or clothing. Beta particles are more penetrating than alpha particles and are stopped by thin sheets of aluminum. Gamma rays are the most penetrating and are stopped by materials such as lead or concrete.

8. Sample answer: You do not expect to find uranium in a spear, because the spear is made of wood. In addition, the half-life of uranium is very long compared with the estimated age of the spear, so very little uranium would have decayed.

9. An alpha particle is made up of two protons and two neutrons. Hydrogen-3 could not undergo alpha decay because it is made up of only two neutrons and one proton, so it does not have enough protons to make an alpha particle.

SECTION
2

Focus

Overview

In this section, students learn about nuclear fission and nuclear fusion. They also learn some of the pros and cons of each process as an alternative energy source.

🔔 Bellringer

Write the words *fission* and *fusion* on the board. Ask students to write what they think each term means. Discuss some of their ideas with them.

Motivate

Demonstration — GENERAL

Mousetrap Fission Because of the preparation needed, you may want to make a videotape of this demonstration. Set up a closely spaced array of about 20 mechanical mousetraps. Place two table-tennis balls on top of each set mousetrap. Ask students to predict what would happen if you threw a ball into the array. Throw one ball into the array, causing a mousetrap to snap. Point out that this represents an uncontrolled chain reaction. How could a controlled reaction be simulated? (A controlled reaction could be modeled by catching enough balls so that only one ball from each trap sets off another trap.) LS Visual

READING WARM-UP

Objectives
- Describe nuclear fission.
- Identify advantages and disadvantages of fission.
- Describe nuclear fusion.
- Identify advantages and disadvantages of fusion.

Terms to Learn
nuclear fission
nuclear chain reaction
nuclear fusion

READING STRATEGY

Reading Organizer As you read this section, make a table comparing nuclear fission and nuclear fusion.

Energy from the Nucleus

From an early age, you were probably told not to play with fire. But fire itself is neither good nor bad. It simply has benefits and hazards.

Likewise, getting energy from the nucleus of an atom has benefits and hazards. In this section, you will learn about two ways to get energy from the nucleus—fission (FISH uhn) and fusion (FYOO zhuhn). Gaining an understanding of the advantages and disadvantages of fission and fusion is important for people who will make decisions about the use of this energy—people like you!

Nuclear Fission

The nuclei of some atoms decay by breaking into two smaller, more stable nuclei. **Nuclear fission** is the process by which a large nucleus splits into two small nuclei and releases energy.

The nuclei of some uranium atoms, as well as the nuclei of other large atoms, can undergo nuclear fission naturally. Large atoms can also be forced to undergo fission by hitting the atoms with neutrons, as shown by the model in **Figure 1**.

✓ **Reading Check** What happens to a nucleus that undergoes nuclear fission? (*See the Appendix for answers to Reading Checks.*)

Figure 1 Fission of a Uranium-235 Nucleus

Uranium-235

Neutron
Charge: 0

Energy

Neutron
Charge: 0

Barium-142
Charge: 56+

Neutron
Charge: 0

Krypton-91
Charge: 36+

Neutron
Charge: 0

Charge: 92+

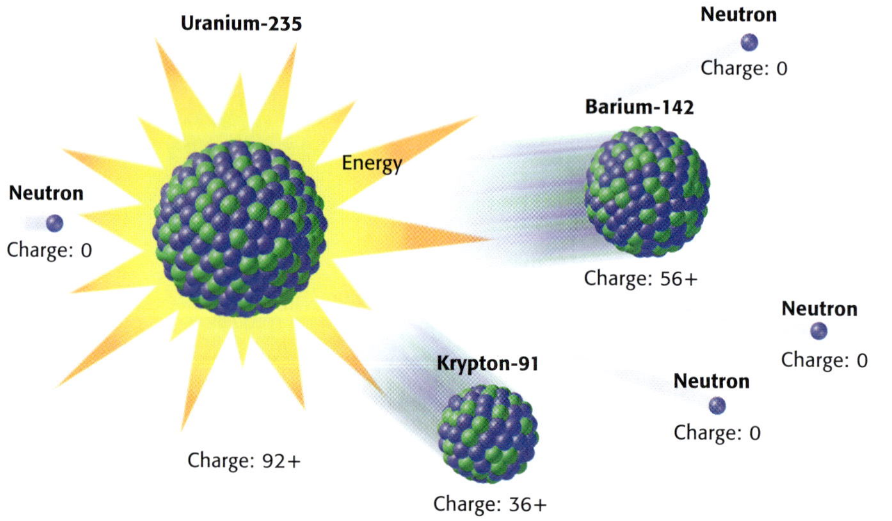

CHAPTER RESOURCES

Chapter Resource File
- Lesson Plan
- Directed Reading A BASIC
- Directed Reading B SPECIAL NEEDS

Technology
- Transparencies
  - Bellringer
  - Fission of a Uranium-235 Nucleus

Answer to Reading Check

A nucleus that undergoes nuclear fission splits into two smaller, more stable nuclei.

Energy from Matter

Did you know that matter can be changed into energy? It's true! If you could find the total mass of the products in **Figure 1** and compare it with the total mass of the reactants, you would find something strange. The total mass of the products is slightly less than the total mass of the reactants. Why are the masses different? Some of the matter was converted into energy.

The amount of energy given off when a single uranium nucleus splits is very small. But this energy comes from a very small amount of matter. The amount of matter converted into energy is only about one-fifth the mass of a hydrogen atom. And hydrogen is the smallest atom that exists! Look at **Figure 2.** The nuclear fission of the uranium nuclei in one fuel pellet releases as much energy as the chemical change of burning about 1,000 kg of coal.

Nuclear Chain Reactions

Look at **Figure 1** again. Suppose that two or three of the neutrons produced split other uranium-235 nuclei. So, energy and more neutrons are given off. And then suppose that two or three of the neutrons that were given off split other nuclei and so on. This example is one type of **nuclear chain reaction,** a continuous series of nuclear fission reactions. A model of an uncontrolled chain reaction is shown in **Figure 3.**

Figure 2 *Each of these small fuel pellets can generate a large amount of energy through the process of nuclear fission.*

nuclear fission the splitting of the nucleus of a large atom into two or more fragments; releases additional neutrons and energy

nuclear chain reaction a continuous series of nuclear fission reactions

Figure 3   An Uncontrolled Nuclear Chain Reaction

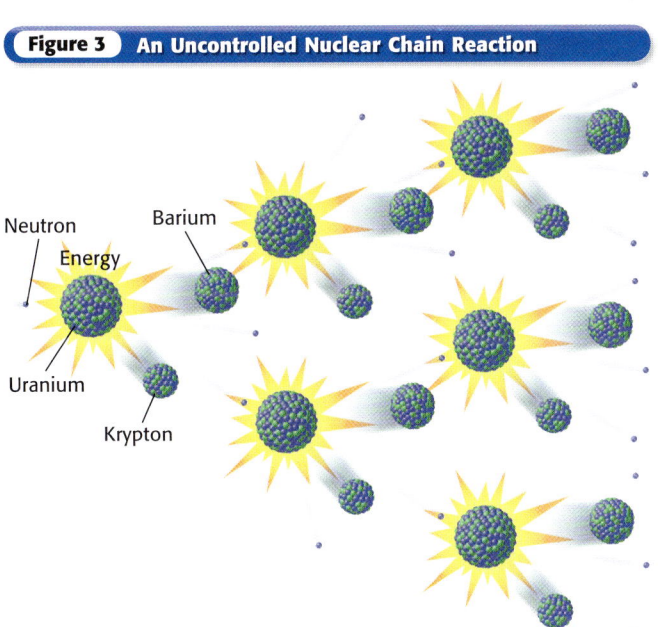

Neutron
Energy
Barium
Uranium
Krypton

BRAIN FOOD

Albert Einstein Although Albert Einstein is considered one of the greatest scientists of the 20th century, he did not begin to speak until the age of 3. In addition, most of his teachers considered him an academic failure because of his apparent lack of interest in classes.

SCIENTISTS AT ODDS

Naming Elements Some scientists proposed that element 105, first synthesized in 1967, be named *hahnium* in honor of Otto Hahn, a German chemist who helped discover nuclear fission. Other scientists proposed other names for the element. The dispute has finally been resolved, and element 105 is now officially known as Dubnium (Db), after the Russian city of Dubna, where the element was first created.

Reactor Types   There are several types of nuclear reactors. The one shown in **Figure 4** is a pressurized water reactor. This reactor uses water under high pressure as a coolant. The pressure allows the water to become superheated without turning into steam. Another type of reactor uses water as a coolant but allows the water to boil and turn to steam. **LS Visual**

CONNECTION ACTIVITY
Environmental Science ———— ADVANCED

Nuclear Plants? Many environmentalists and ecologists share concerns for the flora and fauna in areas around nuclear power plants. Have students research the issues raised by ecologists regarding nuclear power plants. Encourage them to be creative in their presentations. **LS Logical/Verbal**

Debate ———— GENERAL

Space Case Have students do research and hold a debate about the issue of using nuclear energy in spacecraft and satellites. Students on one side can defend the use of nuclear energy in space. The other side can point out the potential hazards associated with using nuclear energy in space. **LS Verbal/Logical**

Energy from a Chain Reaction

In an *uncontrolled chain reaction,* huge amounts of energy are given off very quickly. For example, the tremendous energy of an atomic bomb is the result of an uncontrolled chain reaction. On the other hand, nuclear power plants use *controlled chain reactions.* The energy released from the nuclei in the uranium fuel within the nuclear power plants is used to generate electrical energy. **Figure 4** shows how a nuclear power plant works.

Advantages and Disadvantages of Fission

Every form of energy has advantages and disadvantages. To make informed decisions about energy use, you need to know both sides. For example, burning wood to keep warm on a cold night could save your life. But a spark from the fire could start a forest fire. Nuclear fission has advantages and disadvantages that you should think about.

Figure 4   **How a Nuclear Power Plant Works**

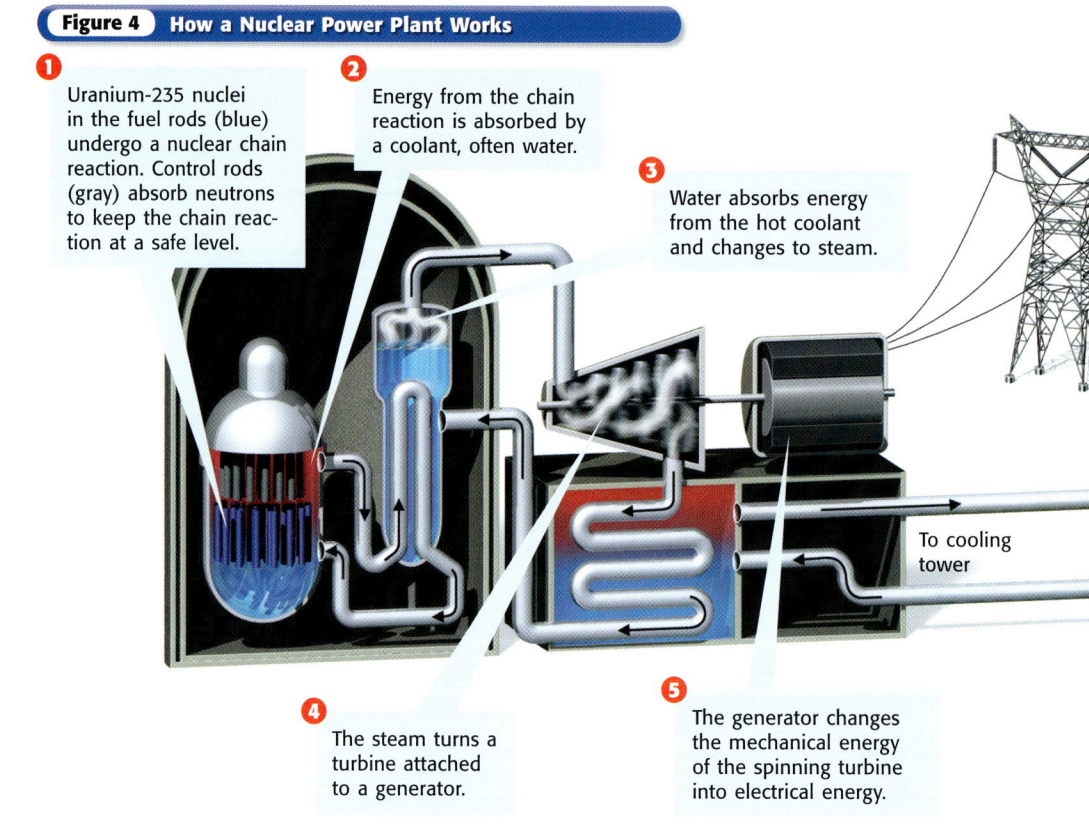

1 Uranium-235 nuclei in the fuel rods (blue) undergo a nuclear chain reaction. Control rods (gray) absorb neutrons to keep the chain reaction at a safe level.

2 Energy from the chain reaction is absorbed by a coolant, often water.

3 Water absorbs energy from the hot coolant and changes to steam.

To cooling tower

4 The steam turns a turbine attached to a generator.

5 The generator changes the mechanical energy of the spinning turbine into electrical energy.

CHAPTER RESOURCES

Technology

 Transparencies
• How a Nuclear Power Plant Works

Q: Why did the student bring a rod and reel to class?

A: The teacher said they would be studying fission.

Accidents

A concern that many people have about nuclear power is the risk of an accident. In Chernobyl, Ukraine, on April 26, 1986, an accident happened, as shown in **Figure 5.** An explosion put large amounts of radioactive uranium fuel and waste products into the atmosphere. The cloud of radioactive material spread over most of Europe and Asia. It reached as far as North America.

What Waste!

Another concern about nuclear power is nuclear waste. This waste includes used fuel rods, chemicals used to process uranium, and even shoe covers and overalls worn by workers. Controlled fission has been carried out for only about 50 years. But the waste will give off high levels of radiation for thousands of years. The rate of radioactive decay cannot be changed. So, the nuclear waste must be stored until it becomes less radioactive. Most of the used fuel rods are stored in huge pools of water. Some of the liquid wastes are stored in underground tanks. However, scientists continue to look for better ideas for long-term storage of nuclear waste.

Nuclear Versus Fossil Fuel

Nuclear power plants cost more to build than power plants that use fossil fuels. But nuclear power plants often cost less to run than plants that use fossil fuels because less fuel is needed. Also, nuclear power plants do not release gases, such as carbon dioxide, into the atmosphere. The use of fission allows our supply of fossil fuels to last longer. However, the supply of uranium is limited.

 Reading Check What are two advantages of using nuclear fission to generate electrical energy?

Figure 5 *During a test at the Chernobyl nuclear power plant, the emergency protection system was turned off. The reactor overheated, which resulted in an explosion.*

CONNECTION TO Language Arts

WRITING SKILL **Storage Site** The government of the United States is required by law to build underground storage for nuclear waste. The waste must be stored for a very long time and cannot escape into the environment. In your **science journal,** write a one-page paper describing the characteristics of a good location for these underground storage sites.

Gone Fission

1. Make two paper balls from a **sheet of paper.**
2. Stand in a group with your classmates. Make sure you are an arm's length from your other classmates.
3. Your teacher will gently toss a paper ball at the group. If you are touched by a ball, gently toss your paper balls at the group.
4. Explain how this activity is a model of a chain reaction. Be sure to explain what the students and the paper balls represent.

The first nuclear-powered submarine was the USS *Nautilus*. The *Nautilus* made its first sea run on January 17, 1955. Because a nuclear generator requires no oxygen, a nuclear-powered submarine can remain underwater for very long periods of time.

Is That a Fact!

The United States has the largest number of operational nuclear power plants in the world (more than 100). The country that comes closest to the United States is France, which has 56 reactors; however, France is only about one-seventeenth the area of the United States.

Figure 6 **Nuclear Fusion of Hydrogen**

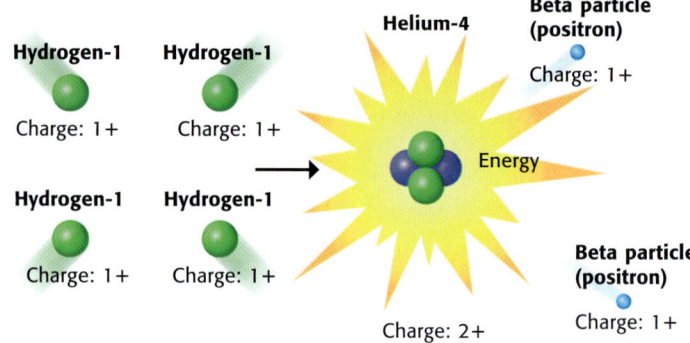

Close

Clay Models Have students model fission by breaking a piece of putty into smaller pieces and fusion by combining small pieces of putty to make one larger piece. **LS** Visual

1. Define *nuclear fission.* (the splitting of the nucleus of a large atom into two or more fragments)

2. Define *nuclear fusion.* (the combination of the nuclei of small atoms to form a larger nucleus)

3. What is the main product of nuclear fusion in the sun? (helium nuclei)

Concept Mapping Ask students to create concept maps to explain fission and fusion. Encourage students to label their diagrams carefully. Students should include where each type of reaction occurs. (fission: nuclear reactor, atomic bomb; fusion: stars, sun, hydrogen bomb) **LS** Visual

Answer to Reading Check

In nuclear fusion, two or more nuclei that have small masses combine to form a larger nucleus. During fusion, energy is released.

---

nuclear fusion the combination of the nuclei of small atoms to form a larger nucleus; releases energy

---

CONNECTION TO Astronomy

Elements of the Stars
Hydrogen is not the only fuel that stars use for fusion. Research other elements that stars can use as fuels and the fusion reactions that make these elements. Make a poster showing what you learn.

ACTIVITY

Nuclear Fusion

Fusion is another nuclear reaction in which matter is converted into energy. In **nuclear fusion,** two or more nuclei that have small masses combine, or fuse, to form a larger nucleus.

Plasma Needed

In order for fusion to happen, the repulsion between positively charged nuclei must be overcome. Very high temperatures are needed—more than 100,000,000°C! At these high temperatures, matter is a plasma. *Plasma* is the state of matter in which electrons have been removed from atoms. So, plasma is made up of ions and electrons. One place that has such temperatures is the sun. In the sun's core, hydrogen nuclei fuse to form a helium nucleus, as shown in the model in **Figure 6.**

✔ **Reading Check** Describe the process of nuclear fusion.

Advantages and Disadvantages of Fusion

Energy for your home cannot yet be generated using nuclear fusion. First, very high temperatures are needed. Second, more energy is needed to make and hold the plasma than is generated by fusion. But scientists predict that fusion will provide electrical energy in the future—maybe in your lifetime!

Less Accident Prone

The concern about an accident such as the one at Chernobyl is much lower for fusion reactors. If an explosion of a fusion reactor happened, very little radioactive material would be released. The hydrogen-3 used for fuel in experimental fusion reactors is much less radioactive than the uranium used in fission reactors.

Science Bloopers

Cold Fusion In 1989, two chemists at the University of Utah, B. Stanley Pons and Martin Fleischmann, claimed that they had produced nuclear fusion at room temperature. However, no one could duplicate their results. Most scientists have concluded that the reports were incorrect. Others still try to explain the results, which cannot be explained by current understanding.

Oceans of Fuel

Scientists studying fusion use hydrogen-2 and hydrogen-3 in their work. Hydrogen-1 is much more common than these isotopes. But there is enough of them in Earth's waters to provide fuel for millions of years. Also, a fusion reaction releases more energy per gram of fuel than a fission reaction does. So, fusion saves more resources than fission does, as shown in **Figure 7**.

Less Waste

The products of fusion reactions are not radioactive. So, fusion is a "cleaner" source of energy than fission is. There would be much less radioactive waste. But to have the benefits of fusion, scientists need money to pay for research.

Figure 7 *Fusing the hydrogen-2 in 3.8 L of water would release about the same amount of energy as burning 1,140 L of gasoline!*

SECTION Review

Summary

- In nuclear fission, a massive nucleus breaks into two nuclei.
- In nuclear fusion, two or more nuclei combine to form a larger nucleus.
- Nuclear fission is used in power plants to generate electrical energy. A limited fuel supply and radioactive waste products are disadvantages of fission.
- Nuclear fusion cannot yet be used as an energy source, but plentiful fuel and little waste are advantages of fusion.

Using Key Terms

Complete each of the following sentences by choosing the correct term from the word bank.

  nuclear fission
  nuclear fusion
  nuclear chain reaction

1. During ___, small nuclei combine.

2. During ___, nuclei split one after another.

Understanding Key Ideas

3. Which of the following is an advantage nuclear fission has over fossil fuels?

  a. unlimited supply of fuel
  b. less radioactive waste
  c. fewer building expenses
  d. less released carbon dioxide

4. Which kind of nuclear reaction is currently used to generate electrical energy?

5. Which kind of nuclear reaction is the source of the sun's energy?

6. What particle is needed to begin a nuclear chain reaction?

7. In both fission and fusion, what is converted into energy?

Math Skills

8. Imagine that a uranium nucleus splits and releases three neutrons and that each neutron splits another nucleus. If the first split occurs in stage 1, how many nuclei will split during stage 4?

Critical Thinking

9. **Making Comparisons** Compare nuclear fission with nuclear fusion.

10. **Analyzing Processes** The floor of a room is covered in mousetraps that each hold two table-tennis balls. One ball is dropped onto a trap. The trap snaps shut, and the balls on it fly into the air and fall on other traps. What nuclear process is modeled here? Explain your answer.

SCLINKS.

NSTA
Developed and maintained by the
National Science Teachers Association

For a variety of links related to this chapter, go to www.scilinks.org

Topic: Nuclear Fission; Nuclear Fusion
SciLinks code: HSM1048; HSM1050

CONNECTION to Earth Science —— GENERAL

Fusion Use the transparency "Fusion of Hydrogen in the Sun" to show students a fusion reaction of other hydrogen isotopes. **Visual**

CHAPTER RESOURCES

Chapter Resource File

- **Section Quiz** GENERAL
- **Section Review** GENERAL
- **Vocabulary and Section Summary** GENERAL
- **Reinforcement Worksheet** BASIC
- **Critical Thinking** ADVANCED
- **Datasheet for Quick Lab**

Technology

Transparencies
- Fusion of Hydrogen-1 Nuclei
- **LINK TO EARTH SCIENCE** Fusion of Hydrogen in the Sun

Domino Chain Reactions

Teacher's Notes

Time Required

One or two 45-minute class periods

Lab Ratings

EASY ———————————→ HARD

Teacher Prep 🜊
Student Set-Up 🜊🜊
Concept Level 🜊🜊🜊
Clean Up 🜊

Lab Notes

In this chapter, images of fission reactions show three neutrons produced when a uranium-235 nucleus splits. However, there are fissions that release only two neutrons. In this activity, each "fission" releases only two neutrons, each of which is represented by a domino.

Model-Making Lab

OBJECTIVES

Build models to represent controlled and uncontrolled nuclear chain reactions.

Compare models of controlled and uncontrolled nuclear chain reactions.

MATERIALS

- dominoes (15)
- stopwatch

Domino Chain Reactions

Fission of uranium-235 is a process that relies on neutrons. When a uranium-235 nucleus splits into two smaller nuclei, it releases two or three neutrons that can cause neighboring nuclei to undergo fission. This fission can result in a nuclear chain reaction. In this lab, you will build two models of nuclear chain reactions, using dominoes.

Procedure

1 For the first model, set up the dominoes as shown below. When pushed over, each domino should hit two dominoes in the next row.

2 Measure the time it takes for all the dominoes to fall. To do this, start the stopwatch as you tip over the front domino. Stop the stopwatch when the last domino falls. Record this time.

3 If some of the dominoes do not fall, repeat steps 1 and 2. You may have to adjust the setup a few times.

4 For the second model, set up the dominoes as shown at left. The domino in the first row should hit both of the dominoes in the second row. Beginning with the second row, only one domino from each row should hit both of the dominoes in the next row.

5 Repeat step 2. Again, you may have to adjust the setup a few times to get all the dominoes to fall.

Larry Tackett
Andrew Jackson Middle School
Cross Lanes, West Virginia

CHAPTER RESOURCES

Chapter Resource File

📁 • Datasheet for Chapter Lab
   • Lab Notes and Answers

Technology

💿 **Classroom Videos**
   • Lab Video

Analyze the Results

1 **Classifying** Which model represents an uncontrolled chain reaction? Which represents a controlled chain reaction? Explain your answers.

2 **Analyzing Results** Imagine that each domino releases a certain amount of energy as it falls. Compare the total amount of energy released in the two models.

3 **Analyzing Data** Compare the time needed to release the energy in the models. Which model took longer to release its energy?

Draw Conclusions

4 **Evaluating Models** In a nuclear power plant, a chain reaction is controlled by using a material that absorbs neutrons. Only enough neutrons to continue the chain reaction are allowed to continue splitting uranium-235 nuclei. Explain how your model of a controlled nuclear chain reaction modeled this process.

5 **Applying Conclusions** Why must uranium nuclei be close to each other in order for a nuclear chain reaction to happen? (Hint: What would happen in your model if the dominoes were too far apart?)

Analyze the Results

1. The first model represents an uncontrolled chain reaction. The second model represents a controlled chain reaction. The number of fission reactions doubles in each step of the first model and would quickly get out of control. Only one fission reaction occurs in each step of the second model, so the reaction happens at a steady, controlled rate.

2. The same amount of energy was released in each model because the same number of dominoes fell in each model.

3. The second model took longer to release its energy.

Draw Conclusions

4. In the second model, the dominoes were set up so that only one of each falling pair would knock down another pair of dominoes. So, only a few dominoes were used to continue the "reaction," just as only some neutrons may continue a chain reaction.

5. The nuclei must be close to each other so that a neutron produced in one fission has a greater chance of hitting another nucleus and causing it to split.

Chapter Review

Assignment Guide

SECTION	QUESTIONS
1	2–5, 7–8, 14–17, 19–22
2	1, 6, 9–13, 18

ANSWERS

Using Key Terms

1. Nuclear fission
2. half-life
3. Nuclear fusion
4. mass numbers

Understanding Key Ideas

5. d
6. b
7. a
8. c
9. d
10. c
11. Two dangers associated with nuclear fission are the potential for an accident that could release radioactive material and the potential of radioactive waste leaking into the environment.

USING KEY TERMS

The statements below are false. For each statement, replace the underlined term to make a true statement.

1. Nuclear fusion involves splitting a nucleus.

2. During one beta decay, half of a radioactive sample will decay.

3. Radioactivity involves the joining of nuclei.

4. Isotopes of an element have different atomic numbers.

UNDERSTANDING KEY IDEAS

Multiple Choice

5. Which of the following is a use of radioactive material?
   a. detecting smoke
   b. locating defects in materials
   c. generating electrical energy
   d. All of the above

6. Which particle both begins and is produced by a nuclear chain reaction?
   a. positron   c. alpha particle
   b. neutron   d. beta particle

7. Which nuclear radiation can be stopped by paper?
   a. alpha particles   c. gamma rays
   b. beta particles   d. None of the above

8. The half-life of a radioactive atom is 2 months. If you start with 1 g of the element, how much will remain after 6 months?
   a. One-half of a gram will remain.
   b. One-fourth of a gram will remain.
   c. One-eighth of a gram will remain.
   d. None of the sample will remain.

9. The waste products of nuclear fission
   a. are harmless.
   b. are safe after 20 years.
   c. can be destroyed by burning them.
   d. remain radioactive for thousands of years.

10. Which statement about nuclear fusion is false?
    a. Nuclear fusion happens in the sun.
    b. Nuclear fusion is the joining of the nuclei of atoms.
    c. Nuclear fusion is currently used to generate electrical energy.
    d. Nuclear fusion can use hydrogen as fuel.

Short Answer

11. What are two dangers associated with nuclear fission?

12. What are two of the problems that need to be solved in order to make nuclear fusion a usable energy source?

13. In fission, the products have less mass than the starting materials do. Explain why this happens.

12. Two problems that need to be solved to make nuclear fusion a practical energy source are finding a way to contain the plasma and finding a process that generates more electrical energy than is needed to heat and contain the plasma.

13. Some matter is converted into energy during fission, so the mass of the products is less than the mass of the starting materials.

Math Skills

14 A scientist used 10 g of phosphorus-32 in a test on plant growth but forgot to record the date. When measured some time later, only 2.5 g of phosphorus-32 remained. If phosphorus-32 has a half-life of 14 days, how many days ago did the experiment begin?

CRITICAL THINKING

15 **Concept Mapping** Use the following terms to create a concept map: *radioactive decay, alpha particle, beta particle, gamma ray,* and *nuclear radiation.*

16 **Expressing Opinions** Smoke detectors often use americium-243 to detect smoke particles in the air. Americium-243 undergoes alpha decay. Do you think that these smoke detectors are safe to have in your home if used properly? Explain. (Hint: Think about how penetrating alpha particles are.)

17 **Applying Concepts** How can radiation cause cancer?

18 **Analyzing Processes** Explain why nuclei of carbon, oxygen, and iron can be found in stars.

19 **Making Inferences** If you could block all radiation from sources outside your body, explain why you would still be exposed to some radiation.

INTERPRETING GRAPHICS

20 The image below was made in a manner similar to that of Becquerel's original experiment. What conclusions can be drawn from this image about the penetrating power of radiation?

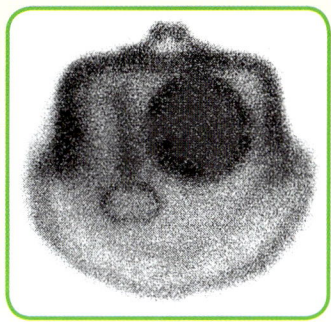

Use the graph below to answer the questions that follow.

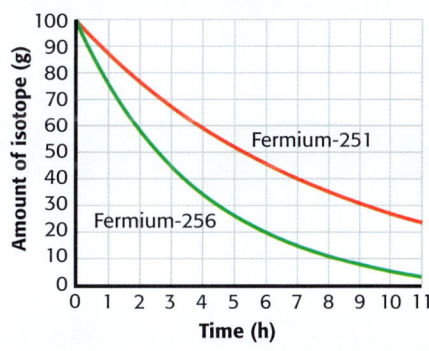

21 What is the half-life of fermium-256?

22 What is the half-life of fermium-251?

Standardized Test Preparation

Read each of the passages below. Then, answer the questions that follow each passage.

Passage 1 Have you noticed that your forks, knives, and spoons don't tarnish easily? Most metal utensils are made of stainless steel. Because it doesn't tarnish, stainless steel is also used in nuclear reactors. Some scientists study radiation's effects on metals and other substances. An important focus of their studies is radiation's effect on the structure of stainless steel. The damage to stainless steel is caused mainly by neutron and heavy ion radiation inside nuclear reactors. The radiation causes stress in the metal. The stress leads to corrosion and finally to cracking. Clearly, this feature is not desirable in parts of a nuclear reactor! Scientists hope that by studying the way radiation affects the atoms of metals, they can find a way to use the incoming radiation to make the surface stronger instead of weaker.

1. Which of the following happens last as stainless steel is damaged in a nuclear reactor?
   A The steel corrodes.
   B The steel is exposed to radiation.
   C The steel cracks.
   D The steel is stressed.

2. Which of the following is a goal of the scientists?
   F to use radiation to strengthen stainless steel
   G to keep stainless steel from tarnishing
   H to keep spoons and forks from cracking
   I to prevent stainless steel from absorbing radiation

3. Why is stainless steel a good metal to use in a nuclear reactor?
   A It is made stronger by radiation.
   B It does not tarnish easily.
   C It cracks under stress.
   D It is not affected by radiation.

Passage 2 A space probe takes about 7 years to reach Saturn. What could supply energy for the cameras and equipment after that time in space? The answer is the radioactive element plutonium! The nuclei (singular, *nucleus*) of plutonium atoms are <u>radioactive</u>, so the nuclei are unstable. They become stable by giving off radiation in the form of particles and rays. This process heats the materials surrounding the plutonium, and the thermal energy of the materials is converted into electrical energy by a radioisotope thermoelectric generator (RTG). Spacecraft such as *Voyager*, *Galileo*, and *Ulysses* depended on RTGs for electrical energy. Because an RTG can generate electrical energy for 10 or more years by using one sample of plutonium, an RTG provides energy longer than any battery can. In fact, the RTGs on *Voyager* were still providing energy after 20 years!

1. Plutonium in an RTG can be expected to provide energy for how long?
   A much less than 7 years
   B about 7 years
   C 10 years or more
   D 20 years

2. Which of the following terms has the most similar meaning to the term *radioactive*?
   F thermoelectric
   G plutonium
   H radiation
   I unstable

3. What is the final form of energy provided by RTGs?
   A thermal
   B electrical
   C particles and rays
   D nuclear

READING

Passage 1
1. C
2. F
3. B

TEST DOCTOR

Question 2: Answer G describes a normal property of stainless steel, so it cannot be a goal of the scientists. Answer H seems reasonable because the scientists want to strengthen stainless steel, which is used to make spoons and forks. Answer I seems like a reasonable goal because the stainless steel would not become damaged by radiation if it never absorbed radiation.

Passage 2
1. C
2. I
3. B

TEST DOCTOR

Question 1: A rapid scan of the passage might cause students to answer this question incorrectly. Students might choose answer A or answer B because of the first line stating that it takes about 7 years to reach Saturn. Students might choose answer D because the last sentence of the passage highlights RTGs that have operated for 20 years.

The table below shows the half-lives of some radioactive isotopes. Use the table below to answer the questions that follow.

Examples of Half-Lives

Isotope	Half-life
Hydrogen-3	12.3 years
Nitrogen-13	10 min
Oxygen-21	3.14 s
Calcium-36	0.1 s
Polonium-210	138 days
Uranium-238	4.5 billion years

1. Half of a sample of which of the following isotopes would take the longest to decay?
   - **A** uranium-238
   - **B** hydrogen-3
   - **C** polonium-210
   - **D** calcium-36

2. How old is an artifact if only one-fourth of the hydrogen-3 in the sample remains?
   - **F** 3.075 years
   - **G** 6.15 years
   - **H** 12.3 years
   - **I** 24.6 years

3. How many days will it take for three-fourths of a sample of radioactive polonium-210 to decay?
   - **A** 69 days
   - **B** 103.5 days
   - **C** 138 days
   - **D** 276 days

4. How many isotopes have shorter half-lives than polonium-210?
   - **F** two
   - **G** three
   - **H** four
   - **I** five

MATH

Read each question below, and choose the best answer.

1. The Butterfly Society spent 1.5 h planting a butterfly garden on Saturday and twice as many hours on Sunday. Which equation could be used to find the total number of hours they spent planting on those 2 days?
   - **A** $n = 2(1.5)$
   - **B** $n = 1.5 + 2(1.5)$
   - **C** $n = 1.5 + 1.5 + 2$
   - **D** $n = 2 \times 2 \times 1.5$

2. How many half-lives have passed if one-eighth of a sample of radioactive carbon-14 remains?
   - **F** two
   - **G** three
   - **H** four
   - **I** eight

3. Which of the following shows the correct fraction of the original sample of radioactive isotope that remains after four half-lives?
   - **A** $4(1/2)$
   - **B** $(1/2)(1/4)$
   - **C** $4\ 1/2$
   - **D** $(1/2)(1/2)(1/2)(1/2)$

4. To find the area of a circle, use the equation $area = \pi r^2$. If the radius of circle A is doubled, how will the area of the circle change?

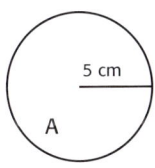

5 cm

A

   - **F** The area will be 1/4 as large.
   - **G** The area will be 1/2 as large.
   - **H** The area will be 2 times larger.
   - **I** The area will be 4 times larger.

Standardized Test Preparation

1. A
2. I
3. D
4. G

✚ TEST DOCTOR

Question 2: Students would choose answer F if they took one-fourth of the half-life of hydrogen-3. Students would choose answer G if they took one-half of the isotope's half-life. Students would choose answer H if they merely looked up the half-life in the table.

MATH

1. B
2. G
3. D
4. I

✚ TEST DOCTOR

Question 3: Students who choose answer A might think that the amount is calculated by multiplying one-half by the number of half-lives that have passed. Students might choose answer B if they think that you multiply one-fourth and one-half to find the amount that remains after four half-lives have passed.

CHAPTER RESOURCES

Chapter Resource File

- Standardized Test Preparation **GENERAL**

State Resources

For specific resources for your state, visit **go.hrw.com** and type in the keyword **HSMSTR**.

Science, Technology, and Society

Background

The FDA approved irradiation as a method to control mold in wheat flour in 1963. Since then, the process has been approved for many types of foods. The dose of ionizing radiation depends on the food and the organism being targeted. The doses are expressed in units of Grays (Gy). The energy added to food irradiated at 1 kGy is equal to the energy given off in 1 s by a 100 W light bulb. This energy raises the temperature of the food less than 1°C. The lowest doses (about 0.15 kGy) are used to prevent white potatoes from sprouting, while the highest doses (30 kGy) are used to sterilize dry spices and seasonings. The dose used to kill bacteria in meat is 4.50 kGy. Tests have shown that many foods can be irradiated without changing their properties. But some foods undergo significant changes. Irradiated alfalfa seeds do not sprout as well as untreated seeds, and meats with a high fat content develop bad odors. Because of these effects and the possibility of changes to the nutritional and chemical makeup of the foods, some experts have recommended that individual foods be extensively tested before they are approved for irradiation.

Science in Action

Science, Technology, and Society

Irradiated Food

One way to help keep food fresh for longer periods of time is to irradiate it. Exposing food to radiation can kill organisms such as mold or bacteria that cause food to spoil. In addition, irradiated potatoes and onions can be stored for a longer time without sprouting. Radiation can even be used to control pests such as beetles that could cause a lot of damage to stored grains.

Social Studies ACTIVITY

WRITING SKILL Food preservation is an important development of history. Write a one-page report that compares methods that you use to keep food from spoiling with methods used in the late 1800s.

Weird Science

Nuclear-Powered Bacteria

Deep under Earth's surface, there is no light. Temperatures are high, water is scarce, and oxygen is difficult to find. For many years, scientists thought that nothing could live under these extreme conditions. But in 1989, a team of scientists found bacteria living in rocks that are 500 m below Earth's surface. Since then, bacteria have been found living in rocks that are as deep as 3.5 km below Earth's surface! Scientists wondered what these bacteria use for food. These bacteria seem to get their food from an unusual source—the radioactive decay of uranium. The idea that radioactive decay can be a food source is new to science and is changing the way that scientists think about life.

Math ACTIVITY

How deep is 3.5 km? To help you imagine this depth, calculate how many Statues of Liberty could be stacked in a hole that is 3.5 km deep. The Statue of Liberty in New York is about 46 m tall.

Answer to Social Studies Activity

Modern preservation techniques include irradiation, refrigeration, freezing, and canning. Preservation methods used in the late 1800s include canning, salting, and drying.

Answer to Math Activity

3.5 km = 3,500 m

3,500 m ÷ 46 m = 76 Statues of Liberty

Marie and Pierre Curie

A Great Team You may have heard the saying "Two heads are better than one." For scientific discoveries, this saying is quite true. The husband and wife team Pierre and Marie Curie put their heads together and discovered the elements radium and polonium. Their work also helped them describe radioactivity.

Working side by side for long hours under poor conditions, Marie and Pierre Curie studied the mysterious rays given off by the element uranium. They processed huge amounts of an ore called *pitchblende* to collect the uranium from it. Strangely, the leftover material was more active than uranium. They spent several more months working with the material and discovered an element that was 300 times more active than uranium. Marie called it *polonium* in honor of Poland, which was the country in which she was born. For their research on radiation, the Curies were awarded the Nobel Prize in physics in 1903.

Language Arts ACTiViTy

WRITING SKILL Think of a time that you and a friend solved a problem together that neither of you could solve alone. Write a one-page story about how you each helped solve the problem.

To learn more about these Science in Action topics, visit **go.hrw.com** and type in the keyword **HP5RADF.**

Current Science

Check out Current Science® articles related to this chapter by visiting **go.hrw.com**. Just type in the keyword **HP5CS16.**

Answer to Language Arts Activity
Accept all reasonable responses.

Weird Science

Background

With the help of the National Science Foundation's Life in Extreme Environments Program, Dr. T. C. Onstott from Princeton University is establishing a long-term research center in South African gold mines, the deepest mines in the world. Scientists reach the research site by taking an elevator 3.5 km into the mine. At this depth, humidity is 100%, and the temperature of the rock reaches 60°C (140°F). Scientists think that life may exist as far as 4 km below Earth's surface and up to 7 km into the oceanic crust. Researchers hope that samples from the South African mines will help them understand how old the underground bacteria are, how common they are, and how they are related to life-forms on Earth's surface. Based on early research, Thomas Gold of Cornell University has calculated that the mass of all organisms living underground could be as great as the mass of all organisms living on Earth's surface.

People in Science

Discussion ——— GENERAL

The discoveries of Marie and Pierre Curie were very important to the scientific world at the time they were made. However, the greatest gains that would come from their work came later. What are some of the uses of radioactivity today? (radiation therapy for cancer, treatment of food, nuclear power, and nuclear weapons)

Finding a Balance

Teacher's Notes

Time Required

One 45-minute class period

Lab Ratings

EASY —————→ HARD

Teacher Prep 🧪🧪
Student Set-Up 🧪
Concept Level 🧪🧪
Clean Up 🧪

MATERIALS

Create at least two envelopes per group. Label each envelope with an unbalanced chemical equation. In each envelope, place one paper arrow, some reactant molecules, and some product molecules. Include extra reactants and products so students do not just put the reactants on one side of the arrow and the products on the other. To create molecule models, draw squares onto a sheet of paper, and label them. Color the squares so that each element has a unique color. (Hint: Draw the molecule models on a sheet of paper, make enough photocopies for all your groups, color the squares, laminate the pages, and then cut out the squares.)

Sample equations:

$Na + Cl_2 \rightarrow NaCl$

$C_2H_4 + O_2 \rightarrow CO_2 + H_2O$

$Fe + O_2 \rightarrow FeO$

$Al + CuSO_4 \rightarrow Al_2(SO_4)_3 + Cu$

$Ba(CN)_2 + H_2SO_4 \rightarrow BaSO_4 + HCN$

Model-Making Lab

Finding a Balance

Usually, balancing a chemical equation involves just writing. But in this activity, you will use models to practice balancing chemical equations, as shown below. By following the rules, you will soon become an expert equation balancer!

MATERIALS

• envelopes, each labeled with an unbalanced equation

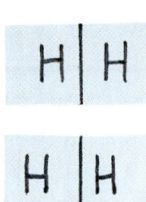 **Example**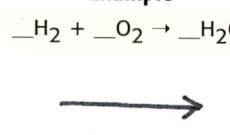

$\_H_2 + \_O_2 \rightarrow \_H_2O$

Balanced Equation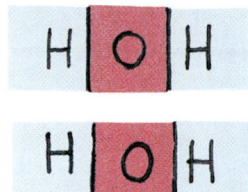
$2H_2 + O_2 \rightarrow 2H_2O$

Procedure

1. The rules are as follows:
   a. Reactant-molecule models may be placed only to the left of the arrow.
   b. Product-molecule models may be placed only to the right of the arrow.
   c. You may use only complete molecule models.
   d. At least one of each of the reactant and product molecules shown in the equation must be included in the model when you are finished.

2. Select one of the labeled envelopes. Copy the unbalanced equation written on the envelope.

3. Open the envelope, and pull out the molecule models and the arrow. Place the arrow in the center of your work area.

4. Put one model of each molecule that is a reactant on the left side of the arrow and one model of each product on the right side.

5. Add one reactant-molecule or product-molecule model at a time until the number of each of the different-colored squares on each side of the arrow is the same. Remember to follow the rules.

6. When the equation is balanced, count the number of each of the molecule models you used. Write these numbers as coefficients, as shown in the balanced equation above.

7. Select another envelope, and repeat the steps until you have balanced all of the equations.

Analyze the Results

1. The rules specify that you are allowed to use only complete molecule models. How are these rules similar to what occurs in a real chemical reaction?

2. In chemical reactions, energy is either released or absorbed. Devise a way to improve the model to show energy being released or absorbed.

CHAPTER RESOURCES

Chapter Resource File

• Datasheet for LabBook
• Lab Notes and Answers

Analyze the Results

1. Chemical reactions cannot involve partial molecules.

2. Answers may vary. Sample answer: Create a symbol for energy that can be used with the reaction models.

Laura Fleet
Alice B. Landrum Middle School
Ponte Vedra Beach, Florida

Skills Practice Lab

Cata-what? Catalyst!

Catalysts increase the rate of a chemical reaction without being changed during the reaction. In this experiment, hydrogen peroxide, H_2O_2, decomposes into oxygen, O_2, and water, H_2O. An enzyme present in liver cells acts as a catalyst for this reaction. You will investigate the relationship between the amount of the catalyst and the rate of the decomposition reaction.

Ask a Question

1 How does the amount of a catalyst affect reaction rate?

Form a Hypothesis

2 Write a statement that answers the question above. Explain your reasoning.

Test the Hypothesis

3 Put a small piece of masking tape near the top of each test tube, and label the tubes "1," "2," and "3."

4 Create a hot-water bath by filling the beaker half full with hot water.

5 Using the funnel and graduated cylinder, measure 5 mL of the hydrogen peroxide solution into each test tube. Place the test tubes in the hot-water bath for 5 min.

6 While the test tubes warm up, grind one liver cube with the mortar and pestle.

7 After 5 min, use the tweezers to place the cube of liver in test tube 1. Place the ground liver in test tube 2. Leave test tube 3 alone.

8 Observe the reaction rate (the amount of bubbling) in all three test tubes, and record your observations.

Analyze the Results

1 Does liver appear to be a catalyst? Explain your answer.

2 Which type of liver (whole or ground) produced a faster reaction? Why?

3 What is the purpose of test tube 3?

MATERIALS

- beaker, 600 mL
- funnel
- graduated cylinder, 10 mL
- hydrogen peroxide, 3% solution
- liver cubes, small (2)
- mortar and pestle
- tape, masking
- test tubes, 10 mL (3)
- tweezers
- water, hot

SAFETY

Draw Conclusions

4 How do your results support or disprove your hypothesis?

5 Why was a hot-water bath used? (Hint: Look in your book for a definition of *activation energy*.)

CHAPTER RESOURCES

Chapter Resource File

- **Datasheet for LabBook**
- **Lab Notes and Answers**

Rodney A. Sandefur
Naturita Middle School
Naturita, Colorado

Disposal Information

Solutions may be washed down the sink if your school drains are connected to a sanitary sewer system with a treatment plant. Students should clean the lab area and wash their hands thoroughly.

Skills Practice Lab

Cata-what? Catalyst!

Teacher's Notes

Time Required

One 45-minute class period

Lab Ratings

EASY ——————→ HARD

Teacher Prep 🧪🧪🧪
Student Set-Up 🧪
Concept Level 🧪🧪🧪
Clean Up 🧪🧪

MATERIALS

Materials listed are for each group of 2–3 students. Liver cubes should be about 1 cm³. Hot water can be obtained from the tap.

Safety Caution

Remind students to review all safety cautions and icons before beginning this lab activity. Use hydrogen peroxide solutions with concentrations of no more than 3%.

Analyze the Results

1. yes; The more vigorous bubbling in the test tubes with liver indicates a faster reaction rate.

2. ground liver; Grinding released more catalyst (enzyme) from the liver cells.

3. Test tube 3 is a control test tube. It is used to compare the rate of bubbling with liver to the rate without liver.

Draw Conclusions

4. Accept all reasonable answers.

5. The bath provides activation energy to start the reaction. The higher temperature also allows the reaction to happen faster.

Putting Elements Together

Teacher's Notes

Time Required

One to two 45-minute class periods

Lab Ratings

EASY ————————→ HARD

Teacher Prep 🧪🧪
Student Set-Up 🧪🧪
Concept Level 🧪🧪
Clean Up 🧪

Safety Caution

Remind students to review all safety cautions and icons before beginning this lab activity. Caution students to be careful around the open flame. All loose hair or clothing should be tied back. Students should wear protective gloves when working with the copper powder and use tongs properly when working near the flame. Caution students not to touch the hot evaporating dish with their bare hands.

Procedure Notes

Set up the ring stand so that the ring will be about 5 cm above the flame.

Putting Elements Together

A synthesis reaction is a reaction in which two or more substances combine to form a single compound. The resulting compound has different chemical and physical properties than the substances from which it is composed. In this activity, you will synthesize, or create, copper(II) oxide from the elements copper and oxygen.

Procedure

❶ Copy the table below.

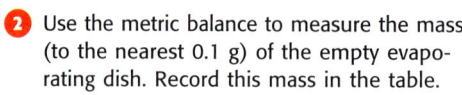

Data Collection Table	
Object	Mass (g)
Evaporating dish	
Copper powder	*DO NOT WRITE IN BOOK*
Copper + evaporating dish after heating	
Copper(II) oxide	

❷ Use the metric balance to measure the mass (to the nearest 0.1 g) of the empty evaporating dish. Record this mass in the table.

❸ Place a piece of weighing paper on the metric balance, and measure approximately 10 g of copper powder. Record the mass (to the nearest 0.1 g) in the table. **Caution:** Wear protective gloves when working with copper powder.

❹ Use the weighing paper to place the copper powder in the evaporating dish. Spread the powder over the bottom and up the sides as much as possible. Discard the weighing paper.

- balance, metric
- Bunsen burner (or portable burner)
- copper powder
- evaporating dish
- gauze, wire
- gloves, protective
- igniter
- paper, weighing
- ring stand and ring
- tongs

SAFETY

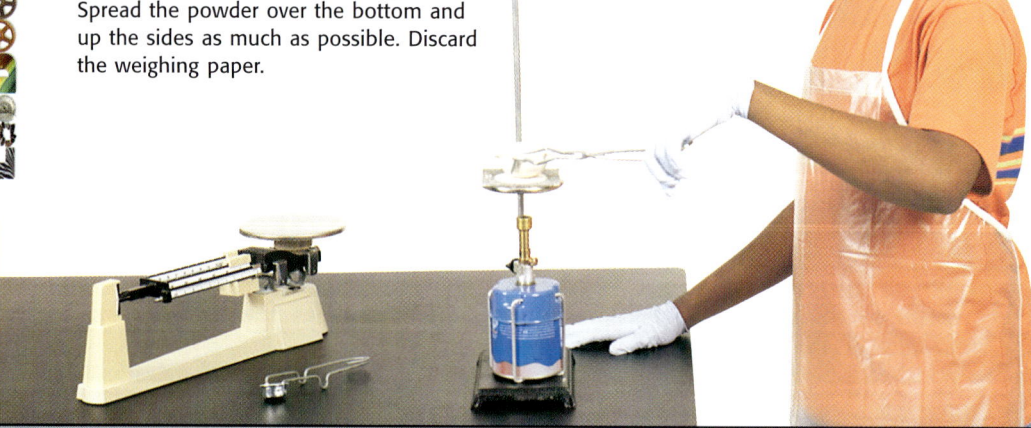

CLASSROOM TESTED & APPROVED

Paul Boyle
Perry Heights Middle School
Evansville, Indiana

CHAPTER RESOURCES

Chapter Resource File

- Datasheet for LabBook
- Lab Notes and Answers

5 Set up the ring stand and ring. Place the wire gauze on top of the ring. Carefully place the evaporating dish on the wire gauze.

6 Place the Bunsen burner under the ring and wire gauze. Use the igniter to light the Bunsen burner. **Caution:** Use extreme care when working near an open flame.

7 Heat the evaporating dish for 10 min.

8 Turn off the burner, and allow the evaporating dish to cool for 10 min. Use tongs to remove the evaporating dish and to place it on the balance to determine the mass. Record the mass in the table.

9 Determine the mass of the reaction product—copper(II) oxide—by subtracting the mass of the evaporating dish from the mass of the evaporating dish and copper powder after heating. Record this mass in the table.

Analyze the Results

1 What evidence of a chemical reaction did you observe after the copper was heated?

2 Explain why there was a change in mass.

3 How does the change in mass support the idea that this reaction is a synthesis reaction?

Draw Conclusions

4 Why was powdered copper used rather than a small piece of copper? (Hint: How does surface area affect the rate of the reaction?)

5 Why was the copper heated? (Hint: Look in your book for the discussion of activation energy.)

6 The copper bottoms of cooking pots can turn black when used. How is that similar to the results you obtained in this lab?

> ### Applying Your Data
>
> Rust, shown above, is iron(III) oxide—the product of a synthesis reaction between iron and oxygen. How does painting a car help prevent this type of reaction?

Analyze the Results

1. The copper changed color, and the mass changed.

2. A change in mass occurred because the copper combined with oxygen from the air. The resulting copper(II) oxide has more mass than the original copper alone.

3. A synthesis reaction is one in which two or more substances join to form a new substance. The mass of the copper(II) oxide is greater than the mass of the copper alone, so a synthesis reaction, in which the copper combined with oxygen from the air, must have occurred, resulting in the change in mass.

Draw Conclusions

4. Powdered copper has a larger surface area than a piece of copper. More surface area increases the rate of the reaction because more copper is exposed to oxygen.

5. The copper was heated because the formation of copper(II) oxide requires a large activation energy.

6. The copper(II) oxide synthesized in this experiment is the same black powder that appears on copper pots.

Applying Your Data

Sample answer: Painting a car helps prevent rust from forming on it by creating a barrier between the iron of the car and the oxygen. If the iron and oxygen are not in contact, they cannot react to form rust.

Disposal Information

Any leftover copper powder can be thrown in the trash. Students should wash their hands thoroughly after completing this lab. Dispose of the copper(II) oxide by letting it cool thoroughly, wrapping it in newspaper or paper towels, and then putting it in the trash.

Making Salt

Teacher's Notes

Time Required

One 45-minute class period, plus 10 minutes the following day

Lab Ratings

EASY ———————————→ HARD

Teacher Prep
Student Set-Up
Concept Level
Clean Up

Safety Caution

Review all proper safety precautions with your students. Students should wear safety goggles, protective gloves, and an apron. In case of an acid or a base spill, first dilute the spill with water. Then, while wearing disposable plastic gloves, mop up the spill with wet cloths designated for spill cleanup. A wet cloth mop can be rinsed out a few times and used until it falls apart. Work with another person nearby who can call for help in case of an emergency, and work near (no more than a few seconds away from) a safety shower and eyewash station known to be in operating condition.

Procedure Notes

You may wish to do this lab as a demonstration or class activity if time or materials are limited.

Making Salt

A neutralization reaction between an acid and a base produces water and a salt. In this lab, you will react an acid with a base and then let the water evaporate. You will then examine what is left for properties that tell you that it is indeed a salt.

Ask a Question

1 Write a question about reactions between acids and bases.

Form a Hypothesis

2 Write a hypothesis that may answer the question you asked in the step above.

Test the Hypothesis

3 Put on protective gloves. Carefully measure 25 mL of hydrochloric acid in a graduated cylinder, and then pour it into the beaker. Carefully rinse the graduated cylinder with distilled water to clean out any leftover acid. **Caution:** Hydrochloric acid is corrosive. If any should spill on you, immediately flush the area with water, and notify your teacher.

4 Add 3 drops of phenolphthalein indicator to the acid in the beaker. You will not see anything happen yet because this indicator won't show its color unless too much base is present.

5 Measure 20 mL of sodium hydroxide (base) in the graduated cylinder, and add it slowly to the beaker with the acid. Use the stirring rod to mix the substances completely. **Caution:** Sodium hydroxide is also corrosive. If any should spill on you, immediately flush the area with water, and notify your teacher.

6 Use an eyedropper to add more base, a few drops at a time, to the acid-base mixture in the beaker. Be sure to stir the mixture after each few drops. Continue adding drops of base until the mixture remains colored after stirring.

MATERIALS

- beaker, 100 mL
- eyedroppers (2)
- evaporating dish
- gloves, protective
- graduated cylinder, 100 mL
- hydrochloric acid
- magnifying lens
- phenolphthalein solution in a dropper bottle
- stirring rod, glass
- sodium hydroxide
- water, distilled

SAFETY

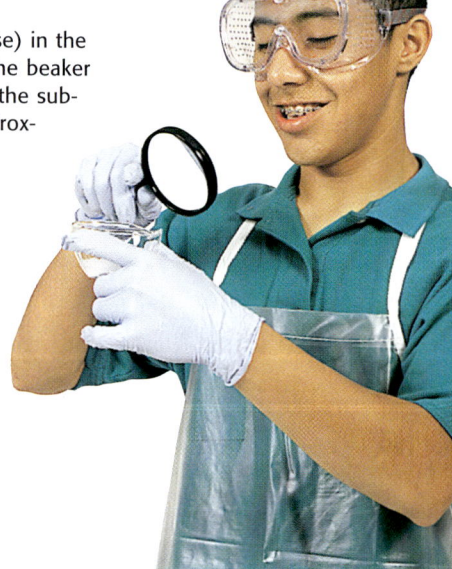

Safety Caution

Hydrochloric Acid   Use only concentrations of hydrochloric acid below 1.0 M. Students should not handle concentrated solutions. Avoid contact with skin and eyes, and avoid breathing vapors. When making a solution, it is important always to add the acid to the water so that if something splashes out, it will most likely be water.

Safety Caution

Sodium Hydroxide   Use only concentrations of sodium hydroxide below 1.0 M. Students should not handle concentrated solutions. Avoid contact with skin and eyes. You should wear goggles, a face shield, impermeable gloves, and a lab apron if you must prepare a solution of NaOH.

7 Use another eyedropper to add acid to the beaker, 1 drop at a time, until the color just disappears after stirring.

8 Pour the mixture carefully into an evaporating dish, and place the dish where your teacher tells you to allow the water to evaporate overnight.

9 The next day, examine your evaporating dish, and with a magnifying lens, study the crystals that were left. Identify the color, shape, and other properties of the crystals.

Analyze the Results

1 The following equation is for the reaction that occured in this experiment:

$$HCl + NaOH \longrightarrow H_2O + NaCl$$

NaCl is ordinary table salt and forms very regular cubic crystals that are white. Did you find white cubic crystals?

2 The phenolphthalein indicator changes color in the presence of a base. Why did you add more acid in step 7 until the color disappeared?

Applying Your Data

Another neutralization reaction occurs between hydrochloric acid and potassium hydroxide, KOH. The equation for this reaction is as follows:

$$HCl + KOH \longrightarrow H_2O + KCl$$

What are the products of this neutralization reaction? How do they compare with those you discovered in this experiment?

Safety Caution

Phenolphthalein Students should use only pre-mixed solutions (2 g in 100 mL 95% ethanol; add 100 mL water). Phenolphthalein solutions are flammable, and the vapors can explode when mixed with air. Ensure that there are no flames or sources of ignition, such as sparks, when you are using the phenolphthalein solution. Restrict the amount of phenolphthalein in the room to 100 mL.

Caution students not to taste the salt they create—it will have phenolphthalein in it.

Disposal Information

Hydrochloric Acid Titrate with 0.1 M NaOH as required until the pH is between 6 and 8, and then pour down the drain.

Sodium Hydroxide Titrate with 0.1 M HCl as required until the pH is between 5 and 9, and then pour down the drain.

Phenolphthalein Set out a container for any used indicator solutions that are left over at the end of the procedure. Titrate the mixture with 0.1 M HCl or 0.1 M NaOH as required until the pH is between 6 and 8, and then pour down the drain. Unused indicators should be tightly covered and returned to the storage shelf.

Analyze the Results

1. Students should observe white cubic crystals.

2. The phenolphthalein changing color in step 6 meant that too much base was present. Acid was added to bring the solution back to neutral.

Applying Your Data

The products are water and a salt, KCl (potassium chloride).

Contents

Appendix

 **Answers**

Chapter 1 Chemical Bonding

Section 1

Page 5: Most atoms form bonds only with their valence electrons.

Page 6: Atoms in Group 18 (the noble gases) rarely form chemical bonds.

Section 2

Page 7: Atoms are neutral because the number of protons in an atom always equals the number of electrons in the atom.

Page 9: Atoms in Group 17 give off the most energy when forming negative ions.

Section 3

Page 12: A covalent bond is a bond that forms when atoms share one or more pairs of electrons.

Page 14: There are two atoms in a diatomic molecule.

Page 16: Ductility is the ability to be drawn into wires.

Chapter 2 Chemical Reactions

Section 1

Page 29: A precipitate is a solid substance that is formed in a solution.

Page 30: In a chemical reaction, the chemical bonds in the starting substances break, and then new bonds form to make new substances.

Section 2

Page 33: Ionic compounds are made up of a metal and a nonmetal.

Page 34: Reactants are the starting substances in a chemical reaction, and products are the substances that are formed.

Page 36: 4

Section 3

Page 38: A synthesis reaction is a reaction in which two or more substances combine to form one new compound.

Page 39: In a decomposition reaction, a substance breaks down into simpler substances. In a synthesis reaction, two or more substances combine to form one new compound.

Page 40: In a single-displacement reaction, an element may replace another element if the replacing element is more reactive than the original element.

Section 4

Page 43: An endothermic reaction is a chemical reaction in which energy is taken in.

Page 44: Activation energy is the energy that is needed to start a chemical reaction.

Page 46: A high concentration of reactants allows the particles of the reactants to run into each other more often, so the reaction proceeds at a faster rate.

Chapter 3 Chemical Compounds

Section 1

Page 59: Ionic solutions conduct an electric current because the ions in the solution are charged and are able to move past each other easily.

Page 60: Most covalent compounds will not dissolve in water because the attraction of the water molecules to each other is much stronger than their attraction to the compound.

Section 2

Page 62: A hydronium ion forms when a hydrogen ion bonds to a water molecule in a water solution.

Page 64: Sulfuric acid is used in car batteries to conduct electric current. Hydrochloric acid is used as an algaecide in swimming pools. Nitric acid is used to make fertilizers.

Page 67: Bases can be used at home in the form of soap, oven cleaner, or antacid.

Section 3

Page 68: In a strong acid, all of the molecules of the acid break apart when the acid is dissolved in water. In a weak acid, only a few of the acid molecules break apart when the acid is dissolved in water.

Page 70: Indicators turn different colors at different pH levels. The color on the pH strip can be compared with the colors on the indicator scale to determine the pH of the solution being tested.

Section 4

Page 72: Structural formulas show how atoms in a molecule are connected.

Page 75: Proteins are made of building blocks called *amino acids.*

Page 76: Nucleic acids store genetic information and build proteins.

Chapter 4   Atomic Energy

Section 1

Page 89: mass number and charge

Page 91: fatigue, loss of appetite, and hair loss

Page 93: 5,730 years

Page 94: A tracer is a radioactive element whose path can be followed through a process or reaction.

Section 2

Page 96: A nucleus that undergoes nuclear fission splits into two smaller, more stable nuclei.

Page 99: Sample answer: Using nuclear fission to generate electrical energy can help our supply of fossil fuels last longer, can help protect the environment because gases such as carbon dioxide are not released during fission, and can save money because nuclear power plants often cost less to run than power plants that use fossil fuels.

Page 100: In nuclear fusion, two or more nuclei that have small masses combine to form a larger nucleus. During fusion, energy is released.

Appendix

Study Skills

FoldNote Instructions

 Have you ever tried to study for a test or quiz but didn't know where to start? Or have you read a chapter and found that you can remember only a few ideas? Well, FoldNotes are a fun and exciting way to help you learn and remember the ideas you encounter as you learn science!

FoldNotes are tools that you can use to organize concepts. By focusing on a few main concepts, FoldNotes help you learn and remember how the concepts fit together. They can help you see the "big picture." Below you will find instructions for building 10 different FoldNotes.

Pyramid

1. Place a sheet of paper in front of you. Fold the lower left-hand corner of the paper diagonally to the opposite edge of the paper.

2. Cut off the tab of paper created by the fold (at the top).

3. Open the paper so that it is a square. Fold the lower right-hand corner of the paper diagonally to the opposite corner to form a triangle.

4. Open the paper. The creases of the two folds will have created an X.

5. Using scissors, cut along one of the creases. Start from any corner, and stop at the center point to create two flaps. Use tape or glue to attach one of the flaps on top of the other flap.

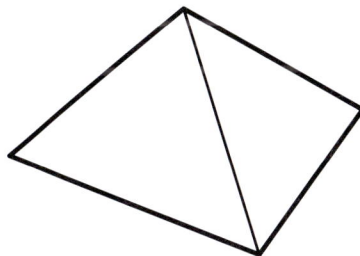

Double Door

1. Fold a sheet of paper in half from the top to the bottom. Then, unfold the paper.

2. Fold the top and bottom edges of the paper to the crease.

Booklet

1. Fold a sheet of paper in half from left to right. Then, unfold the paper.

2. Fold the sheet of paper in half again from the top to the bottom. Then, unfold the paper.

3. Refold the sheet of paper in half from left to right.

4. Fold the top and bottom edges to the center crease.

5. Completely unfold the paper.

6. Refold the paper from top to bottom.

7. Using scissors, cut a slit along the center crease of the sheet from the folded edge to the creases made in step 4. Do not cut the entire sheet in half.

8. Fold the sheet of paper in half from left to right. While holding the bottom and top edges of the paper, push the bottom and top edges together so that the center collapses at the center slit. Fold the four flaps to form a four-page book.

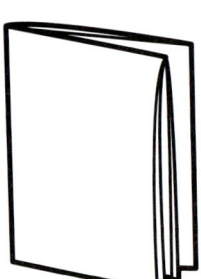

Layered Book

1. Lay one sheet of paper on top of another sheet. Slide the top sheet up so that 2 cm of the bottom sheet is showing.

2. Hold the two sheets together, fold down the top of the two sheets so that you see four 2 cm tabs along the bottom.

3. Using a stapler, staple the top of the FoldNote.

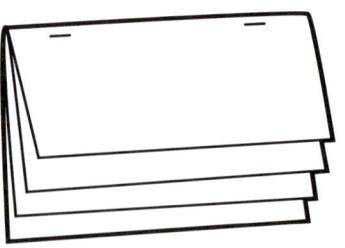

Key-Term Fold

1. Fold a sheet of lined notebook paper in half from left to right.

2. Using scissors, cut along every third line from the right edge of the paper to the center fold to make tabs.

Four-Corner Fold

1. Fold a sheet of paper in half from left to right. Then, unfold the paper.

2. Fold each side of the paper to the crease in the center of the paper.

3. Fold the paper in half from the top to the bottom. Then, unfold the paper.

4. Using scissors, cut the top flap creases made in step 3 to form four flaps.

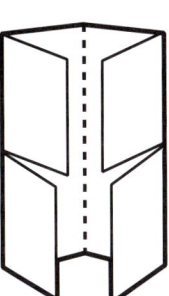

Three-Panel Flip Chart

1. Fold a piece of paper in half from the top to the bottom.

2. Fold the paper in thirds from side to side. Then, unfold the paper so that you can see the three sections.

3. From the top of the paper, cut along each of the vertical fold lines to the fold in the middle of the paper. You will now have three flaps.

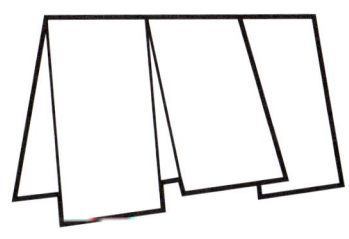

Appendix

Table Fold

1. Fold a piece of paper in half from the top to the bottom. Then, fold the paper in half again.

2. Fold the paper in thirds from side to side.

3. Unfold the paper completely. Carefully trace the fold lines by using a pen or pencil.

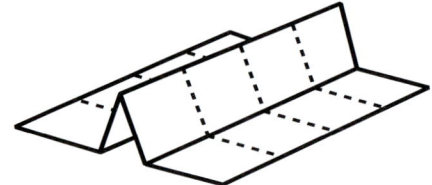

Two-Panel Flip Chart

1. Fold a piece of paper in half from the top to the bottom.

2. Fold the paper in half from side to side. Then, unfold the paper so that you can see the two sections.

3. From the top of the paper, cut along the vertical fold line to the fold in the middle of the paper. You will now have two flaps.

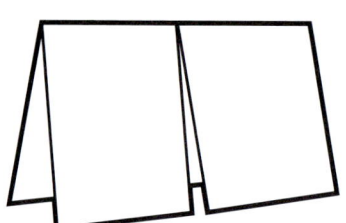

Tri-Fold

1. Fold a piece a paper in thirds from the top to the bottom.

2. Unfold the paper so that you can see the three sections. Then, turn the paper sideways so that the three sections form vertical columns.

3. Trace the fold lines by using a pen or pencil. Label the columns "Know," "Want," and "Learn."

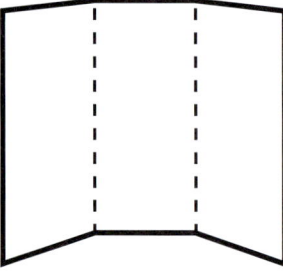

Appendix

Graphic Organizer Instructions

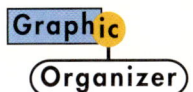 Have you ever wished that you could "draw out" the many concepts you learn in your science class? Sometimes, being able to *see* how concepts are related really helps you remember what you've learned. Graphic Organizers do just that! They give you a way to draw or map out concepts.

All you need to make a Graphic Organizer is a piece of paper and a pencil. Below you will find instructions for four different Graphic Organizers designed to help you organize the concepts you'll learn in this book.

Spider Map

1. Draw a diagram like the one shown. In the circle, write the main topic.

2. From the circle, draw legs to represent different categories of the main topic. You can have as many categories as you want.

3. From the category legs, draw horizontal lines. As you read the chapter, write details about each category on the horizontal lines.

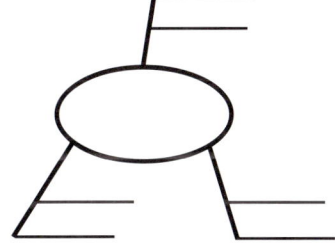

Comparison Table

1. Draw a chart like the one shown. Your chart can have as many columns and rows as you want.

2. In the top row, write the topics that you want to compare.

3. In the left column, write characteristics of the topics that you want to compare. As you read the chapter, fill in the characteristics for each topic in the appropriate boxes.

Appendix

Chain-of-Events-Chart

1. Draw a box. In the box, write the first step of a process or the first event of a timeline.

2. Under the box, draw another box, and use an arrow to connect the two boxes. In the second box, write the next step of the process or the next event in the timeline.

3. Continue adding boxes until the process or timeline is finished.

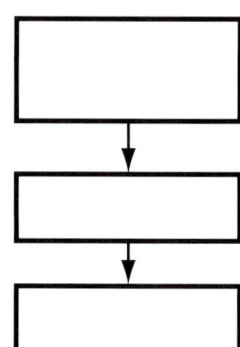

Concept Map

1. Draw a circle in the center of a piece of paper. Write the main idea of the chapter in the center of the circle.

2. From the circle, draw other circles. In those circles, write characteristics of the main idea. Draw arrows from the center circle to the circles that contain the characteristics.

3. From each circle that contains a characteristic, draw other circles. In those circles, write specific details about the characteristic. Draw arrows from each circle that contains a characteristic to the circles that contain specific details. You may draw as many circles as you want.

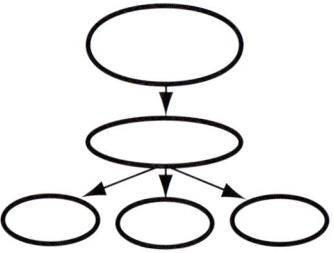

SI Measurement

The International System of Units, or SI, is the standard system of measurement used by many scientists. Using the same standards of measurement makes it easier for scientists to communicate with one another.

SI works by combining prefixes and base units. Each base unit can be used with different prefixes to define smaller and larger quantities. The table below lists common SI prefixes.

SI Prefixes

Prefix	Symbol	Factor	Example
kilo-	k	1,000	kilogram, 1 kg = 1,000 g
hecto-	h	100	hectoliter, 1 hL = 100 L
deka-	da	10	dekameter, 1 dam = 10 m
		1	meter, liter, gram
deci-	d	0.1	decigram, 1 dg = 0.1 g
centi-	c	0.01	centimeter, 1 cm = 0.01 m
milli-	m	0.001	milliliter, 1 mL = 0.001 L
micro-	μ	0.000 001	micrometer, 1 μm = 0.000 001 m

SI Conversion Table

SI units	From SI to English	From English to SI
Length		
kilometer (km) = 1,000 m	1 km = 0.621 mi	1 mi = 1.609 km
meter (m) = 100 cm	1 m = 3.281 ft	1 ft = 0.305 m
centimeter (cm) = 0.01 m	1 cm = 0.394 in.	1 in. = 2.540 cm
millimeter (mm) = 0.001 m	1 mm = 0.039 in.	
micrometer (μm) = 0.000 001 m		
nanometer (nm) = 0.000 000 001 m		
Area		
square kilometer (km^2) = 100 hectares	1 km^2 = 0.386 mi^2	1 mi^2 = 2.590 km^2
hectare (ha) = 10,000 m^2	1 ha = 2.471 acres	1 acre = 0.405 ha
square meter (m^2) = 10,000 cm^2	1 m^2 = 10.764 ft^2	1 ft^2 = 0.093 m^2
square centimeter (cm^2) = 100 mm^2	1 cm^2 = 0.155 in.^{2}	1 in.^{2} = 6.452 cm^2
Volume		
liter (L) = 1,000 mL = 1 dm^3	1 L = 1.057 fl qt	1 fl qt = 0.946 L
milliliter (mL) = 0.001 L = 1 cm^3	1 mL = 0.034 fl oz	1 fl oz = 29.574 mL
microliter (μL) = 0.000 001 L		
Mass		
kilogram (kg) = 1,000 g	1 kg = 2.205 lb	1 lb = 0.454 kg
gram (g) = 1,000 mg	1 g = 0.035 oz	1 oz = 28.350 g
milligram (mg) = 0.001 g		
microgram (μg) = 0.000 001 g		

Temperature Scales

Temperature can be expressed by using three different scales: Fahrenheit, Celsius, and Kelvin. The SI unit for temperature is the kelvin (K).

Although 0 K is much colder than 0°C, a change of 1 K is equal to a change of 1°C.

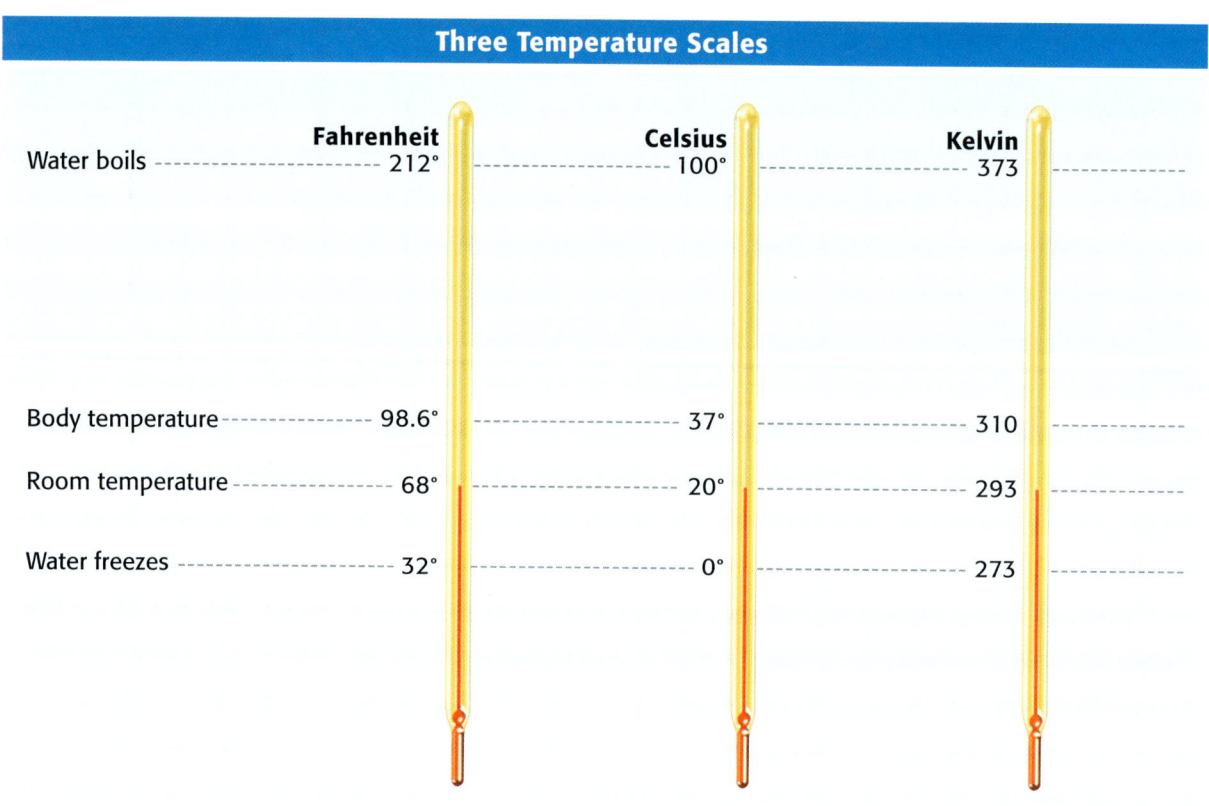

Three Temperature Scales

	Fahrenheit	Celsius	Kelvin
Water boils	212°	100°	373
Body temperature	98.6°	37°	310
Room temperature	68°	20°	293
Water freezes	32°	0°	273

Temperature Conversions Table		
To convert	**Use this equation:**	**Example**
Celsius to Fahrenheit °C → °F	$°F = \left(\dfrac{9}{5} \times °C \right) + 32$	Convert 45°C to °F. $°F = \left(\dfrac{9}{5} \times 45°C \right) + 32 = 113°F$
Fahrenheit to Celsius °F → °C	$°C = \dfrac{5}{9} \times (°F - 32)$	Convert 68°F to °C. $°C = \dfrac{5}{9} \times (68°F - 32) = 20°C$
Celsius to Kelvin °C → K	$K = °C + 273$	Convert 45°C to K. $K = 45°C + 273 = 318 \ K$
Kelvin to Celsius K → °C	$°C = K - 273$	Convert 32 K to °C. $°C = 32K - 273 = -241°C$

Measuring Skills

Using a Graduated Cylinder

When using a graduated cylinder to measure volume, keep the following procedures in mind:

1. Place the cylinder on a flat, level surface before measuring liquid.

2. Move your head so that your eye is level with the surface of the liquid.

3. Read the mark closest to the liquid level. On glass graduated cylinders, read the mark closest to the center of the curve in the liquid's surface.

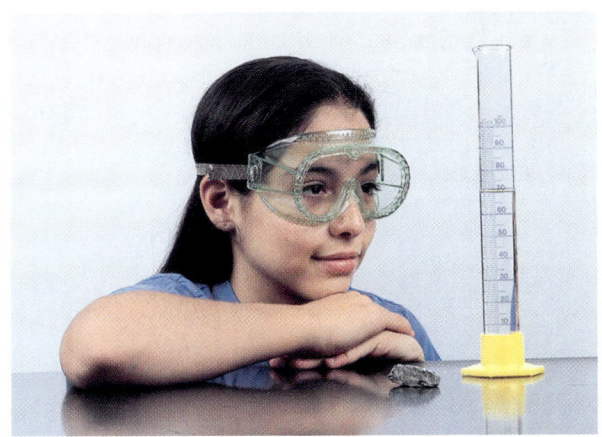

Using a Meterstick or Metric Ruler

When using a meterstick or metric ruler to measure length, keep the following procedures in mind:

1. Place the ruler firmly against the object that you are measuring.

2. Align one edge of the object exactly with the 0 end of the ruler.

3. Look at the other edge of the object to see which of the marks on the ruler is closest to that edge. (Note: Each small slash between the centimeters represents a millimeter, which is one-tenth of a centimeter.)

Using a Triple-Beam Balance

When using a triple-beam balance to measure mass, keep the following procedures in mind:

1. Make sure the balance is on a level surface.

2. Place all of the countermasses at 0. Adjust the balancing knob until the pointer rests at 0.

3. Place the object you wish to measure on the pan. **Caution:** Do not place hot objects or chemicals directly on the balance pan.

4. Move the largest countermass along the beam to the right until it is at the last notch that does not tip the balance. Follow the same procedure with the next-largest countermass. Then, move the smallest countermass until the pointer rests at 0.

5. Add the readings from the three beams together to determine the mass of the object.

6. When determining the mass of crystals or powders, first find the mass of a piece of filter paper. Then, add the crystals or powder to the paper, and remeasure. The actual mass of the crystals or powder is the total mass minus the mass of the paper. When finding the mass of liquids, first find the mass of the empty container. Then, find the combined mass of the liquid and container. The mass of the liquid is the total mass minus the mass of the container.

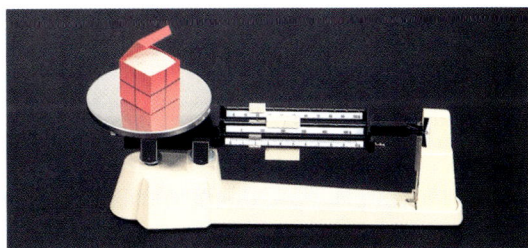

Scientific Methods

The ways in which scientists answer questions and solve problems are called **scientific methods.** The same steps are often used by scientists as they look for answers. However, there is more than one way to use these steps. Scientists may use all of the steps or just some of the steps during an investigation. They may even repeat some of the steps. The goal of using scientific methods is to come up with reliable answers and solutions.

Six Steps of Scientific Methods

 **Ask a Question** Good questions come from careful **observations.** You make observations by using your senses to gather information. Sometimes, you may use instruments, such as microscopes and telescopes, to extend the range of your senses. As you observe the natural world, you will discover that you have many more questions than answers. These questions drive investigations.

Questions beginning with *what, why, how,* and *when* are important in focusing an investigation. Here is an example of a question that could lead to an investigation.

Question: How does acid rain affect plant growth?

 **Form a Hypothesis** After you ask a question, you need to form a **hypothesis.** A hypothesis is a clear statement of what you expect the answer to your question to be. Your hypothesis will represent your best "educated guess" based on what you have observed and what you already know. A good hypothesis is testable. Otherwise, the investigation can go no further. Here is a hypothesis based on the question, "How does acid rain affect plant growth?"

Hypothesis: Acid rain slows plant growth.

The hypothesis can lead to predictions. A prediction is what you think the outcome of your experiment or data collection will be. Predictions are usually stated in an if-then format. Here is a sample prediction for the hypothesis that acid rain slows plant growth.

Prediction: If a plant is watered with only acid rain (which has a pH of 4), then the plant will grow at half its normal rate.

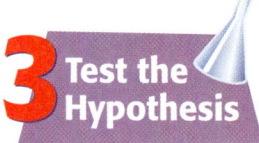 **Test the Hypothesis** After you have formed a hypothesis and made a prediction, your hypothesis should be tested. One way to test a hypothesis is with a controlled experiment. A **controlled experiment** tests only one factor at a time. In an experiment to test the effect of acid rain on plant growth, the **control group** would be watered with normal rain water. The **experimental group** would be watered with acid rain. All of the plants should receive the same amount of sunlight and water each day. The air temperature should be the same for all groups. However, the acidity of the water will be a variable. In fact, any factor that is different from one group to another is a **variable.** If your hypothesis is correct, then the acidity of the water and plant growth are *dependant variables.* The amount a plant grows is dependent on the acidity of the water. However, the amount of water each plant receives and the amount of sunlight each plant receives are *independent variables.* Either of these factors could change without affecting the other factor.

Sometimes, the nature of an investigation makes a controlled experiment impossible. For example, the Earth's core is surrounded by thousands of meters of rock. Under such circumstances, a hypothesis may be tested by making detailed observations.

 **Analyze the Results** After you have completed your experiments, made your observations, and collected your data, you must analyze all the information you have gathered. Tables and graphs are often used in this step to organize the data.

5 Draw Conclusions

After analyzing your data, you can determine if your results support your hypothesis. If your hypothesis is supported, you (or others) might want to repeat the observations or experiments to verify your results. If your hypothesis is not supported by the data, you may have to check your procedure for errors. You may even have to reject your hypothesis and make a new one. If you cannot draw a conclusion from your results, you may have to try the investigation again or carry out further observations or experiments.

6 Communicate Results

After any scientific investigation, you should report your results. By preparing a written or oral report, you let others know what you have learned. They may repeat your investigation to see if they get the same results. Your report may even lead to another question and then to another investigation.

Scientific Methods in Action

Scientific methods contain loops in which several steps may be repeated over and over again. In some cases, certain steps are unnecessary. Thus, there is not a "straight line" of steps. For example, sometimes scientists find that testing one hypothesis raises new questions and new hypotheses to be tested. And sometimes, testing the hypothesis leads directly to a conclusion. Furthermore, the steps in scientific methods are not always used in the same order. Follow the steps in the diagram, and see how many different directions scientific methods can take you.

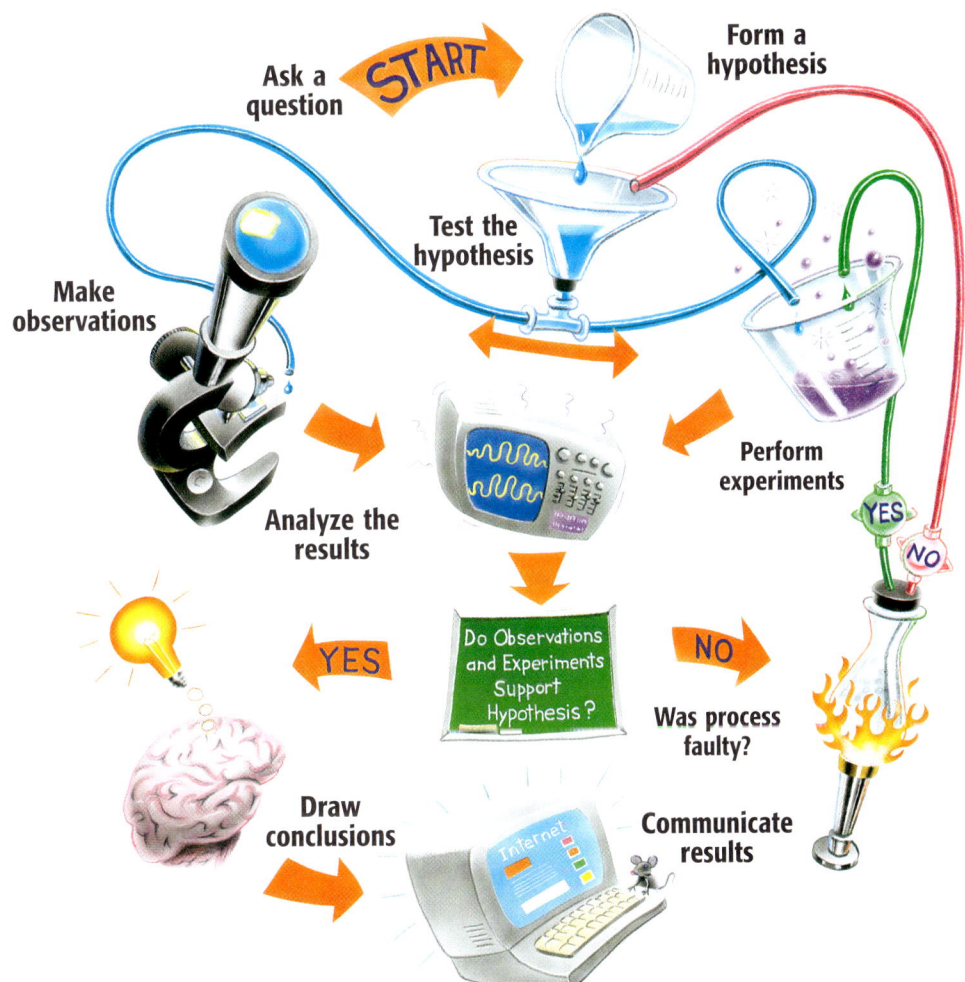

Periodic Table of the Elements

Each square on the table includes an element's name, chemical symbol, atomic number, and atomic mass.

The color of the chemical symbol indicates the physical state at room temperature. Carbon is a solid.

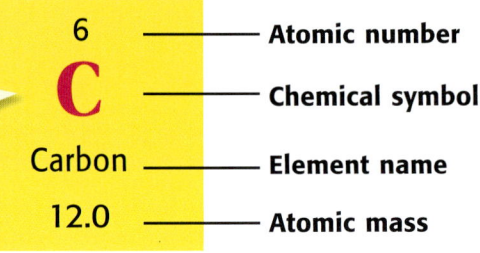

6	Atomic number
C	Chemical symbol
Carbon	Element name
12.0	Atomic mass

The background color indicates the type of element. Carbon is a nonmetal.

Background
- Metals
- Metalloids
- Nonmetals

Chemical symbol
- Solid
- Liquid
- Gas

Period 1

1
H
Hydrogen
1.0

	Group 1	Group 2
Period 2	3 **Li** Lithium 6.9	4 **Be** Beryllium 9.0
Period 3	11 **Na** Sodium 23.0	12 **Mg** Magnesium 24.3

	Group 3	Group 4	Group 5	Group 6	Group 7	Group 8	Group 9
Period 4	21 **Sc** Scandium 45.0	22 **Ti** Titanium 47.9	23 **V** Vanadium 50.9	24 **Cr** Chromium 52.0	25 **Mn** Manganese 54.9	26 **Fe** Iron 55.8	27 **Co** Cobalt 58.9
Period 5	39 **Y** Yttrium 88.9	40 **Zr** Zirconium 91.2	41 **Nb** Niobium 92.9	42 **Mo** Molybdenum 95.9	43 **Tc** Technetium (98)	44 **Ru** Ruthenium 101.1	45 **Rh** Rhodium 102.9
Period 6	57 **La** Lanthanum 138.9	72 **Hf** Hafnium 178.5	73 **Ta** Tantalum 180.9	74 **W** Tungsten 183.8	75 **Re** Rhenium 186.2	76 **Os** Osmium 190.2	77 **Ir** Iridium 192.2
Period 7	89 **Ac** Actinium (227)	104 **Rf** Rutherfordium (261)	105 **Db** Dubnium (262)	106 **Sg** Seaborgium (263)	107 **Bh** Bohrium (264)	108 **Hs** Hassium (265)†	109 **Mt** Meitnerium (268)†

Period 4 Group 1: 19 **K** Potassium 39.1; Group 2: 20 **Ca** Calcium 40.1
Period 5 Group 1: 37 **Rb** Rubidium 85.5; Group 2: 38 **Sr** Strontium 87.6
Period 6 Group 1: 55 **Cs** Cesium 132.9; Group 2: 56 **Ba** Barium 137.3
Period 7 Group 1: 87 **Fr** Francium (223); Group 2: 88 **Ra** Radium (226)

† Estimated from currently available IUPAC data.

A row of elements is called a *period*.

A column of elements is called a *group* or *family*.

Values in parentheses are of the most stable isotope of the element.

These elements are placed below the table to allow the table to be narrower.

Lanthanides

58 **Ce** Cerium 140.1	59 **Pr** Praseodymium 140.9	60 **Nd** Neodymium 144.2	61 **Pm** Promethium (145)	62 **Sm** Samarium 150.4

Actinides

90 **Th** Thorium 232.0	91 **Pa** Protactinium 231.0	92 **U** Uranium 238.0	93 **Np** Neptunium (237)	94 **Pu** Plutonium (244)

Appendix

Topic: **Periodic Table**
Go To: **go.hrw.com**
Keyword: **HN0 PERIODIC**
Visit the HRW Web site for
updates on the periodic table.

This zigzag line reminds you where the metals, nonmetals, and metalloids are.

Group 18

2 **He** Helium 4.0

Group 13	Group 14	Group 15	Group 16	Group 17
5 **B** Boron 10.8	6 **C** Carbon 12.0	7 **N** Nitrogen 14.0	8 **O** Oxygen 16.0	9 **F** Fluorine 19.0	10 **Ne** Neon 20.2
13 **Al** Aluminum 27.0	14 **Si** Silicon 28.1	15 **P** Phosphorus 31.0	16 **S** Sulfur 32.1	17 **Cl** Chlorine 35.5	18 **Ar** Argon 39.9

Group 10	Group 11	Group 12					
28 **Ni** Nickel 58.7	29 **Cu** Copper 63.5	30 **Zn** Zinc 65.4	31 **Ga** Gallium 69.7	32 **Ge** Germanium 72.6	33 **As** Arsenic 74.9	34 **Se** Selenium 79.0	35 **Br** Bromine 79.9	36 **Kr** Krypton 83.8
46 **Pd** Palladium 106.4	47 **Ag** Silver 107.9	48 **Cd** Cadmium 112.4	49 **In** Indium 114.8	50 **Sn** Tin 118.7	51 **Sb** Antimony 121.8	52 **Te** Tellurium 127.6	53 **I** Iodine 126.9	54 **Xe** Xenon 131.3
78 **Pt** Platinum 195.1	79 **Au** Gold 197.0	80 **Hg** Mercury 200.6	81 **Tl** Thallium 204.4	82 **Pb** Lead 207.2	83 **Bi** Bismuth 209.0	84 **Po** Polonium (209)	85 **At** Astatine (210)	86 **Rn** Radon (222)
110 **Dm** Darmstadtium (269)†	111 **Uuu** Unununium (272)†	112 **Uub** Ununbium (277)†		114 **Uuq** Ununquadium (285)†				

The names and three-letter symbols of elements are temporary. They are based on the atomic numbers of the elements. Official names and symbols will be approved by an international committee of scientists.

63 **Eu** Europium 152.0	64 **Gd** Gadolinium 157.2	65 **Tb** Terbium 158.9	66 **Dy** Dysprosium 162.5	67 **Ho** Holmium 164.9	68 **Er** Erbium 167.3	69 **Tm** Thulium 168.9	70 **Yb** Ytterbium 173.0	71 **Lu** Lutetium 175.0
95 **Am** Americium (243)	96 **Cm** Curium (247)	97 **Bk** Berkelium (247)	98 **Cf** Californium (251)	99 **Es** Einsteinium (252)	100 **Fm** Fermium (257)	101 **Md** Mendelevium (258)	102 **No** Nobelium (259)	103 **Lr** Lawrencium (262)

Appendix

Making Charts and Graphs

Pie Charts

A pie chart shows how each group of data relates to all of the data. Each part of the circle forming the chart represents a category of the data. The entire circle represents all of the data. For example, a biologist studying a hardwood forest in Wisconsin found that there were five different types of trees. The data table at right summarizes the biologist's findings.

Wisconsin Hardwood Trees	
Type of tree	Number found
Oak	600
Maple	750
Beech	300
Birch	1,200
Hickory	150
Total	3,000

How to Make a Pie Chart

1 To make a pie chart of these data, first find the percentage of each type of tree. Divide the number of trees of each type by the total number of trees, and multiply by 100.

$$\frac{600 \text{ oak}}{3,000 \text{ trees}} \times 100 = 20\%$$

$$\frac{750 \text{ maple}}{3,000 \text{ trees}} \times 100 = 25\%$$

$$\frac{300 \text{ beech}}{3,000 \text{ trees}} \times 100 = 10\%$$

$$\frac{1,200 \text{ birch}}{3,000 \text{ trees}} \times 100 = 40\%$$

$$\frac{150 \text{ hickory}}{3,000 \text{ trees}} \times 100 = 5\%$$

2 Now, determine the size of the wedges that make up the pie chart. Multiply each percentage by 360°. Remember that a circle contains 360°.

$20\% \times 360° = 72°$ $25\% \times 360° = 90°$

$10\% \times 360° = 36°$ $40\% \times 360° = 144°$

$5\% \times 360° = 18°$

3 Check that the sum of the percentages is 100 and the sum of the degrees is 360.

$20\% + 25\% + 10\% + 40\% + 5\% = 100\%$

$72° + 90° + 36° + 144° + 18° = 360°$

4 Use a compass to draw a circle and mark the center of the circle.

5 Then, use a protractor to draw angles of 72°, 90°, 36°, 144°, and 18° in the circle.

6 Finally, label each part of the chart, and choose an appropriate title.

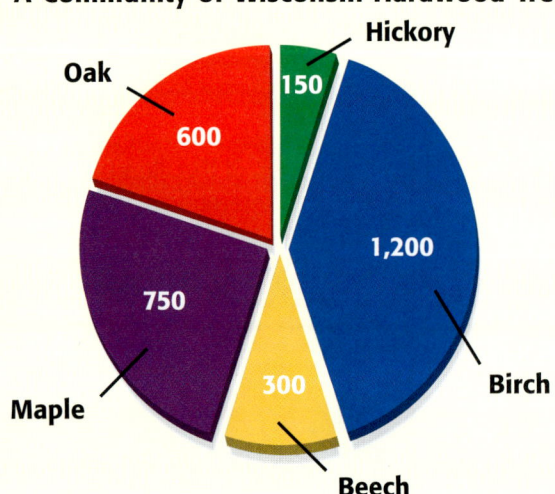

A Community of Wisconsin Hardwood Trees

Line Graphs

Line graphs are most often used to demonstrate continuous change. For example, Mr. Smith's students analyzed the population records for their hometown, Appleton, between 1900 and 2000. Examine the data at right.

Because the year and the population change, they are the *variables*. The population is determined by, or dependent on, the year. Therefore, the population is called the **dependent variable,** and the year is called the **independent variable.** Each set of data is called a **data pair.** To prepare a line graph, you must first organize data pairs into a table like the one at right.

Population of Appleton, 1900–2000	
Year	**Population**
1900	1,800
1920	2,500
1940	3,200
1960	3,900
1980	4,600
2000	5,300

How to Make a Line Graph

1 Place the independent variable along the horizontal (*x*) axis. Place the dependent variable along the vertical (*y*) axis.

2 Label the *x*-axis "Year" and the *y*-axis "Population." Look at your largest and smallest values for the population. For the *y*-axis, determine a scale that will provide enough space to show these values. You must use the same scale for the entire length of the axis. Next, find an appropriate scale for the *x*-axis.

3 Choose reasonable starting points for each axis.

4 Plot the data pairs as accurately as possible.

5 Choose a title that accurately represents the data.

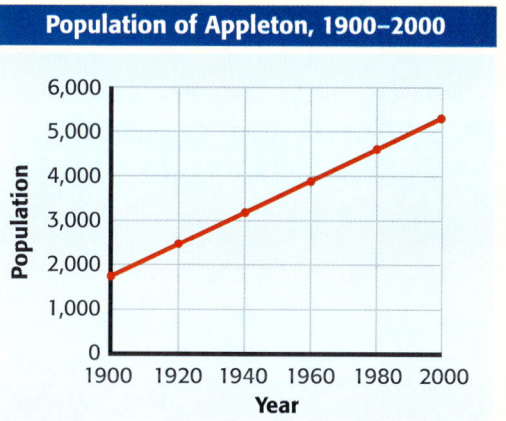

How to Determine Slope

Slope is the ratio of the change in the *y*-value to the change in the *x*-value, or "rise over run."

1 Choose two points on the line graph. For example, the population of Appleton in 2000 was 5,300 people. Therefore, you can define point *a* as (2000, 5,300). In 1900, the population was 1,800 people. You can define point *b* as (1900, 1,800).

2 Find the change in the *y*-value.
(*y* at point *a*) − (*y* at point *b*) =
5,300 people − 1,800 people =
3,500 people

3 Find the change in the *x*-value.
(*x* at point *a*) − (*x* at point *b*) =
2000 − 1900 = 100 years

4 Calculate the slope of the graph by dividing the change in *y* by the change in *x*.

$$slope = \frac{change\ in\ y}{change\ in\ x}$$

$$slope = \frac{3,500\ people}{100\ years}$$

$$slope = 35\ people\ per\ year$$

In this example, the population in Appleton increased by a fixed amount each year. The graph of these data is a straight line. Therefore, the relationship is **linear.** When the graph of a set of data is not a straight line, the relationship is **nonlinear.**

Using Algebra to Determine Slope

The equation in step 4 may also be arranged to be

$$y = kx$$

where y represents the change in the y-value, k represents the slope, and x represents the change in the x-value.

$$slope = \frac{change\ in\ y}{change\ in\ x}$$

$$k = \frac{y}{x}$$

$$k \times x = \frac{y \times x}{x}$$

$$kx = y$$

Bar Graphs

Bar graphs are used to demonstrate change that is not continuous. These graphs can be used to indicate trends when the data cover a long period of time. A meteorologist gathered the precipitation data shown here for Hartford, Connecticut, for April 1–15, 1996, and used a bar graph to represent the data.

Precipitation in Hartford, Connecticut April 1–15, 1996			
Date	Precipitation (cm)	Date	Precipitation (cm)
April 1	0.5	April 9	0.25
April 2	1.25	April 10	0.0
April 3	0.0	April 11	1.0
April 4	0.0	April 12	0.0
April 5	0.0	April 13	0.25
April 6	0.0	April 14	0.0
April 7	0.0	April 15	6.50
April 8	1.75		

How to Make a Bar Graph

1 Use an appropriate scale and a reasonable starting point for each axis.

2 Label the axes, and plot the data.

3 Choose a title that accurately represents the data.

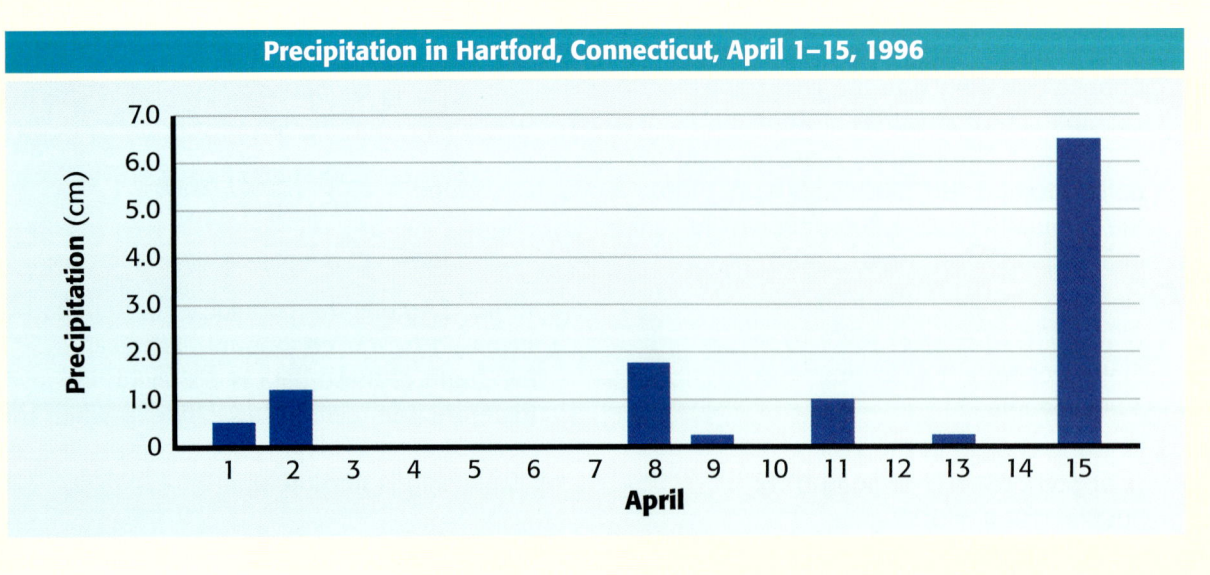

Precipitation in Hartford, Connecticut, April 1–15, 1996

Math Refresher

Science requires an understanding of many math concepts. The following pages will help you review some important math skills.

Averages

An **average,** or **mean,** simplifies a set of numbers into a single number that *approximates* the value of the set.

> **Example:** Find the average of the following set of numbers: 5, 4, 7, and 8.

Step 1: Find the sum.

$$5 + 4 + 7 + 8 = 24$$

Step 2: Divide the sum by the number of numbers in your set. Because there are four numbers in this example, divide the sum by 4.

$$\frac{24}{4} = 6$$

The average, or mean, is **6.**

Ratios

A **ratio** is a comparison between numbers, and it is usually written as a fraction.

> **Example:** Find the ratio of thermometers to students if you have 36 thermometers and 48 students in your class.

Step 1: Make the ratio.

$$\frac{36 \text{ thermometers}}{48 \text{ students}}$$

Step 2: Reduce the fraction to its simplest form.

$$\frac{36}{48} = \frac{36 \div 12}{48 \div 12} = \frac{3}{4}$$

The ratio of thermometers to students is **3 to 4,** or $\frac{3}{4}$. The ratio may also be written in the form 3:4.

Proportions

A **proportion** is an equation that states that two ratios are equal.

$$\frac{3}{1} = \frac{12}{4}$$

To solve a proportion, first multiply across the equal sign. This is called *cross-multiplication*. If you know three of the quantities in a proportion, you can use cross-multiplication to find the fourth.

> **Example:** Imagine that you are making a scale model of the solar system for your science project. The diameter of Jupiter is 11.2 times the diameter of the Earth. If you are using a plastic-foam ball that has a diameter of 2 cm to represent the Earth, what must the diameter of the ball representing Jupiter be?

$$\frac{11.2}{1} = \frac{x}{2 \text{ cm}}$$

Step 1: Cross-multiply.

$$\frac{11.2}{1} \diagup\!\!\!\!\diagdown \frac{x}{2}$$

$$11.2 \times 2 = x \times 1$$

Step 2: Multiply.

$$22.4 = x \times 1$$

Step 3: Isolate the variable by dividing both sides by 1.

$$x = \frac{22.4}{1}$$

$$x = 22.4 \text{ cm}$$

You will need to use a ball that has a diameter of **22.4** cm to represent Jupiter.

Percentages

A **percentage** is a ratio of a given number to 100.

Example: What is 85% of 40?

Step 1: Rewrite the percentage by moving the decimal point two places to the left.

0.85

Step 2: Multiply the decimal by the number that you are calculating the percentage of.

0.85 × 40 = 34

85% of 40 is **34.**

Decimals

To **add** or **subtract decimals,** line up the digits vertically so that the decimal points line up. Then, add or subtract the columns from right to left. Carry or borrow numbers as necessary.

Example: Add the following numbers: 3.1415 and 2.96.

Step 1: Line up the digits vertically so that the decimal points line up.

```
  3.1415
+ 2.96
```

Step 2: Add the columns from right to left, and carry when necessary.

```
  1 1
  3.1415
+ 2.96
-------
  6.1015
```

The sum is **6.1015.**

Fractions

Numbers tell you how many; **fractions** tell you *how much of a whole.*

Example: Your class has 24 plants. Your teacher instructs you to put 5 plants in a shady spot. What fraction of the plants in your class will you put in a shady spot?

Step 1: In the denominator, write the total number of parts in the whole.

$$\frac{?}{24}$$

Step 2: In the numerator, write the number of parts of the whole that are being considered.

$$\frac{5}{24}$$

So, $\frac{5}{24}$ of the plants will be in the shade.

Reducing Fractions

It is usually best to express a fraction in its simplest form. Expressing a fraction in its simplest form is called *reducing* a fraction.

Example: Reduce the fraction $\frac{30}{45}$ to its simplest form.

Step 1: Find the largest whole number that will divide evenly into both the numerator and denominator. This number is called the *greatest common factor* (GCF).

Factors of the numerator 30:

1, 2, 3, 5, 6, 10, **15,** 30

Factors of the denominator 45:

1, 3, 5, 9, **15,** 45

Step 2: Divide both the numerator and the denominator by the GCF, which in this case is 15.

$$\frac{30}{45} = \frac{30 \div 15}{45 \div 15} = \frac{2}{3}$$

Thus, $\frac{30}{45}$ reduced to its simplest form is $\frac{2}{3}$.

Adding and Subtracting Fractions

To **add** or **subtract fractions** that have the **same denominator,** simply add or subtract the numerators.

Examples:

$$\frac{3}{5} + \frac{1}{5} = ? \text{ and } \frac{3}{4} - \frac{1}{4} = ?$$

Step 1: Add or subtract the numerators.

$$\frac{3}{5} + \frac{1}{5} = \frac{4}{} \text{ and } \frac{3}{4} - \frac{1}{4} = \frac{2}{}$$

Step 2: Write the sum or difference over the denominator.

$$\frac{3}{5} + \frac{1}{5} = \frac{4}{5} \text{ and } \frac{3}{4} - \frac{1}{4} = \frac{2}{4}$$

Step 3: If necessary, reduce the fraction to its simplest form.

$\frac{4}{5}$ cannot be reduced, and $\frac{2}{4} = \frac{1}{2}$.

To **add** or **subtract fractions** that have **different denominators,** first find the least common denominator (LCD).

Examples:

$$\frac{1}{2} + \frac{1}{6} = ? \text{ and } \frac{3}{4} - \frac{2}{3} = ?$$

Step 1: Write the equivalent fractions that have a common denominator.

$$\frac{3}{6} + \frac{1}{6} = ? \text{ and } \frac{9}{12} - \frac{8}{12} = ?$$

Step 2: Add or subtract the fractions.

$$\frac{3}{6} + \frac{1}{6} = \frac{4}{6} \text{ and } \frac{9}{12} - \frac{8}{12} = \frac{1}{12}$$

Step 3: If necessary, reduce the fraction to its simplest form.

The fraction $\frac{4}{6} = \frac{2}{3}$, and $\frac{1}{12}$ cannot be reduced.

Multiplying Fractions

To **multiply fractions,** multiply the numerators and the denominators together, and then reduce the fraction to its simplest form.

Example:

$$\frac{5}{9} \times \frac{7}{10} = ?$$

Step 1: Multiply the numerators and denominators.

$$\frac{5}{9} \times \frac{7}{10} = \frac{5 \times 7}{9 \times 10} = \frac{35}{90}$$

Step 2: Reduce the fraction.

$$\frac{35}{90} = \frac{35 \div 5}{90 \div 5} = \frac{7}{18}$$

Dividing Fractions

To **divide fractions,** first rewrite the divisor (the number you divide by) upside down. This number is called the *reciprocal* of the divisor. Then multiply and reduce if necessary.

Example:

$$\frac{5}{8} \div \frac{3}{2} = ?$$

Step 1: Rewrite the divisor as its reciprocal.

$$\frac{3}{2} \rightarrow \frac{2}{3}$$

Step 2: Multiply the fractions.

$$\frac{5}{8} \times \frac{2}{3} = \frac{5 \times 2}{8 \times 3} = \frac{10}{24}$$

Step 3: Reduce the fraction.

$$\frac{10}{24} = \frac{10 \div 2}{24 \div 2} = \frac{5}{12}$$

Appendix

Scientific Notation

Scientific notation is a short way of representing very large and very small numbers without writing all of the place-holding zeros.

> **Example:** Write 653,000,000 in scientific notation.

Step 1: Write the number without the place-holding zeros.

653

Step 2: Place the decimal point after the first digit.

6.53

Step 3: Find the exponent by counting the number of places that you moved the decimal point.

6.53000000

The decimal point was moved eight places to the left. Therefore, the exponent of 10 is positive 8. If you had moved the decimal point to the right, the exponent would be negative.

Step 4: Write the number in scientific notation.

$$\mathbf{6.53 \times 10^8}$$

Area

Area is the number of square units needed to cover the surface of an object.

Formulas:

area of a square = side × side
area of a rectangle = length × width
area of a triangle = $\frac{1}{2}$ × base × height

Examples: Find the areas.

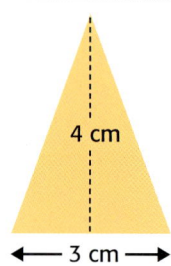

Triangle

$area = \frac{1}{2} \times base \times height$

$area = \frac{1}{2} \times 3\ cm \times 4\ cm$

$area = \mathbf{6\ cm^2}$

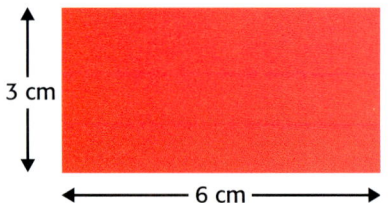

Rectangle

$area = length \times width$

$area = 6\ cm \times 3\ cm$

$area = \mathbf{18\ cm^2}$

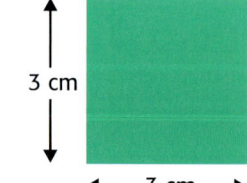

Square

$area = side \times side$

$area = 3\ cm \times 3\ cm$

$area = \mathbf{9\ cm^2}$

Volume

Volume is the amount of space that something occupies.

Formulas:

volume of a cube =
side × side × side

volume of a prism =
area of base × height

Examples:

Find the volume of the solids.

Cube

$volume = side \times side \times side$

$volume = 4\ cm \times 4\ cm \times 4\ cm$

$volume = \mathbf{64\ cm^3}$

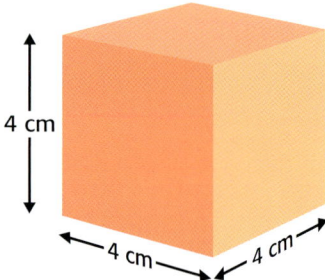

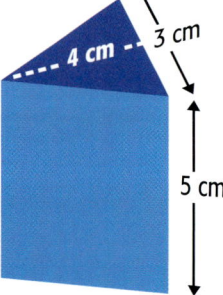

Prism

$volume = area\ of\ base \times height$

$volume = (area\ of\ triangle) \times height$

$volume = (\frac{1}{2} \times 3\ cm \times 4\ cm) \times 5\ cm$

$volume = 6\ cm^2 \times 5\ cm$

$volume = \mathbf{30\ cm^3}$

Appendix

Physical Science Laws and Principles

Law of Conservation of Energy

The law of conservation of energy states that energy can be neither created nor destroyed.

The total amount of energy in a closed system is always the same. Energy can be changed from one form to another, but all of the different forms of energy in a system always add up to the same total amount of energy no matter how many energy conversions occur.

Law of Universal Gravitation

The law of universal gravitation states that all objects in the universe attract each other by a force called *gravity*. The size of the force depends on the masses of the objects and the distance between objects.

The first part of the law explains why a bowling ball is much harder to lift than a table-tennis ball. Because the bowling ball has a much larger mass than the table-tennis ball does, the amount of gravity between the Earth and the bowling ball is greater than the amount of gravity between the Earth and the table-tennis ball.

The second part of the law explains why a satellite can remain in orbit around the Earth. The satellite is carefully placed at a distance great enough to prevent the Earth's gravity from immediately pulling the satellite down but small enough to prevent the satellite from completely escaping the Earth's gravity and wandering off into space.

Newton's Laws of Motion

Newton's first law of motion states that an object at rest remains at rest and an object in motion remains in motion at constant speed and in a straight line unless acted on by an unbalanced force.

The first part of the law explains why a football will remain on a tee until it is kicked off or until a gust of wind blows it off.

The second part of the law explains why a bike rider will continue moving forward after the bike comes to an abrupt stop. Gravity and the friction of the sidewalk will eventually stop the rider.

Newton's second law of motion states that the acceleration of an object depends on the mass of the object and the amount of force applied.

The first part of the law explains why the acceleration of a 4 kg bowling ball will be greater than the acceleration of a 6 kg bowling ball if the same force is applied to both.

The second part of the law explains why the acceleration of a bowling ball will be larger if a larger force is applied to the bowling ball.

The relationship of acceleration (a) to mass (m) and force (F) can be expressed mathematically by the following equation:

$$acceleration = \frac{force}{mass}, \text{ or } a = \frac{F}{m}$$

This equation is often rearranged to the form

$$force = mass \times acceleration$$
$$or$$
$$F = m \times a$$

Newton's third law of motion states that whenever one object exerts a force on a second object, the second object exerts an equal and opposite force on the first.

This law explains that a runner is able to move forward because of the equal and opposite force that the ground exerts on the runner's foot after each step.

Law of Reflection

The **law of reflection** states that the angle of incidence is equal to the angle of reflection. This law explains why light reflects off a surface at the same angle that the light strikes the surface.

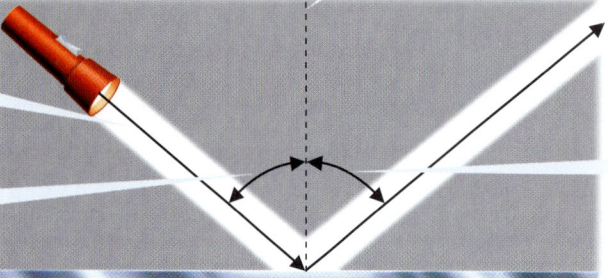

A line perpendicular to the mirror's surface is called the *normal*.

The beam of light reflected off the mirror is called the *reflected beam*.

The beam of light traveling toward the mirror is called the *incident beam*.

The angle between the incident beam and the normal is called the *angle of incidence*.

The angle between the reflected beam and the normal is called the *angle of reflection*.

Charles's Law

Charles's law states that for a fixed amount of gas at a constant pressure, the volume of the gas increases as the temperature of the gas increases. Likewise, the volume of the gas decreases as the temperature of the gas decreases.

If a basketball that was inflated indoors is left outside on a cold winter day, the air particles inside the ball will move more slowly. They will hit the sides of the basketball less often and with less force. The ball will get smaller as the volume of the air decreases.

Boyle's Law

Boyle's law states that for a fixed amount of gas at a constant temperature, the volume of a gas increases as the pressure of the gas decreases. Likewise, the volume of a gas decreases as its pressure increases.

If an inflated balloon is pulled down to the bottom of a swimming pool, the pressure of the water on the balloon increases. The pressure of the air particles inside the balloon must increase to match that of the water outside, so the volume of the air inside the balloon decreases.

Pascal's Principle

Pascal's principle states that a change in pressure at any point in an enclosed fluid will be transmitted equally to all parts of that fluid.

When a mechanic uses a hydraulic jack to raise an automobile off the ground, he or she increases the pressure on the fluid in the jack by pushing on the jack handle. The pressure is transmitted equally to all parts of the fluid-filled jacking system. As fluid presses the jack plate against the frame of the car, the car is lifed off the ground.

Archimedes' Principle

Archimedes' principle states that the buoyant force on an object in a fluid is equal to the weight of the volume of fluid that the object displaces.

A person floating in a swimming pool displaces 20 L of water. The weight of that volume of water is about 200 N. Therefore, the buoyant force on the person is 200 N.

Bernoulli's Principle

Bernoulli's principle states that as the speed of a moving fluid increases, the fluid's pressure decreases.

The lift on an airplane wing or on a Frisbee® can be explained in part by using Bernoulli's principle. Because of the shape of the Frisbee, the air moving over the top of the Frisbee must travel farther than the air below the Frisbee in the same amount of time. In other words, the air above the Frisbee is moving faster than the air below it. This faster-moving air above the Frisbee exerts less pressure than the slower-moving air below it does. The resulting increased pressure below exerts an upward force and pushes the Frisbee up.

Useful Equations

Average speed

$$average\ speed = \frac{total\ distance}{total\ time}$$

Example: A bicycle messenger traveled a distance of 136 km in 8 h. What was the messenger's average speed?

$$\frac{136\ km}{8\ h} = 17\ km/h$$

The messenger's average speed was **17 km/h.**

Average acceleration

$$\frac{average}{acceleration} = \frac{final\ velocity - starting\ velocity}{time\ it\ takes\ to\ change\ velocity}$$

Example: Calculate the average acceleration of an Olympic 100 m dash sprinter who reaches a velocity of 20 m/s south at the finish line. The race was in a straight line and lasted 10 s.

$$\frac{20\ m/s - 0\ m/s}{10s} = 2\ m/s/s$$

The sprinter's average acceleration is **2 m/s/s south.**

Net force

Forces in the Same Direction
When forces are in the same direction, add the forces together to determine the net force.

Example: Calculate the net force on a stalled car that is being pushed by two people. One person is pushing with a force of 13 N northwest, and the other person is pushing with a force of 8 N in the same direction.

$$13\ N + 8\ N = 21\ N$$

The net force is **21 N northwest.**

Forces in Opposite Directions
When forces are in opposite directions, subtract the smaller force from the larger force to determine the net force. The net force will be in the direction of the larger force.

Example: Calculate the net force on a rope that is being pulled on each end. One person is pulling on one end of the rope with a force of 12 N south. Another person is pulling on the opposite end of the rope with a force of 7 N north.

$$12\ N - 7\ N = 5\ N$$

The net force is **5 N south.**

Work

Work is done by exerting a force through a distance. Work has units of joules (J), which are equivalent to Newton-meters.

$$Work = F \times d$$

Example: Calculate the amount of work done by a man who lifts a 100 N toddler 1.5 m off the floor.

$Work$ = 100 N × 1.5 m = 150 N•m = 150 J

The man did **150 J** of work.

Power

Power is the rate at which work is done. Power is measured in watts (W), which are equivalent to joules per second.

$$P = \frac{Work}{t}$$

Example: Calculate the power of a weight-lifter who raises a 300 N barbell 2.1 m off the floor in 1.25 s.

$Work$ = 300 N × 2.1 m = 630 N•m = 630 J

$$P = \frac{630 \text{ J}}{1.25 \text{ s}} = \frac{504 \text{ J}}{\text{s}} = 504 \text{ W}$$

The weightlifter has **504 W** of power.

Pressure

Pressure is the force exerted over a given area. The SI unit for pressure is the pascal (Pa).

$$pressure = \frac{force}{area}$$

Example: Calculate the pressure of the air in a soccer ball if the air exerts a force of 25,000 N over an area of 0.15 m².

$$pressure = \frac{25,000 \text{ N}}{0.15 \text{ m}^2} = \frac{167,000 \text{ N}}{\text{m}^2} = 167,000 \text{ Pa}$$

The pressure of the air inside the soccer ball is **167,000 Pa.**

Density

$$density = \frac{mass}{volume}$$

Example: Calculate the density of a sponge that has a mass of 10 g and a volume of 40 cm³.

$$\frac{10 \text{ g}}{40 \text{ cm}^3} = \frac{0.25 \text{ g}}{\text{cm}^3}$$

The density of the sponge is $\frac{0.25 \text{ g}}{\text{cm}^3}$.

Concentration

$$concentration = \frac{mass \text{ of } solute}{volume \text{ of } solvent}$$

Example: Calculate the concentration of a solution in which 10 g of sugar is dissolved in 125 mL of water.

$$\frac{10 \text{ g of sugar}}{125 \text{ mL of water}} = \frac{0.08 \text{ g}}{\text{mL}}$$

The concentration of this solution is $\frac{0.08 \text{ g}}{\text{mL}}$.

Glossary

A

acid any compound that increases the number of hydronium ions when dissolved in water (62)

activation energy the minimum amount of energy required to start a chemical reaction (44)

B

base any compound that increases the number of hydroxide ions when dissolved in water (65)

C

carbohydrate a class of energy-giving nutrients that includes sugars, starches, and fiber; contains carbon, hydrogen, and oxygen (74)

catalyst (KAT uh LIST) a substance that changes the rate of a chemical reaction without being used up or changed very much (47)

chemical bond an interaction that holds atoms or ions together (4, 58)

chemical bonding the combining of atoms to form molecules or ionic compounds (4)

chemical equation a representation of a chemical reaction that uses symbols to show the relationship between the reactants and the products (34)

chemical formula a combination of chemical symbols and numbers to represent a substance (32)

chemical reaction the process by which one or more substances change to produce one or more different substances (28)

covalent bond (koh VAY luhnt BAHND) a bond formed when atoms share one or more pairs of electrons (12)

covalent compound a chemical compound that is formed by the sharing of electrons (420)

crystal lattice (KRIS tuhl LAT is) the regular pattern in which a crystal is arranged (11)

D

decomposition reaction a reaction in which a single compound breaks down to form two or more simpler substances (39)

double-displacement reaction a reaction in which a gas, a solid precipitate, or a molecular compound forms from the exchange of ions between two compounds (41)

E

endothermic reaction a chemical reaction that requires heat (43)

exothermic reaction a chemical reaction in which heat is released to the surroundings (42)

H

half-life the time needed for half of a sample of a radioactive substance to undergo radioactive decay (93)

hydrocarbon an organic compound composed only of carbon and hydrogen (73)

I

indicator a compound that can reversibly change color depending on conditions such as pH (63)

inhibitor a substance that slows down or stops a chemical reaction (46)

ion a charged particle that forms when an atom or group of atoms gains or loses one or more electrons (8)

ionic bond (ie AHN ik BAHND) a bond that forms when electrons are transferred from one atom to another, which results in a positive ion and a negative ion (8)

ionic compound a compound made of oppositely charged ions (58)

isotope (IE suh TOHP) an atom that has the same number of protons (or the same atomic number) as other atoms of the same element do but that has a different number of neutrons (and thus a different atomic mass) (90)

L

law of conservation of energy the law that states that energy cannot be created or destroyed but can be changed from one form to another (43)

law of conservation of mass the law that states that mass cannot be created or destroyed in ordinary chemical and physical changes (35)

lipid a type of biochemical that does not dissolve in water; fats and steroids are lipids (75)

M

mass number the sum of the numbers of protons and neutrons in the nucleus of an atom (89)

metallic bond a bond formed by the attraction between positively charged metal ions and the electrons around them (15)

molecule (MAHL i KYOOL) the smallest unit of a substance that keeps all of the physical and chemical properties of that substance (13)

N

neutralization reaction (NOO truhl i ZA shuhn ree AK shuhn) the reaction of an acid and a base to form a neutral solution of water and a salt (69)

nuclear chain reaction a continuous series of nuclear fission reactions (97)

nuclear fission (NOO klee uhr FISH uhn) the splitting of the nucleus of a large atom into two or more fragments; releases additional neutrons and energy (96)

nuclear fusion (NOO klee uhr FYOO zhuhn) the combination of the nuclei of small atoms to form a larger nucleus; releases energy (100)

nucleic acid (noo KLEE ik AS id) a molecule made up of subunits called *nucleotides* (76)

O

organic compound a covalently bonded compound that contains carbon (72)

P

pH a value that is used to express the acidity or basicity (alkalinity) of a system (69)

precipitate (pree SIP uh TAYT) a solid that is produced as a result of a chemical reaction in solution (29)

product a substance that forms in a chemical reaction (34)

protein a molecule that is made up of amino acids and that is needed to build and repair body structures and to regulate processes in the body (75)

R

radioactivity the process by which an unstable nucleus gives off nuclear radiation (88)

reactant (ree AK tuhnt) a substance or molecule that participates in a chemical reaction (34)

S

salt an ionic compound that forms when a metal atom replaces the hydrogen of an acid (71)

single-displacement reaction a reaction in which one element takes the place of another element in a compound (39)

synthesis reaction (SIN thuh sis ree AK shuhn) a reaction in which two or more substances combine to form a new compound (38)

V

valence electron (VAY luhns ee LEK TRAHN) an electron that is found in the outermost shell of an atom and that determines the atom's chemical properties (5)

Spanish Glossary

A

acid/ácido cualquier compuesto que aumenta el número de iones de hidrógeno cuando se disuelve en agua (62)

activation energy/energía de activación la cantidad mínima de energía que se requiere para iniciar una reacción química (44)

B

base/base cualquier compuesto que aumenta el número de iones de hidróxido cuando se disuelve en agua (65)

C

carbohydrate/carbohidrato una clase de nutrientes que proporcionan energía; incluye los azúcares, los almidones y las fibras; contiene carbono, hidrógeno y oxígeno (74)

catalyst/catalizador una substancia que cambia la tasa de una reacción química sin consumirse ni cambiar demasiado (47)

chemical bond/enlace químico una interacción que mantiene unidos los átomos o los iones (4, 58)

chemical bonding/formación de un enlace químico la combinación de átomos para formar moléculas o compuestos iónicos (4)

chemical equation/ecuación química una representación de una reacción química que usa símbolos para mostrar la relación entre los reactivos y los productos (34)

chemical formula/fórmula química una combinación de símbolos químicos y números que se usan para representar una substancia (32)

chemical reaction/reacción química el proceso por medio del cual una o más substancia cambian para producir una o más substancias distintas (28)

covalent bond/enlace covalente un enlace formado cuando los átomos comparten uno más pares de electrones (12)

covalent compound/compuesto covalente un compuesto químico que se forma al compartir electrones (420)

crystal lattice/red cristalina el patrón regular en el que un cristal está ordenado (11)

D

decomposition reaction/reacción de descomposición una reacción en la que un solo compuesto se descompone para formar dos o más substancias más simples (39)

double-displacement reaction/reacción de doble desplazamiento una reacción en la que se forma un gas, un precipitado sólido o un compuesto molecular a partir del intercambio de iones entre dos compuestos (41)

E

endothermic reaction/reacción endotérmica una reacción química que necesita calor (43)

exothermic reaction/reacción exotérmica una reacción química en la que se libera calor a los alrededores (42)

H

half-life/vida media el tiempo que tarda la mitad de la muestra de una substancia radiactiva en desintegrarse por desintegración radiactiva (93)

hydrocarbon/hidrocarburo un compuesto orgánico compuesto únicamente por carbono e hidrogeno (73)

I

indicator/indicador un compuesto que puede cambiar de color de forma reversible dependiendo de pH de la solución o de otro cambio químico (63)

inhibitor/inhibidor una substancia que desacelera o detiene una reacción química (46)

ion/ion una partícula cargada que se forma cuando un átomo o grupo de átomos gana o pierde uno o más electrones (8)

ionic bond/enlace iónico un enlace que se forma cuando los electrones se transfieren de un átomo a otro, y que produce un ion positivo y uno negativo (8)

ionic compound/compuesto iónico un compuesto formado por iones con cargas opuestas (58)

isotope/isótopo un átomo que tiene el mismo número de protones (o el mismo número atómico) que otros átomos del mismo elemento, pero que tiene un número diferente de neutrones (y, por lo tanto, otra masa atómica) (90)

L

law of conservation of energy/ley de la conservación de la energía la ley que establece que la energía ni se crea ni se destruye, sólo se transforma de una forma a otra (43)

law of conservation of mass/ley de la conservación de la masa la ley que establece que la masa no se crea ni se destruye por cambios químicos o físicos comunes (35)

lipid/lípido un tipo de substancia bioquímica que no se disuelve en agua; las grasas y los esteroides son lípidos (75)

M

mass number/número de masa la suma de los números de protones y neutrones que hay en el núcleo de un átomo (89)

metallic bond/enlace metálico un enlace formado por la atracción entre iones metálicos cargados positivamente y los electrones que los rodean (15)

molecule/molécula la unidad más pequeña de una substancia que conserva todas las propiedades físicas y químicas de esa substancia (13)

N

neutralization reaction/reacción de neutralización la reacción de un ácido y una base que forma una solución neutra de agua y una sal (69)

nuclear chain reaction/reacción nuclear en cadena una serie continua de reacciones nucleares de fisión (97)

nuclear fission/fisión nuclear la partición del núcleo de un átomo grande en dos o más fragmentos; libera neutrones y energía adicionales (96)

nuclear fusion/fusión nuclear combinación de los núcleos de átomos pequeños para formar un núcleo más grande; libera energía (100)

nucleic acid/ácido nucleico una molécula formada por subunidades llamadas nucleótidos (76)

O

organic compound/compuesto orgánico un compuesto enlazado de manera covalente que contiene carbono (72)

P

pH/pH un valor que expresa la acidez o la basicidad (alcalinidad) de un sistema (69)

precipitate/precipitado un sólido que se produce como resultado de una reacción química en una solución (29)

product/producto una substancia que se forma en una reacción química (34)

protein/proteína una molécula formada por aminoácidos que es necesaria para construir y reparar estructuras corporales y para regular procesos del cuerpo (75)

R

radioactivity/radiactividad el proceso por medio del cual un núcleo inestable emite radiación nuclear (88)

reactant/reactivo una substancia o molécula que participa en una reacción química (34)

S

salt/sal un compuesto iónico que se forma cuando un átomo de un metal reemplaza el hidrógeno de un ácido (71)

single-displacement reaction/reacción de sustitución simple una reacción en la que un elemento toma el lugar de otro elemento en un compuesto (39)

synthesis reaction/reacción de síntesis una reacción en la que dos o más sustancias se combinan para formar un compuesto nuevo (38)

V

valence electron/electrón de valencia un electrón que se encuentra en el orbital más externo de un átomo y que determina las propiedades químicas del átomo (5)

Spanish Glossary

Index

Index

Index

Index

Credits

Abbreviations used: (t) top, (c) center, (b) bottom, (l) left, (r) right, (bkgd) background

PHOTOGRAPHY

Front Cover PhotoLink/Getty Images

Skills Practice Lab Teens Sam Dudgeon/HRW

Connection to Astrology Corbis Images; **Connection to Biology** David M. Phillips/Visuals Unlimited; **Connection to Chemistry** Digital Image copyright © 2005 PhotoDisc; **Connection to Environment** Digital Image copyright © 2005 PhotoDisc; **Connection to Geology** Letraset Phototone; **Connection to Language Arts** Digital Image copyright © 2005 PhotoDisc; **Connection to Meteorology** Digital Image copyright © 2005 PhotoDisc; **Connection to Oceanography** © ICONOTEC; **Connection to Physics** Digital Image copyright © 2005 PhotoDisc

Table of Contents iv (cl), © Konrad Wothe/Minden Pictures; iv (b), Victoria Smith/HRW; v (t), Victoria Smith/HRW; v (cl), ©Bob Thomason/Getty Images; vi–vii, Victoria Smith/HRW; x (bl), Sam Dudgeon/HRW; xi (tl), John Langford/HRW; xi (b), Sam Dudgeon/HRW; xii (tl), Victoria Smith/HRW; xii (bl), Stephanie Morris/HRW; xii (br), Sam Dudgeon/HRW; xiii (tl), Patti Murray/Animals, Animals; xiii (tr), Jana Birchum/HRW; xiii (b), Peter Van Steen/HRW

Chapter One 2–3 (all), © Doug Struthers/Getty Images; 4 (bl), © Charles Gupton/CORBIS; 8 (br), © Konrad Wothe/Minden Pictures; 11 (cl), Paul Silverman/Fundamental Photographs; 14 (tr), Sam Dudgeon/HRW; 15 (tc), Sam Dudgeon/HRW; 15 (bl), © Jonathan Blair/CORBIS; 16 (tr), Victoria Smith/HRW; 17 (tr), John Langford/HRW; 19 (b), Sam Dudgeon/HRW; 20 (br), Victoria Smith/HRW; 21 (cr, br), Sam Dudgeon/HRW; 21 (tc), © Konrad Wothe/Minden Pictures; 24 (tr), Peter Oxford/Nature Picture Library; 24 (tl), Diaphor Agency/Index Stock Imagery, Inc.; 25 (cr), Steve Fischbach/HRW; 25 (bl), W. & D. McIntyre/Photo Researchers, Inc.

Chapter Two 26–27 (all), Corbis Images; 28 (bl), Rob Matheson/The Stock Market; 28 (br), Sam Dudgeon/HRW; 29 (cl, cr), Richard Megna/Fundamental Photographs, New York; 29 (br), Scott Van Osdol/HRW; 29 (bl), J.T. Wright/Bruce Coleman Inc./Picture Quest; 30 (all), Charlie Winters; 31 (br), Charlie Winters/HRW; 34 (tl), John Langford/HRW; 34 (bl), Richard Haynes/HRW; 35 (tr), Charles D. Winters/Photo Researchers, Inc.; 35 (tc), John Langford/HRW; 35 (tl), © Ingram Publishing; 40 (tl), Peticolas/Megna/Fundamental Photographs; 40 (tr), Richard Megna/Fundamental Photographs; 42 (bl), Victoria Smith/HRW; 42 (bc), Peter Van Steen/HRW; 42 (br), © Tom Stewart/The Stock Market; 43 (br), © David Stoecklein/CORBIS; 44 (t), Michael Newman/PhotoEdit; 45 (cr), Richard Megna/Fundamental Photographs; 46 (t), Sam Dudgeon/HRW; 47 (tr), Dorling Kindersley Limited courtesy of the Science Museum, London/CORBIS; 47 (bl), Victoria Smith/HRW; 48 (b), Victoria Smith/HRW; 50 (tr), Richard Megna/Fundamental Photographs; 51 (cr), Richard Megna/Fundamental Photographs; 51 (br), Rob Matheson/The Stock Market; 54 (tr), Tony Freeman/PhotoEdit; 54 (tl), Henry Bargas/Amarillo Globe–News/AP/Wide World Photos; 55 (all), Bob Parker/Austin Fire Investigation

Chapter Three 56–57 (all), © Dr. Dennis Kunkel/Visuals Unlimited; 58 (bl), © Andrew Syred/Getty Images; 59 (all), Richard Megna/Fundamental Photographs; 60 (b), Victoria Smith/HRW; 61 (tr), Richard Megna/Fundamental Photographs; 62 (br), Jack Newkirk/HRW; 63 (br), Charles D. Winters/Timeframe Photography, Inc.; 63 (tl, tr), Peter Van Steen/HRW ; 64 (br), Tom Tracy/The Stock Shop/Medichrome ; 65 (tc), Victoria Smith/HRW; 65 (tr), © Peter Cade/Getty Images; 65 (tl), © Bob Thomason/Getty Images; 66 (all), Peter Van Steen/HRW ; 67 (tr), Peter Van Steen/HRW ; 70 (bl), Digital Image copyright © 2005 PhotoDisc; 70 (tl), Victoria Smith/HRW; 70 (tc, tr), Scott Van Osdol/HRW; 71 (tr), Miro Vinton/Stock Boston/PictureQuest; 73 (tl), Sam Dudgeon/HRW; 73 (tc), John Langford/HRW; 73 (tr), Charles D. Winters/Timeframe Photography, Inc.; 74 (tc), Digital Image copyright © 2005 PhotoDisc; 75 (bl), Sam Dudgeon/HRW; 76 (tl), Hans Reinhard/Bruce Coleman, Inc. ; 77 (tr), CORBIS Images/HRW; 78 (b), Sam Dudgeon/HRW; 80 (tr), Peter Van Steen/HRW ; 81 (all), Digital Image copyright © 2005 PhotoDisc; 84 (tr), Dan Loh/AP/Wide World Photos; 84 (tl), Sygma; 85 (tr), Nicole Guglielmo; 85 (bl), Corbis Images

Chapter Four 86–87 (all), GJLP/CNRI/PhotoTake; 88 (br), Henri Becquerel/The Granger Collection; 88 (bl), Roberto De Gugliemo/Science Photo Library/Photo Researchers, Inc.; 89 (tr), Digital Image copyright © 2005 PhotoDisc; 92 (br), Sygma; 94 (br), Tim Wright/CORBIS; 94 (bl), Custom Medical Stock Photo; 95 (tr), Roberto De Gugliemo/Science Photo Library/Photo Researchers, Inc.; 97 (tr), Emory Kristof/National Geographic Society Image Collection; 99 (tr), © Shone/Gamma; 101 (tc), Sam Dudgeon/HRW; 101 (tr), John Langford/HRW; 102 (all), Sam Dudgeon/HRW; 103 (b), Sam Dudgeon/HRW; 104 (tl), Tim Wright/CORBIS; 105 (bl), John Langford/HRW; 105 (cr), Science Photo Library/Photo Researchers, Inc.; 108 (tl), Courtesy USDA; 108 (tr), SABA Press Photos, Inc.; 109 (cr), © Underwood & Underwood/CORBIS; 109 (bl), © The Nobel Foundation

Lab Book/Appendix "LabBook Header", "L", Corbis Images; "a", Letraset Phototone; "b", and "B", HRW; "o", and "k", images ©2006 PhotoDisc/HRW; 110 (all), Sam Dudgeon/HRW; 111 (br), Sam Dudgeon/HRW; 112 (b), Sam Dudgeon/HRW; 113 (br), Rob Boudreau/Getty Images; 114 (br), Victoria Smith/HRW; 115 (cr), John Langford/HRW; 120 (br), Victoria Smith; 121 (br), Victoria Smith; 127 (tr), Peter Van Steen/HRW; 127 (br), Sam Dudgeon/HRW; 141 (tr), Sam Dudgeon/HRW

TEACHER EDITION CREDITS

1E (cl), © Charles Gupton/CORBIS; 1E (br), Paul Silverman/Fundamental Photographs, New York; 1F (cl), Sam Dudgeon/HRW; 1F (tr), © Jonathan Blair/CORBIS; 25E (cl), © David Stoecklein/CORBIS; 25F (cr), © Tom Stewart/The Stock Market; 55E (cl), Richard Megna/Fundamental Photographs, New York; 55E (br), Victoria Smith/HRW; 55F (tl), Miro Vinton/Stock Boston/PictureQuest; 55F (cr), Charles D. Winters/Timeframe Photography, Inc.; 85E (cl), Henri Becquerel/The Granger Collection; 85E (br), Roberto De Gugliemo/Science Photo Library/Photo Researchers, Inc.; 85F (cl), © Shone/Gamma

Answers to Concept Mapping Questions

The following pages contain sample answers to all of the concept mapping questions that appear in the Chapter Reviews. Because there is more than one way to do a concept map, your students' answers may vary.

CHAPTER **1** **Chemical Bonding**

16.

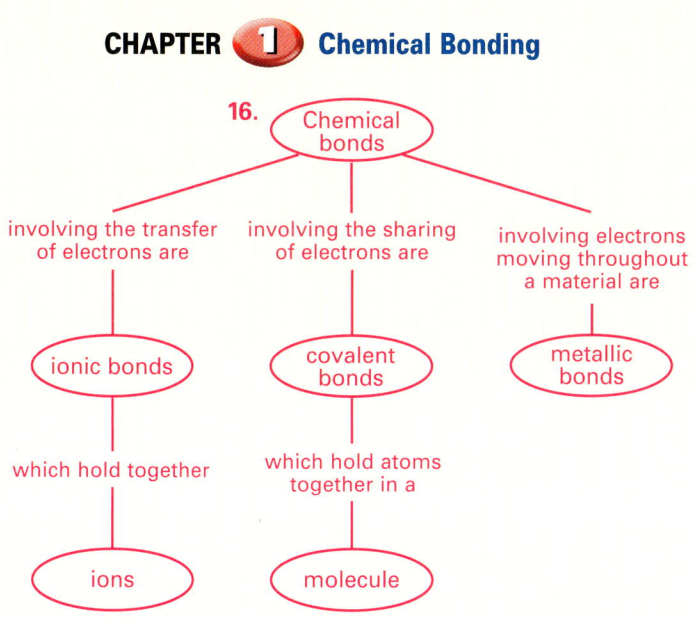

CHAPTER **2** **Chemical Reactions**

16.

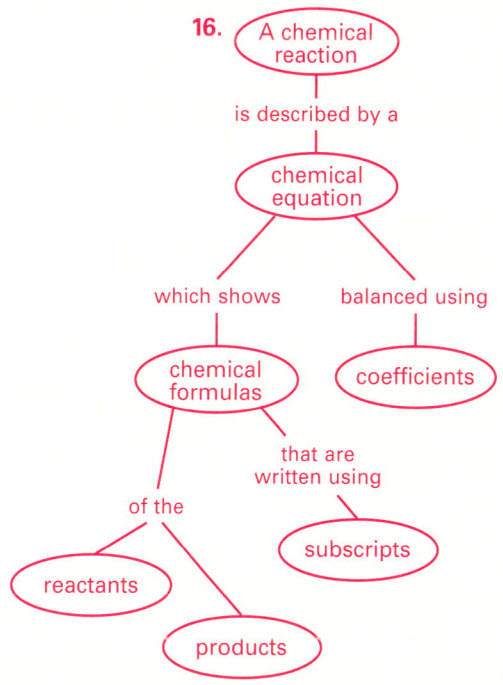

CHAPTER 3 Chemical Compounds

15.

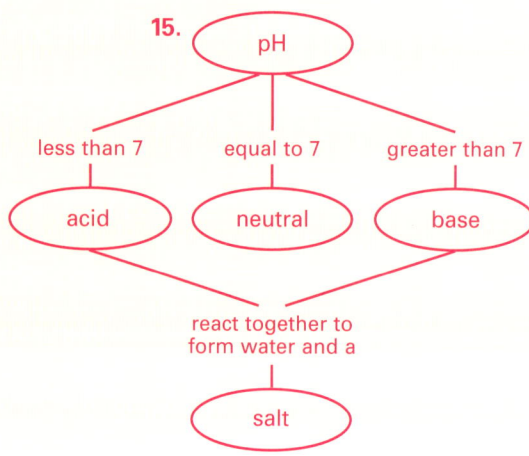

pH
- less than 7 → acid
- equal to 7 → neutral
- greater than 7 → base

acid and base react together to form water and a → salt

CHAPTER 4 Atomic Energy

15.

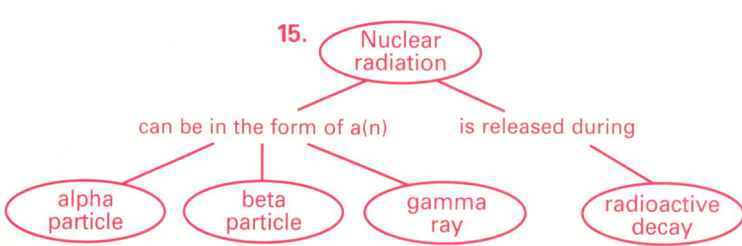

Nuclear radiation
- can be in the form of a(n): alpha particle, beta particle, gamma ray
- is released during: radioactive decay